AF726552

Lecture Notes in Computational Science and Engineering

30

Editors

Timothy J. Barth, Moffett Field, CA
Michael Griebel, Bonn
David E. Keyes, Norfolk
Risto M. Nieminen, Espoo
Dirk Roose, Leuven
Tamar Schlick, New York

Springer
Berlin
Heidelberg
New York
Hong Kong
London
Milan
Paris
Tokyo

Lorenz T. Biegler
Omar Ghattas
Matthias Heinkenschloss
Bart van Bloemen Waanders
Editors

Large-Scale PDE-Constrained Optimization

With 84 Figures, 11 in Color

Springer

Editors

Lorenz T. Biegler

Department of Chemical Engineering
Carnegie Mellon University
Pittsburgh, PA 15213, USA
e-mail: lb01@andrew.cmu.edu

Matthias Heinkenschloss

Department of Computational
and Applied Mathematics
Rice University
6100 S. Main Street - MS 134
Houston, TX 77005, USA
e-mail: heinken@caam.rice.edu

Omar Ghattas

Department of Biomedical Engineering
and Civil & Environmental Engineering
Carnegie Mellon University
Pittsburgh, PA 15213, USA
e-mail: oghattas@cs.cmu.edu

Bart van Bloemen Waanders

Optimization & Uncertainty Estimation Dept.
Sandia National Laboratories - MS 0847
P.O. Box 5800
Albuquerque, NM 87185, USA
e-mail: bartv@sandia.gov

Cataloging-in-Publication Data applied for

A catalog record for this book is available from the Library of Congress.

Bibliographic information published by Die Deutsche Bibliothek
Die Deutsche Bibliothek lists this publication in the Deutsche Nationalbibliografie;
detailed bibliographic data is available in the Internet at <http://dnb.ddb.de>.

Mathematics Subject Classification (2000):
34-XX, 35-XX, 49-XX, 65-XX, 68-XX, 74-XX, 76-XX, 90-XX

ISSN 1439-7358
ISBN 3-540-05045-0 Springer-Verlag Berlin Heidelberg New York

This work is subject to copyright. All rights are reserved, whether the whole or part of the material is concerned, specifically the rights of translation, reprinting, reuse of illustrations, recitation, broadcasting, reproduction on microfilm or in any other way, and storage in data banks. Duplication of this publication or parts thereof is permitted only under the provisions of the German Copyright Law of September 9, 1965, in its current version, and permission for use must always be obtained from Springer-Verlag. Violations are liable for prosecution under the German Copyright Law.

Springer-Verlag Berlin Heidelberg New York
a member of BertelsmannSpringer Science + Business Media GmbH

http://www.springer.de

© Springer-Verlag Berlin Heidelberg 2003
Printed in Germany

The use of general descriptive names, registered names, trademarks, etc. in this publication does not imply, even in the absence of a specific statement, that such names are exempt from the relevant protective laws and regulations and therefore free for general use.

Cover Design: Friedhelm Steinen-Broo, Estudio Calamar, Spain
Cover production: *design & production*
Typeset by the authors using a Springer TEX macro package

Printed on acid-free paper   46/3142/LK - 5 4 3 2 1 0

Table of Contents

I Introduction

Large-Scale PDE-Constrained Optimization: An Introduction 3
Lorenz T. Biegler, Omar Ghattas, Matthias Heinkenschloss, Bart van Bloemen Waanders

II Large-Scale CFD Applications

Nonlinear Elimination in Aerodynamic Analysis and Design Optimization 17
D. P. Young, W. P. Huffman, R. G. Melvin, C. L. Hilmes, F. T. Johnson

Optimization of Large-Scale Reacting Flows using MPSalsa and Sequential Quadratic Programming 45
A.G. Salinger, R.P. Pawlowski, J.N. Shadid, B. van Bloemen Waanders, R. Bartlett, G.C. Itle, L. Biegler

III Multifidelity Models and Inexactness

First-Order Approximation and Model Management in Optimization .. 63
Natalia M. Alexandrov, Robert Michael Lewis

Multifidelity Global Optimization Using DIRECT 80
Steven E. Cox, Raphael T. Haftka

Inexactness Issues in the Lagrange-Newton-Krylov-Schur Method for PDE-constrained Optimization 93
George Biros, Omar Ghattas

IV Sensitivities for PDE-based Optimization

Solution Adapted Mesh Refinement and Sensitivity Analysis for Parabolic Partial Differential Equation Systems 117
Shengtai Li, Linda R. Petzold, James M. Hyman

Challenges and Opportunities in Using Automatic Differentiation with Object-Oriented Toolkits for Scientific Computing 133
Paul Hovland, Steven Lee, Lois McInnes, Boyana Norris, Barry Smith

Piggyback Differentiation and Optimization 148
Andreas Griewank, Christèle Faure

V NLP Algorithms and Inequality Constraints

Assessing the Potential of Interior Methods for Nonlinear Optimization 167
José Luis Morales, Jorge Nocedal, Richard A. Waltz, Guanghui Liu, Jean-Pierre Goux

An Interior-Point Algorithm for Large Scale Optimization 184
John T. Betts, Samuel K. Eldersveld, Paul D. Frank, John G. Lewis

SQP SAND Strategies that Link to Existing Modeling Systems 199
Lorenz T. Biegler and Andreas Wächter

Interior Methods For a Class of Elliptic Variational Inequalities 218
Randolph E. Bank, Philip E. Gill, Roummel F. Marcia

Hierarchical Control of a Linear Diffusion Equation 236
Anthony J. Kearsley, Jon W. Tolle, Paul T. Boggs

VI Time-Dependent Problems

A Sequential Quadratic Programming Method for Nonlinear Model Predictive Control 253
Kenneth R. Muske, James W. Howse

Reduced Order Modelling Approaches to PDE-Constrained Optimization Based on Proper Orthogonal Decomposition 268
Marco Fahl, Ekkehard W. Sachs

Adaptive Simulation, the Adjoint State Method, and Optimization 281
Mark S. Gockenbach, William W. Symes

VII Frameworks for PDE-Constrained Optimization

The SIERRA Framework for Developing Advanced Parallel Mechanics Applications.. 301
James R. Stewart, H. Carter Edwards

rSQP++ : An Object-Oriented Framework for Successive Quadratic Programming.. 316
Roscoe A. Bartlett, Lorenz T. Biegler

Sundance Rapid Prototyping Tool for Parallel PDE Optimization 331
Kevin R. Long

Color Plates **343**

Part I

Introduction

Large-Scale PDE-Constrained Optimization: An Introduction

Lorenz T. Biegler[1], Omar Ghattas[1], Matthias Heinkenschloss[2], and Bart van Bloemen Waanders[3]

[1] Carnegie Mellon University, Pittsburgh, PA 15213 USA
[2] Rice University, Houston, TX 77005 USA
[3] Sandia National Laboratories[†], Albuquerque, NM 87185 USA

Abstract. Optimal design, optimal control, and parameter estimation of systems governed by partial differential equations (PDE) give rise to a class of problems known as PDE-constrained optimization. The size and complexity of the discretized PDEs often pose significant challenges for contemporary optimization methods. Recent advances in algorithms, software, and high performance computing systems have resulted in PDE simulations that can often scale to millions of variables, thousands of processors, and multiple physics interactions. As PDE solvers mature, there is increasing interest in industry and the national labs in solving optimization problems governed by such large-scale simulations. This article provides a brief introduction and overview to the *Lecture Notes in Computational Science and Engineering* volume entitled *Large-Scale PDE-Constrained Optimization.*

This volume contains nineteen articles that were initially presented at the *First Sandia Workshop on Large-Scale PDE-Constrained Optimization.* The articles in this volume assess the state-of-the-art in PDE-constrained optimization, identify challenges to optimization presented by modern highly parallel PDE simulation codes and discuss promising algorithmic and software approaches to address them. These contributions represent current research of two strong scientific computing communities, in optimization and PDE simulation. This volume merges perspectives in these two different areas and identifies interesting open questions for further research. We hope that this volume leads to greater synergy and collaboration between these communities.

1 Algorithmic challenges for PDE-constrained optimization

PDE simulation is widespread in science and engineering applications. Moreover, with increasing development and application of supercomputing hardware and advances in numerical methods, very large-scale and detailed simulations can now be considered. An essential sequel to simulation is its application in design, control, data assimilation, and inversion. Most of these tasks are naturally stated as continuous variable optimization problems (i.e.,

[†] Sandia is a multiprogram laboratory operated by Sandia Corporation, a Lockheed-Martin Company, for the United States Department of Energy under Contract DE-AC04-94AL85000.

nonlinear programs) and it is clear that superior designs and estimates result with efficient, high performance optimization strategies.

Research in optimization algorithms and numerical methods for PDEs has evolved along two parallel paths that are not always compatible with each other. To exploit recent advances in both fields a number of challenges need to be addressed. These can be summarized by the following issues:

- problem size of PDE constrained optimization problems,
- efficiency vs. intrusiveness of nonlinear programming (NLP) algorithms integrated with PDE solvers,
- "physics-based" globalizations and inexact PDE solvers that employ various levels of model fidelity,
- approximate Jacobians used in nonlinear PDE solvers and the application of sensitivity and adjoint methods,
- implicitly-defined PDE residuals, adaptive PDE solvers and nonsmooth solutions,
- treatment of inequality constraints in the NLP, and
- time-dependent PDE optimization.

In the remainder of this section we briefly consider each of these issues. The next section provides a brief overview of the articles in this volume with the perspective of addressing these topics.

1.1 Problem size in PDE-constrained optimization

An essential characteristic of PDE optimization problems is their very large problem size, both in the number of state variables n and decision variables m. These are dictated by the approximation and discretization of the underlying continuous problem. For many production scientific and industrial 3D simulations, the number of state variables and state equations can easily number in the millions. Indeed, current "heroic" simulations are nearing gigascale dimensions (i.e. $\mathcal{O}(10^9)$ state variables and constraints) and terascale memory and work requirements.

On the other hand, the number of decision variables (i.e. design variables, control variables, or inversion parameters) characterizing typical PDE optimization problems varies widely, from just a handful, to on the order of the number of state variables. In optimal design, it is common to parameterize a design with feature-based variables which results in a mesh-independent number of design variables. On the other hand, one can represent the surface design by the location of every grid point, i.e. $m = \mathcal{O}(n^{2/3})$ in three dimensions. Finally, topology optimization methods usually result in $\mathcal{O}(n)$ design variables. In optimal control problems, controls can be applied at points, along surfaces, or throughout a volume. Therefore, the number of controls ranges from $\mathcal{O}(1)$ for point controllers, to $\mathcal{O}(n^{2/3})$ for boundary control, to $\mathcal{O}(n)$ for distributed (volume) control. Inverse problems similarly lead to decision variable parameterizations that range from mesh-independent to mesh-dependent of varying degree. For example, inverting state boundary data to recover parameters of a constitutive equation in a (piecewise) homogeneous medium produces a small number of decision variables, that does not increase as the mesh is refined. On the other hand, the decision space increases as $\mathcal{O}(n^{2/3})$ when inverting for boundary conditions of a PDE. Finally, inverting for initial conditions, or for a mesh-parameterized heterogeneous material field, produces $\mathcal{O}(n)$ decision variables. If we are optimizing for time-dependent decision variables the mesh-dependence expressions above will increase by a factor related to the number of time steps, which for Courant-limited time stepping, scales as $\mathcal{O}(n^{1/3})$.

Large-scale PDE-constrained nonlinear programs are typically solved using variants of Newton's method applied to the optimality (also known as the Karush-Kuhn-Tucker or KKT) conditions of the NLP. This Newton-based approach, extended to deal with inequality constraints, is known as Sequential Quadratic Programming (SQP) and variants that incorporate the elimination of the state variables and linearized equations are known as reduced space SQP, or rSQP methods. For full-space-based optimization of 3D PDE systems, an iterative linear solver is needed as the KKT matrix is too large to be factored. Here preconditioning of this system is essential. As described in the remainder of this section, a number of issues need to be addressed related to the calculation or approximation of first and second derivatives and the choice of linear and nonlinear solution algorithms.

1.2 Efficiency vs. intrusiveness: PDE-solver viewpoint

A major challenge lies in the efficient integration of nonlinear programming algorithms with advanced PDE solvers. To lend further insight into this challenge, we consider four levels of implementation of NLP algorithms with PDE solvers. At the most basic level, we consider a black-box or *Nested Analysis and Design (NAND)* implementation. This interface requires very little interaction of the NLP algorithm with the PDE solver. Gradients for the NLP solver (e.g., SQP or rSQP) are obtained by finite differencing the PDE solver, a nonlinear elimination of the state variables is assumed, and usually only few decision variables can be considered. On the other hand, the NAND implementation can suffer from repeated and time-consuming solution of the PDEs and intermediate convergence failures of the PDE solver.

At the other extreme we can consider a full space SQP method. This implementation allows access to the complete discretization of the PDE model and accurate first and second derivatives that make up the linear KKT system in the space of all of the state and decision variables. To solve this full system, an approximate elimination of states and adjoints is applied as a preconditioner. Therefore, because the iterative solution of the linearized PDE equations is now part of the NLP algorithm, this approach leads to a very fast *SAND (Simultaneous Analysis and Design)* optimization strategy. However, this fully open implementation requires the computation of additional

Jacobians and Hessians that are typically not a part of PDE solvers, thus requiring a rewrite or retrofit of the solver.

One can also consider intermediate implementations that are tailored to the PDE code using rSQP. There are two approaches that are distinguished by whether accurate Jacobian elements of the PDEs can be made available to the NLP algorithm. With the *direct tailored approach*, reduced gradients and search directions for rSQP are calculated directly using Newton steps from the PDE solver system and the "sensitivity" of Newton steps to the decision variables. Nevertheless, these steps require the solution of a linear system with as many right hand sides as decision variables. For large-scale problems, iterative solvers are required, and there is little opportunity to amortize solutions costs over the multiple right hand sides; these approaches are consequently much more expensive than the full space SAND approach. On the other hand, the *adjoint tailored approach* calculates reduced gradients and search directions for rSQP by using both the PDE Jacobian and its transpose. This approach is still more expensive than full space SAND, but it requires only as many iterative linear solutions as there are active constraints, which is typically far fewer than with the direct approach. It also provides multiplier estimates for the NLP solver.

1.3 Physics-based globalizations and inexact solution

In addition to the difficulties PDE constraints present to optimization software due to their large size, they can often exhibit severe nonlinearities along with steep profiles and poorly conditioned solutions. As a result, a number of strategies is needed to promote convergence of PDE solvers, beyond the general-purpose line search and trust region methods commonly used in large-scale optimization software. In particular, a wide variety of PDE physics-based globalizations are used. These include:

- mesh sequencing to reduce the problem nonlinearity, improve the conditioning on a coarse grid, and bootstrap a solution on a fine grid;
- continuation methods on nonlinear parameters or pseudo-time stepping to steady state;
- lower-fidelity precursor models that provide good initializations for the discretized PDEs; and
- approximate Jacobians (e.g. due to dropped terms as in a Picard method) that are known to enlarge the attraction basin of Newton's method.

These globalization strategies lead to considerable deviations from Newton-based equation solving. As a result, they do not lend themselves to straightforward incorporation within simultaneous methods. The challenge here is to find the best ways to integrate these globalizations into SQP methods.

Moreover, large-scale PDE solvers are often inexact. Both the linear and nonlinear solvers applied to the discretized PDEs are iterative, and one is often content only to reduce the (linear, nonlinear) residual by several orders of magnitude. While inexact solvers can be used to advantage far from an optimum, care must be taken in constructing a convergent and efficient optimization strategy that reduces the inexactness as the optimum is neared.

1.4 Approximate Jacobians

The sources of inexactness in PDE solvers are most evident in the construction of approximate Jacobians from the discretized PDE residuals. For convergence and even for successful performance, PDE solvers do not require exact Jacobians of the residual, and most production PDE solvers do not form the exact Jacobian for expense or difficulty of implementation reasons. For PDE optimization, on the other hand, the exact PDE Jacobian is required to compute the reduced gradients of the objective (and constraint functions) and the KKT conditions for termination.

Approximate Jacobians in PDE solvers result from a number of sources. Often, derivative terms are dropped due to difficulties in differentiating complex terms in the residual, as in complicated constitutive and subgrid scale models, as well as stabilization, shock capturing, and moving mesh schemes. Approximate Jacobians appear also in coupled problems under the framework of operator-split or staggered solution methods. These techniques often amount to ignoring the Jacobians of coupling terms, and are widely accepted as the method of choice for many multiphysics problems. Also popular is the use of a lower-order discretization for constructing the "Jacobian" than is used for the residual, which makes it approximate.

Finally, it should be noted that even when the Jacobian is not too difficult or expensive to form, many modern Krylov-based PDE solvers will still not form it, and instead will approximate the matrix-vector products through directional differencing of the residual. Such "matrix-free" methods still require a preconditioner, but this can invoke any of the Jacobian approximations mentioned above.

The generation of approximate Jacobians has important implications for PDE-based optimization strategies and algorithms. Exact Jacobians and Hessians are essential for the SAND approach but not necessary for black-box NAND. Moreover, approximate Jacobians influence the two intermediate tailored methods in different ways. For the *direct rSQP approach*, reduced gradients are generated by the iterative solution of linear sensitivity equations that have a structure analogous to the state equations. Thus the same matrix-free Krylov solver and preconditioner (represented by the approximate Jacobian) that is applied to the state equations can be used for the sensitivity equations. For the *adjoint rSQP approach*, which is essential for problems with large numbers of decision variables, a matrix-free approach cannot be applied without the ability to evaluate matrix-vector products with the *transpose* of the Jacobian. Hence, exact derivatives are a necessity here.

1.5 Implicitly-defined and nonsmooth PDE residuals

Obtaining accurate Jacobians and higher derivatives for optimization is further complicated by residuals that are nonsmooth or only implicitly defined. Evaluating the PDE residual may require complex internal computations associated with solution adaptivity, discontinuity-capturing and front-tracking. An example of this can be seen in moving-mesh treatment of dynamic interfaces. Model features of PDEs also contribute to difficulties with implicitly-defined residuals. For example, subgrid-scale models, complex constitutive models, contact problems, and plasticity yield conditions may involve "internal" computations that are not exposed in the PDE residuals. In many cases, the global residual may not even be formed, as in code-coupled multiphysics problems.

An additional difficulty, particularly with implicitly-defined residuals, is that the residual may not depend smoothly (or even continuously) on state variables. In some cases, the nonsmoothness is intrinsic to the underlying infinite-dimensional problem (e.g., due to shear banding in a solid or shock formation in a gas flow). In other cases, nonsmoothness is introduced into the residual calculation via advanced numerical devices, for example, through solution-adaptive meshing or time-stepping, or moving mesh schemes. The nonsmoothness can often be mitigated by defeating these advanced features in favor of fixed mesh or fixed order methods. Other nonsmooth effects can possibly be reformulated away by promoting them to level of the optimizer, as with variational inequalities that arise in contact and plasticity models.

1.6 Treatment of inequalities

Many PDE-based optimal design and control problems have bounds on decision variables, or inequalities on state variables, that are expressed pointwise throughout the PDE domain; upon discretization, these can lead to $O(n)$ inequalities, which are often nonlinear. Examples of these include avoiding material failure, reducing vorticity, or limiting fluxes throughout a domain.

In classical SQP and rSQP methods, inequalities are handled through the active set solution of the quadratic programming subproblem. However, determination of the active set of inequalities is combinatorial and may be extremely expensive for large systems. Moreover, in the rSQP method, reduced gradients are required for all of the inequalities as well as for the objective function. While this information is provided for free with the direct rSQP method, the calculation of these reduced gradients with the adjoint rSQP method adds a large computational cost, and removes any advantage of the adjoint approach over the direct approach when there are large numbers of active constraints.

Recently, large-scale interior point (or barrier) NLP algorithms have been considered to address these challenges. Here inequality constraints are lumped into a barrier term and the combinatorial problem is transformed to solving a set of nonlinear equations. In addition, a reduced gradient is required only for the barrier function and this is obtained cheaply with an adjoint method. Nevertheless, there are still some open questions with this approach. Barrier problems may be ill-conditioned and more efficient ways to incorporate second order information must be found.

1.7 Time-dependent problems

Because dynamic PDE optimization extends directly from the optimization of steady-state PDE systems, all of the above issues are captured and amplified in time-dependent problems. But even more than in the steady state case, large problem size is a crucial consideration, since the optimization problem "sees" the entire time history of the PDE system. The dynamic structure of these problems needs to be exploited.

Here, SAND methods require the storage of the entire state variable time history; this is clearly not practical for large problems. On the other hand, NAND methods avoid the storage problem with direct sensitivity methods; however, this is workable only for a handful of decision variables. Adjoint techniques can be applied; these involve calculating the reduced gradient by solution of forward-in-time state problems and backward-in-time adjoint problems. To avoid storage of the complete state time history, checkpointing and compression schemes can be used. Nevertheless, there are a number of open questions that remain in design optimization algorithms that best combine the advantages of SAND and NAND.

1.8 Concluding comments

The development of scalable methods for solving optimization problems that are constrained by systems of PDEs presents numerous challenges, a number of which have been outlined above. Nevertheless, the capability for fast and robust PDE optimization will lead to tremendous improvements in the design, control, estimation and analysis of devices and systems in science and engineering. Moreover, tackling the above challenges leads to a rich source of exciting research problems.

The emergence of PDE-constrained optimization comes at a time when many legacy PDE codes are being rewritten to take advantage of distributed memory parallel machines as well as advances in object oriented software development. This therefore provides a window of opportunity to influence design of PDE solvers to make them optimization-friendly. As a result, we hope that these circumstances provide significant opportunities for cross-fertilization between optimization and simulation communities. It is in this spirit that we introduce the following articles in this volume.

2 Overview of Articles

Collectively, the articles in this volume assess the current state of the art in PDE-constrained optimization, and address some of the difficulties expressed above. These include discretization issues, unavailability of exact Jacobians, the expense of computing sensitivities, the need to accommodate approximate solvers and inexact problem information, large sets of inequality constraints that may arise from discretized pointwise constraints, opportunities to exploit models of varying fidelity, large storage problems presented by time-dependent PDEs, and considerable software engineering challenges.

2.1 Large-scale CFD applications

The first two articles deal with issues involved in large-scale CFD optimization, including globalization, inexactness, and interfacing rSQP methods with industrial-strength codes. Applications to compressible and incompressible flow optimization are presented. The first article by *Young et al.* discusses a number of issues related to implementing an SQP optimization strategy to the TRANAIR CFD code at Boeing. Here a nonlinear elimination (NAND) strategy is presented along with a detailed algorithm and examples drawn from airfoil design. The second article by *Salinger et al.* deals with an rSQP tailored optimization implementation on MPSalsa, a reacting flow code at Sandia National Laboratories. This approach is illustrated with the design optimization of a chemical vapor deposition reactor.

2.2 Multifidelity models and inexactness

The next three articles develop optimization strategies that use multifidelity models and inexactness to reduce significantly the cost needed to solve PDE optimization problems. The first article by *Alexandrov and Lewis* deals with a multifidelity model management scheme that enforces first order consistency among the models and guarantees convergence for the high fidelity model. The approach is demonstrated on the optimization of a multi-element airfoil operating in transonic conditions. The second article by *Cox and Haftka* presents a multifidelity extension of the DIRECT global optimization algorithm, a variation of a Lipschitzian method. The article details this approach and describes its extension to search among models with different levels of accuracy, including polynomial interpolants. The approach is illustrated on a small test problem. The third article by *Biros and Ghattas* considers inexactness issues in the Lagrange-Newton-Krylov-Schur (LNKS) method for SAND optimization. The article provides a conceptual derivation of the method, an analysis of tolerance to inexactness in the linear solver, and an application to optimal boundary flow control.

2.3 Sensitivities for PDE-based optimization

The next three articles deal with the computation of sensitivities for PDE-constrained optimization. The first article by *Li et al.* concerns a mesh refinement scheme in space that is combined with backward difference formula solvers in time and adapted to yield sensitivities. The article describes mesh refinement and time integration issues as well as a number of implementation details. The approach is illustrated on a flame propagation problem. The next article by *Hovland et al.* discusses the state of the art for automatic differentiation (AD) in large-scale scientific computing. It also discusses challenges and planned developments in advancing these methods within a number of object oriented toolkits for PDE modeling and optimization. Experimental performance results are presented using the PETSc and SensPVODE toolkits. The third article by *Griewank and Faure* presents new results on properties of automatic differentiation. In particular, they analyze the convergence of AD for fixed point iterations for both direct and adjoint schemes. Moreover, the rapid performance of their AD algorithm is demonstrated on a 2D Euler code. The results of these articles have important implications for obtaining first and second derivatives for PDE-constrained optimization problems.

2.4 NLP algorithms and inequality constraints

The next five articles deal with advances for SQP-type NLP algorithms, particularly with respect to the treatment of inequality constraints. The first article by *Morales et al.* presents an experimental case study that compares two interior point SQP codes, KNITRO and LOQO, with two active set SQP codes, FilterSQP and SNOPT. Applied to unconstrained, equality constrained and general constrained test problems, the authors include a number of observations on these four algorithms and note the advantages of interior point methods. The second article by *Betts et al.* describes recent advances with the SOCS NLP solver at Boeing. Here a primal-dual interior point algorithm is implemented and a filter line search is incorporated for globalization. The approach is illustrated on a highly constrained, high index PDE where the interior point method has significant advantages over the active set strategy. The next article by *Biegler and Wächter* considers a number of options for rSQP strategies including second order information, type of line search strategy and active set vs. interior point methods. A comparison of options is given on discretized optimal control problems and some conclusions are offered for their use with PDE-constrained optimization.

The last two articles in this set address particular classes of PDE optimization problems. The article by *Bank et al.* concerns solution of a class of elliptic variational inequality problems, which result in large bound-constrained optimization problems. In particular, it is shown how a primal-dual interior method can be used to solve a sequence of finite-dimensional approximations of these problems via the multilevel adaptive mesh refinement PDE solver

PLTMG. The article presents several examples involving obstacle and elasto-plastic torsion problems. Finally, the fifth article by *Kearsley et al.* concerns the hierarchical control of a linear diffusion equation. This is posed as a multicriterion optimization problem and reformulated as a highly constrained nonlinear program. Several cases are considered and solved using an SQP approach.

2.5 Time-dependent problems

This set of articles addresses the challenges stemming from optimization of large time-dependent PDE problems. The first article by *Muske and Howse* presents a Lagrangian method for simultaneous nonlinear model predictive control. Here time-dependent advection-diffusion equations are considered and a SAND optimization is proposed using a Newton-Krylov solver and block preconditioner. The next article by *Fahl and Sachs* describes an approach to PDE-constrained optimization based on Proper Orthogonal Decomposition (POD). POD offers a systematic way to simplify the PDE problem but safeguards are needed to ensure convergence for the original model. Here a POD-based trust region framework is proposed. The subsequent algorithm is applied to control of a time-dependent heat conduction problem. The third article by *Gockenbach and Symes* treats optimization problems that depend on adaptive marching simulations in time. Here, the computed states are often nonsmooth even though the underlying problem is well-defined. The authors describe an abstract framework that allows the computation of sensitivities as part of the Hilbert Class Library.

2.6 Frameworks for PDE-constrained optimization

This last section deals with developments and opportunities within computational frameworks for PDE-constrained optimization. The first article by *Stewart and Edwards* details the SIERRA framework developed at Sandia National Labs. The main goal is to integrate mesh management, field variable management and algorithm support services along with a suite of existing application codes within a distributed parallel computing environment. The second article by *Bartlett and Biegler* describes rSQP++, an object-oriented framework for SQP. This framework supports different variations of SQP algorithms as well as specific linear algebra solvers. It leads to algorithm configurations that exploit the structure of PDE-based models as well as the architecture of the computing environment. Finally, the article by *Long* describes Sundance, a rapid prototyping toolkit for parallel PDE solution and optimization. Here the parallel PDE solver can be accessed through a high level problem description. Using a small amount of user-level code, the variational PDE formulation, the discretization method and even operators for Jacobians and Hessians can be specified for use in optimization algorithms.

3 Concluding remarks and acknowledgements

The articles in this volume represent a rich spectrum of research in PDE-constrained optimization. In particular we hope that they will spur future advances in this exciting field, especially through collaborations between the PDE and optimization communities.

Finally, we are pleased to acknowledge the following groups and individuals, without whom this volume would not have been realized. Our thanks go to:

- the authors, reviewers and session chairs for the workshop and their efforts in producing the articles in this volume;
- Ms. Linda Benavides, Ms. Deanna Ceballos and Ms. Barbara Delap at Sandia National Labs for administrative support;
- the Computer Science Research Institute at Sandia National Labs for sponsorship and support of the workshop;
- the National Science Foundation under Grant ACI 0116984 for providing travel support for graduate students and researchers; and
- Dr. Martin Peters, editor of the LNCSE series at Springer Verlag, for his support and assistance related to this volume.

Part II

Large-Scale CFD Applications

Nonlinear Elimination in Aerodynamic Analysis and Design Optimization

D. P. Young, W. P. Huffman, R. G. Melvin, C. L. Hilmes, and F. T. Johnson

The Boeing Company, P. O. Box 3707, M/S 7L-21, Seattle, WA 98124-2207

Abstract. Recent emphasis on reduction of design cycle time and cost in the design of commercial aircraft has sparked a renewed interest in design optimization in aerodynamics, structures, and aeroelastics. The constrained aerodynamic optimization problem is closely related to the problem of solving nonlinear systems of equations. In applying Newton's method to steady-state compressible CFD analysis problems, the nonlinear elimination method has been remarkably successful. In this paper we consider the implications of this experience for design optimization formulations in the general case of state equation equality constraints. This relationship between nonlinear equation solving and design optimization is illustrated by drawing on computational examples from the TRANAIR compressible CFD code. We first discuss various formulations of the PDE constrained optimization problem related to the Lagrange Newton method and the multiplier free version implementation in TRANAIR. We then discuss the nonlinear elimination method and its application to a simple nozzle problem. This method is then applied to derive various globalization methods in design optimization which are illustrated by a computational example in airfoil design. Finally, we discuss some remaining limitations and issues.

1 Introduction

Today, commercial aircraft are routinely designed using a combination of Computational Fluid Dynamics (CFD) and wind tunnel testing [24]. One historically successful design method is the use of repetitive analysis and/or wind tunnel testing to develop engineering understanding. An alternative is the inverse design method, in which the designer specifies desirable flow features, in most cases the pressure distribution on the wing surface, and asks the CFD code to attempt to compute the geometry which produces such flow features. Using inverse design, the many cut-and-try iterations of repetitive analysis are replaced by a systematic computational procedure.

There are, however, still a number of drawbacks of the inverse design approach. The first one is the difficulty of finding "good" pressure distributions for highly three-dimensional flows. A second difficulty is the consideration of off-design performance. The inverse design method is inherently a single point design process, although the designer usually has some knowledge of what pressure architecture at the cruise condition will likely give reasonable off-design characteristics. A third difficulty is the indirect way in which inequality constraints on the geometric surface (arising for example from manufacturing or structural requirements) must be imposed.

These difficulties lead to the desire to develop an optimization methodology for aerodynamic design. Most of the methods for aerodynamic optimization are variations of classical methods for optimal control [3] in which special consideration must be given to computational costs and the mathematical difficulties inherent in compressible viscous analysis and design. A well formulated optimization method may help to quickly achieve a good compromise between aerodynamic or economic cost objectives and the constraints imposed on the geometry by other disciplines such as manufacturing and structures. It will also allow us to consider concurrently multiple critical flight conditions, so that iteration with off-design considerations can be reduced or eliminated. In the past few years, we have developed an optimization capability in the TRANAIR code [1, 16, 19, 27, 28]. TRANAIR is a two- and three-dimensional full potential flow code with a directly coupled boundary layer capable of handling complex geometries through the use of Cartesian grids and solution adaptive local grid refinement.

This paper discusses some of the algorithmic issues dealt with in building the TRANAIR aerodynamic optimization capability. In section 2 we discuss some of the theoretical difficulties encountered in aerodynamic optimization. In section 3 we will give an overview of optimization formulations for PDE constrained problems and in section 4 describe the methods implemented in the TRANAIR design code. In section 5 we will describe various globalization strategies motivated by nonlinear elimination and demonstrate their effectiveness with some examples. In section 6 we will describe how the optimization method is extended to multi-point design. In section 7 we consider some remaining technical issues involving robust convergence and identification of ill-posed problems.

2 Mathematical Difficulties

It has been known for over 40 years that the problem of transonic flow past an airfoil is an ill-posed problem in certain circumstances [22, 23]. Small changes in a nearly flat portion of an airfoil can cause a supersonic zone with a shock wave to appear or disappear. This is usually avoided in engineering practice since the resulting real flow is likely to exhibit undesirable unsteady oscillations. This difficulty is not remedied by using higher order mathematical modeling, e. g., the Euler equations [18]. The Navier-Stokes equations are also well known to exhibit multiple solutions as the Reynolds number is increased. As might be expected, design optimization tends to drive the design towards these impractical configurations unless constrained from doing so. It is also well known that in viscous flow, separation is a highly nonlinear and sensitive phenomena. In a boundary layer, small changes in either the airfoil shape or the outer inviscid flow can result in large changes in the boundary layer. In fact, it is questionable whether steady state solutions exist at all for the Navier-Stokes equations when the flow is separated. Further, turbulence modeling is essential to achieving realistic results for high Reynolds number flows and the dependence of the flow on these models is very nonlinear. In addition, an airframe must have acceptable performance everywhere in the flight envelope to be viable. Once this is realized, it is apparent that the aerodynamic design problem must consider at least a reasonable number of different flight conditions even if only through inequality constraints, e.g., it has been observed that controling the peak Mach number of a wing at cruise flight conditions during design will usually result in acceptable aerodynamic characteristics at higher speeds or lifts. In practice, multi-point optimization is required which increases computational cost and flow time [9].

3 Optimization Formulations

The TRANAIR code is a compressible CFD code designed to model arbitrary geometries robustly [27]. It uses a locally refined Cartesian grid that does not conform to the surface of the body, solution adaptive grid refinement, and a Newton-Krylov solution algorithm. General geometry capability and Newton-Krylov solution methodology have generally been thought to be helpful when constructing a PDE-constrained optimization code. However, it has generally been thought that solution adaptivity is an impediment because it is a source of "noise" in the objective when a black box optimization approach is employed. A further simplifying methodology utilized in TRANAIR is the use of transpiration boundary conditions to simulate the effect of actually moving the surface. This eliminates the necessity of computing sensitivities of the grid itself with respect to the design variables. One of the consequences of the use of Newton's method is the experience with inexactness of derivatives that carries over almost directly to the optimization case [28]. In addition, we will argue below that the use of a sensitivity formulation enables the implicit and cost effective generation of partial second order information.

TRANAIR discretizes and solves the flow problem on a series of grids, $l = 1, 2, \ldots, NG$. We have combined our design and optimization capability with this solution adaptive grid capability by defining a discrete simplified optimization problem on each grid in such a way that the entire process can converge to the solution of the continuous optimal control problem. For simplicity, in the following description of optimization methods, we will deal only with the discrete version of the problem which is posed on a given grid.

The discretized nonlinear state (flow) equations will be denoted by

$$F(X, u) = (F_1(X, u), F_2(X, u), \ldots, F_n(X, u))^T = 0$$

where the state variables, $X = (X_1, X_2, \ldots, X_n)^T$, consist of both the inviscid flow variables and the boundary layer variables. The design parameters, $u = (u_1, u_2, \ldots, u_m)^T$ could represent the geometry shape or flow quantities such as freestream angle of attack or Mach number. Included in $F(X, u)$ is an inhomogeneous Neumann boundary condition $T(X, u)$ used to approximate the effect of boundary motion [16]. One simple form of this condition is $T(X, u) = W \cdot \delta\hat{n}$ where W is the mass flux through the boundary and $\delta\hat{n}$ is the total change in unit normal due to the design parameters u.

Now, consider the problem of minimizing a scalar objective function $I(X, u)$ subject to the constraint that $F(X, u) = 0$. We assume that the Jacobian matrix $\partial F_i/\partial X_k$ is invertible for the values of u of interest and use the notation that

$$\partial I/\partial X = (\partial I/\partial X_1, \partial I/\partial X_2, \ldots, \partial I/\partial X_n)$$

is a row vector.

The necessary conditions for optimality are often formulated by introducing the Lagrange multipliers $\lambda = (\lambda_1, \lambda_2, \ldots, \lambda_n)^T$ as independent variables [3]. The Lagrangian is then defined by

$$L(X, u, \lambda) = I(X, u) + \lambda^T F(X, u).$$

At an optimum, the Lagrange multiplier λ_i is the derivative of the value of I with respect to changes in the value of the constraint F_i, i.e., $\lambda_i = dI/dF_i$. Necessary conditions for an optimum are that the gradient of L be zero,

$$\frac{\partial L}{\partial u} = \frac{\partial I}{\partial u} + \lambda^T \frac{\partial F}{\partial u} = 0 \tag{1}$$

$$\frac{\partial L}{\partial X} = \frac{\partial I}{\partial X} + \lambda^T \frac{\partial F}{\partial X} = 0 \tag{2}$$

$$\frac{\partial L}{\partial \lambda} = F(X, u) = 0. \tag{3}$$

In cases where $\partial F/\partial X$ is invertible for the values of u of interest, the Lagrange multipliers can be eliminated from the formulation. This is done by first solving equation (2) for λ. This yields

$$\lambda = -\left(\frac{\partial F}{\partial X}\right)^{-T} \left(\frac{\partial I}{\partial X}\right)^T. \tag{4}$$

If this is substituted into equation (1), we obtain the following equivalent necessary conditions for optimality:

$$\frac{dI}{du} \equiv \frac{\partial I}{\partial u} - \left(\frac{\partial I}{\partial X}\right)\left(\frac{\partial F}{\partial X}\right)^{-1}\left(\frac{\partial F}{\partial u}\right) = 0 \tag{5}$$

$$F(X, u) = 0.$$

The quantity dI/du defined above is often called the reduced gradient. The necessary conditions can now be formulated either in terms of solving an adjoint problem (4) for λ or directly by solving the linearized state equations using multiple right hand sides to compute each column of the n by m matrix

$$Q \equiv \frac{dX_i}{du_j} = -\sum_{k=1}^{n} \left(\frac{\partial F_k}{\partial X_i}\right)^{-1} \left(\frac{\partial F_k}{\partial u_j}\right).$$

The latter approach is the one implemented in the design and optimization version of TRANAIR. A transpiration boundary condition (an inhomogeneous Neumann boundary condition) is used to approximate $\partial F/\partial u$ for those u's that represent surface geometry, as described in [16]. Equation (5) gives the gradient of I assuming F is held fixed. It does not, however, tell us how to balance decreasing the value of I against making F small. Such a balance can be derived by examining the classical Lagrange-Newton method [11]. Applying Newton's method to equations (1), (2), and (3), we arrive at the following equations for the updates δu to u, δX to X, and $\delta\lambda$ to λ:

$$\begin{bmatrix} L_{u,u} & L_{u,X} & F_u^T \\ L_{X,u} & L_{X,X} & F_X^T \\ F_u & F_X & 0 \end{bmatrix} \begin{bmatrix} \delta u \\ \delta X \\ \hat{\lambda} \end{bmatrix} = - \begin{bmatrix} I_u^T \\ I_X^T \\ F \end{bmatrix} \tag{6}$$

where $\hat{\lambda} = \lambda + \delta\lambda$ and we have used the subscript notation for partial derivatives, e.g., $F_{XX} = \partial^2 F/\partial X_j \partial X_l$. The block entries above can be readily expanded as $L_{u,u} = I_{u,u} + \lambda^T F_{u,u}$, $L_{u,X} = I_{u,X} + \lambda^T F_{u,X}$, etc. Block row operations on (6) yield the following equivalent system:

$$H\delta u = -G^T \tag{7}$$

$$F_u \delta u + F_X \delta X = -F(X, u) \tag{8}$$

$$\hat{\lambda} = F_X^{-T}[-I_X^T - L_{X,u}\delta u - L_{X,X}\delta X] \tag{9}$$

where

$$\begin{aligned} H = {} & I_{uu} - Q^T I_{Xu} - I_{uX}Q + Q^T I_{XX} Q \\ & + \lambda^T F_{uu} - Q^T \lambda^T F_{Xu} \\ & - \lambda^T F_{uX} Q + Q^T \lambda^T F_{XX} Q \end{aligned} \tag{10}$$

and

$$\begin{aligned} G^T = {} & (I_u - I_X F_X^{-1} F_u)^T - (I_{uX} + \lambda^T F_{uX}) F_X^{-1} F \\ & - F_u^T F_X^{-T} (I_{XX} + \lambda^T F_{XX}) F_X^{-1} F. \end{aligned} \tag{11}$$

The matrix $H = d^2 I/du_j du_i$ is often called the reduced Hessian and G is a modified form of the reduced gradient. The first term in the formula for G^T is the transpose of the reduced gradient given by equation (5) above. The last two terms can be neglected if $F_X^{-1}F$, i.e., the Newton step for the state equations, is small. In any case, dropping these terms reduces convergence to two step superlinear [11]. If H is positive definite, solving equation (7) is equivalent to minimizing the quadratic functional

$$QI(\delta u) = \frac{1}{2}(\delta u)^T H(\delta u) + G(\delta u) + I_0. \tag{12}$$

It is possible to impose any inequality or equality constraints involving either X and/or u when solving this quadratic program.

3.1 Block Lagrange Newton Method

One algorithm implementing the Lagrange Newton method is just a block solution method for equation (6). Given the current values for all the variables, first compute H and G from equations (10) and (11) using the current values of the Lagrange multipliers. The step in the design variables δu is obtained by minimizing QI as given by equation (12) subject to any inequality constraints. The step in the state variables is obtained by solving equation (8) for δX. Equation (9) can now be solved for $\hat{\lambda}$, the new value of the Lagrange multipliers. This completes the computation of the Newton direction. This block elimination method gives the same Newton direction as solving the large sparse system (6). Thus, the choice of whether to do the block elimination or solve the large system directly is purely a linear algebra question. The sparse approach has been advocated by Biros and Ghattas (this volume) as well as by Betts (this volume). In the sparse approach, the sensitivities are not required and the Hessian of the Lagrangian can be computed by second differences of L itself, for example. A preconditioning method for the full matrix based on a BFGS estimate of the reduced Hessian has been implemented by Biros and Ghattas [2]. This strategy will only succeed if the large linear system can be solved subject to any inequality constraints. In any Newton method for solving nonlinear systems of equations and in most optimization methods, some form of step length control is required to stabilize the algorithm. Stabilization strategies can range all the way from simple line searches to sophisticated trust region methods. These strategies are called globalizations and can have a dramatic impact on convergence as discussed in section 5.

3.2 Reduced Space Method

The Lagrange Newton method makes no assumptions about the residuals for the state equations except at convergence. This is tied up in the introduction and updating of the Lagrange multipliers. If the state equations are solvable for X for all reasonable values of u, one can implicitly eliminate the state variables from the formulation and merely minimize as a function of u the objective $I(X(u), u)$. In this formulation, the gradient is given by equation (5) and the Hessian is given by equation (10) where λ is given by equation (4). It is assumed that $F(X, u) = 0$, i.e., that X is always updated whenever a new value of u is considered by the optimization algorithm. This reduced space method is a nonlinear elimination method applied to the Lagrange Newton system of equations and its performance will be problem dependent as discussed below.

3.3 Lagrange Newton Method Without Multipliers

It is possible to construct a simplified Lagrange Newton method without explicitly computing or updating the Lagrange multipliers by using certain simplifying assumptions. First, assume that the sensitivities Q are available. Second, if F_{XX}, F_{Xu}, F_{uX}, and F_{uu} can be neglected, the Hessian matrix H can be computed without reference to λ. This would be the case, for example, if the state equations were linear or nearly linear in both the state and control variables. The second and third terms in equation (11) can be neglected degrading convergence from quadratic to two step superlinear. In any case, if our globalization always causes F to be near zero, the terms can be neglected. In either case, G does not explicitly require computation of λ. Now equations (7) and (8) constitute a Newton like method that does not require the explicit computation or updating of the Lagrange multipliers. In TRANAIR, we substitute a more global nonlinear program for the quadratic program and never explicitly compute H, allowing an optimization code to generate an approximation using BFGS. It is an open question whether these simplifying assumptions increase the importance of globalization. Certainly, without multiplier estimates, it is difficult to see how to implement the full Lagrange Newton method. This is a question that deserves further study. Since the question is independent of the linear algebra required to solve the Lagrange Newton system, the question is most naturally studied in the context of a reduced method as given by equations (7), (8), and (9). We suggest an advantage of the sensitivity form over an adjoint form of the reduced gradient in the next section.

4 TRANAIR Solution Adaptive Method

Currently, in TRANAIR a variant of the Lagrange Newton method without multipliers is implemented. If we let superscripts denote the grid number, this method can be described as follows:

1. Given the initial design variables u^1 and an initial guess for the state variables $\hat{X}^0$
2. For each grid $l = 1, 2, \dots, NG$:
   (a) Discretize the state equations (the flow equations) on grid l.
   (b) Solve this discrete problem $F^l(X^l, u^l) = 0$ approximately for X^l using initial values derived from $\hat{X}^{l-1}$ and calculate the sensitivities Q.
   (c) Solve the optimization subproblem discussed below for δu^l. Constraints on either flow or design variables can be applied at this stage.

(d) Determine the steplength ϵ^l using some globalization procedure and update the design variables, $u^{l+1} = u^l + \epsilon^l \delta u^l$.
(e) Update the flow variables by solving equation (8) assuming $F^l = 0$, i.e., $\hat{X}^l = X^l + \epsilon^l (F_X^l)^{-1} F_u^l \delta u^l$. The globalization procedures discussed below in section 5 may give different values of $\hat{X}^l$.
(f) Estimate the discretization error using $\hat{X}^l$.
(g) Use the error estimate to determine grid $l+1$ and go to (a) above.

3. Reloft the geometry using the final value of the design parameters u^{NG}. Determine the new values of the design variables u. If the final transpiration $T^{NG}(X^{NG}, u^{NG})$ is not sufficiently small or if the design space was changed, go to step 1 above.

The outer loop (step 3) enables convergence to the solution of the continuous design problem with some level of artificial dissipation. This outer loop involves relofting the geometry (actually moving the surface), doing a surface discretization (called paneling), and then redesigning until the transpiration error becomes insignificant. In our experience, convergence of this outer loop has taken at most two relofts of the geometry. Step 2 above is an inexact Newton method [7] for solving the necessary conditions for optimality on a fixed grid and allows the incorporation of solution adaptivity into the optimization process. Below we describe the various approximations used for the reduced gradient and Hessian.

Step 2(b) is not strictly needed for the solution of equations (7) and (8) but does enable dropping the $F(X, u)$ right hand side from equation (8) and dropping the last two terms in equation (11) for G. However, as pointed out above, if it is not carried out, it might be advisable to include Lagrange multiplier updates in the method and include the additional terms in our calculation of G and H. For highly nonlinear flow regimes, step 2(b) may be necessary in any case to prevent divergence as discussed in section 5. Step 2(c) above requires linearization of any flow or design constraints, $c(X, u)$. Assuming that the sensitivities dX_i/du_j are known, this is done using the reduced gradient formula given in equation (5):

$$\frac{dc}{du_j} = \frac{\partial c}{\partial u_j} + \sum_{k=1}^{n} \frac{\partial c}{\partial X_k} \frac{dX_k}{du_j} \tag{13}$$

Approximate Hessian information can be introduced in several ways. In the case of a least squares objective function (see section 5.4 below), the normal matrix involves only the sensitivities. Using this matrix to approximate the Hessian gives rise to a linear least squares problem. This method is called the Gauss-Newton method and details are given in [11, 16]. The resulting least squares problem is currently solved with the package LSSOL [13]. In the case of a non-least-squares objective function, the quadratic programming approach given by equations (7),(8), and (9) is not fully implemented.

Instead, an approximate optimization problem is solved which assumes a linearized flow velocity $\boldsymbol{v}$. We assume that I depends on the flow only through $\boldsymbol{v}$ and actually minimize the function J given by

$$J(u) = I\left(\boldsymbol{v}^0 + (u - u^0)\frac{d\boldsymbol{v}}{du}, u\right) \tag{14}$$

where $d\boldsymbol{v}/du$ is computed using equation (13). The gradient of J at u^0 is the same as the gradient of I. However, the Hessian of J is given by the first four terms of equation (10). The minimization is currently accomplished by using the optimization package NPSOL [14]. Since J is inexpensive to evaluate, gradients can be computed by finite differences. The result is that second order information corresponding to the first four terms of equation (10) is generated implicitly in a BFGS update strategy for which the evaluations are very inexpensive allowing a large number of iterations for each subproblem. The final active set and the BFGS approximation to the reduced Hessian for the subproblem can be saved to initialize the optimization on the next grid.

4.1 Algorithmic Issues

The storage for the sensitivities might seem at first glance to be prohibitive for large numbers of design variables. However, most aerodynamic optimization problems involve objective functions and constraints that only depend on the velocity at points on the surface of the configuration. We take advantage of this fact by storing $d\boldsymbol{v}/du$ only for these surface points. The computational cost of the sensitivities can be somewhat ameliorated by using parallelism and by the use of block GMRES [25].

We now comment on the adjoint formulation [17] for optimal control and compare it to the sensitivity method. The two methods arise from the observation that the matrix product given above in equation (5) for the reduced gradient can be computed in two ways. If the multipliers are computed first, the method is usually referred to as an adjoint method. Partial second order information is easily available in the sensitivity method in the form of the first 4 terms of equation (10). In the adjoint method as usually implemented, no second order information is immediately availability and some acceleration strategy such as BFGS is usually relied on to improve convergence over the rate of steepest descent. In fact, if a full Newton method is desired, the apparent cost advantage of the adjoint method disappears since the reduced Hessian H requires both λ and Q. Further, with the sensitivity method, linearizations of all constraints are available immediately.

We have observed reasonably good convergence of our optimal control method especially for least squares objectives. This is probably due to the implicit use of the partial second order information available in a sensitivity method. In engineering practice, TRANAIR design is usually run for 5 to 10 design grids (each with an optimization subproblem). Because of efficient generation of sensitivities, the total CPU cost for up to 400 design variables in three space dimensions is usually less than the cost of 20 analyses. This is at least 2 orders of magnitude less than comparable black box methods.

However, there are still questions about the convergence of the method particularly in challenging cases. It would be of particular interest to determine the effect on convergence of including the curvature of the state equations (the last four terms in the formula for H above) and/or multiplier estimates in the formulation of the optimization subproblem.

A further interesting question concerns the comparison between the reduced space method and the block Lagrange Newton method. In the latter, the Lagrange multipliers must be explicitly computed at each step and there is controversy about whether overall convergence is enhanced by allowing "cutting across the design space."

4.2 Inexactness

The method outlined above with solution adaptive gridding and sensitivities is reliable in the presence of significant inexactness. This is due to the formulation as an inexact Newton method in which each optimization subproblem is consistent to machine precision. In fact, our sensitivities are computed using a block GMRES method in which the residual is only reduced by a factor of 20. This is exactly the convergence criterion used in the analysis version of the code. In running thousands of cases, we have been unable to find a single instance where this level of inexactness has caused a failure to converge either an analysis problem or a design optimization problem. It is much more difficult to quantify the effect of inexactness in methods such as the adjoint method or the nonlinear conjugate gradient method where a consistent subproblem is never defined.

5 Globalization

As mentioned above, the difference between what is usually called the SAND [15] (or the all at once [6] or the Lagrange Newton method [11]) and the NAND (or the discipline feasible or the reduced sequential quadratic programming (RSQP) or the reduced space method) is really globalization. An excellent discussion of many of the algorithmic issues involved is the work of Biros and Ghattas [2]. Typically, the NAND method merely enforces the state equation constraint at every optimization iteration. Complex optimization methods blur this distinction even further. In reality, there is a spectrum of methods that is liable to defy exact classification in any practical implementation [29]. In order to motivate our discussion of globalization methods in optimization, we first examine the role of globalization of Newton's method in compressible viscous fluid dynamics.

5.1 Nonlinear Elimination

When solving nonlinear systems of equations such as those arising in computational fluid dynamics, convergence can in some cases be dramatically accelerated by using the nonlinear elimination method. This method is classical but was analyzed in [20]. Suppose we want to solve the system of nonlinear equations

$$F(X) = 0 \tag{15}$$

and the system can be divided into two sets of equations F_1 and F_2 one of which might be much more linear than the other. Let us further suppose that the variables can be divided into 2 subsets X_1 and X_2 such that $F_2(X_1, X_2) = 0$ is solvable for X_2 given X_1. In this case, we can define the nonlinear function $h(X_1)$ by this latter solution X_2. We can now define a nonlinear system equivalent to (15) by

$$f(X_1) = F_1(X_1, h(X_1)) = 0. \tag{16}$$

Once (16) is solved for X_1, $F_2(X_1, X_2) = 0$ must be solved for X_2. If Newton's method is applied to (16), the Jacobian matrix is $\frac{df}{dX_1} = \frac{\partial F_1}{\partial X_1} - \frac{\partial F_1}{\partial X_2}(\frac{\partial F_2}{\partial X_2})^{-1}\frac{\partial F_2}{\partial X_1}$ and the linear Newton system to be solved would be $\frac{df}{dX_1}\delta X_1 = -f$. As can be seen by using block elimination on equation (17), the resulting δX_1 is exactly the δX_1 that would result if the Newton step for the entire system is computed by solving

$$\begin{bmatrix} \frac{\partial F_1}{\partial X_1} & \frac{\partial F_1}{\partial X_2} \\ \frac{\partial F_2}{\partial X_1} & \frac{\partial F_2}{\partial X_2} \end{bmatrix} \begin{bmatrix} \delta X_1 \\ \delta X_2 \end{bmatrix} = - \begin{bmatrix} F_1(X) \\ F_2(X) \end{bmatrix} \tag{17}$$

Thus, the Newton direction for the reduced system (16) is just the X_1 component of the Newton direction for the full system (15). However, the globalization for the two systems is different. Given the current iterate X_1^n, a standard line search for (16) would be to choose the step length ϵ to minimize $\parallel F_1(X_1^{n+1}, X_2^{n+1}) \parallel$ where $X_1^{n+1} = X_1^n + \epsilon\delta X_1^n$ and $X_2^{n+1} = h(X_1^{n+1})$ is determined by solving exactly $F_2(X_1^n + \epsilon\delta X_1^n, X_2^{n+1}) = 0$.

The maximum computational benefit of the method is achieved when F_2 is much more nonlinear than F_1, the dimension of X_2 is small compared to X, and $F_2(X_1, X_2) = 0$ is relatively easy to solve for X_2 because of special features. An example of the last characteristic is solving the boundary layer equations given the velocity at the edge of the layer. These equations are parabolic in the streamwise direction and hence can be solved rather easily by space marching. They are also very nonlinear especially near separation. Another example of this type of situation is the compressible fluid flow equations near a strong shock wave. This is the situation discussed at length below for a simple one dimensional converging diverging nozzle problem.

5.2 Nonlinear Elimination Applied to Nozzle Flow

Differential Equation Consider a one-dimensional nozzle problem described by the differential equation

$$[A(x)\rho(u)u]_x = 0 \tag{18}$$

where $0 \le x \le L$. $A(x)$ is the cross sectional area of the nozzle (assumed to be converging/diverging), $u(x)$ is the velocity, and $\rho(u)$ is the density. This single equation embodies conservation of mass in the nozzle. We assume that the velocity is given as the gradient of a scalar potential, i.e., $u = \phi_x$ and that the density is given by the isentropic formula $\rho = \left[1 + \frac{1}{2}(\gamma - 1)(1 - u^2)\right]^{\frac{1}{\gamma-1}}$ where $\gamma = 1.4$. The boundary conditions are that $\phi(0) = 0$ and $\phi(L) = K$ where K is given. A typical nozzle area distribution is $A(x) = 0.6(x-1)^2+0.4$ where $L = 2.0$. In this formulation, we have the following identities and definitions for the pressure p, the local speed of sound c, and the local Mach number $\mathcal{M}$: $p = \rho^\gamma/\gamma$, $c^2 = \gamma p/\rho = \rho^{\gamma-1}$, and $\mathcal{M} = u/c$.

A small value of K will give a totally subsonic and symmetric (about the throat) velocity. In this formulation, $\rho = c = 1$ when $u = 1$, $u < 1$ corresponds to subsonic flow, and $u > 1$ corresponds to supersonic flow. A simple calculation shows that equation (18) is equivalent to

$$A_x \rho u - A\rho u_x(1 - \mathcal{M}^2) = 0$$

At the throat, $A_x = 0$ so that either $u_x = 0$ or $\mathcal{M} = 1$. In subsonic flow, it turns out that $u_x = 0$ at the throat whereas in transonic flow, $\mathcal{M} = 1$ at the throat. For K sufficiently large, the flow will be subsonic in the converging part of the nozzle, sonic at the throat, and supersonic downstream of the throat. The supersonic zone terminates in a shock wave followed by subsonic flow to the exit. The larger K is, the stronger the shock will be.

Discretization If we consider a mesh of points $x_i, i = 1 \ldots NX$ and let $h_i = x_{i+1} - x_i$, $x_{i+\frac{1}{2}} = 0.5(x_{i+1} + x_i)$, $u_{i+\frac{1}{2}} = (\phi(x_{i+1}) - \phi(x_i))/(x_{i+1} - x_i)$, and $\rho_{i+\frac{1}{2}} = \rho(u_{i+\frac{1}{2}})$, a typical centered difference formula would be

$$\frac{A(x_{i+\frac{1}{2}})\rho_{i+\frac{1}{2}}u_{i+\frac{1}{2}} - A(x_{i-\frac{1}{2}})\rho_{i-\frac{1}{2}}u_{i-\frac{1}{2}}}{0.5(h_{i-1} + h_i)} = 0. \tag{19}$$

The boundary conditions are $\phi(x_1) = 0$ and $\phi(x_{NX}) = K$.

First Order Density Biasing A standard form of stabilization (which is required in regions of supersonic flow to rule out expansion shocks) is upwind biasing of the density. We define the switching function μ by

$$\mu = \text{MAX}\left[0, 1 - \frac{\mathcal{M}_c^2}{\mathcal{M}^2}\right]$$

where $\mathcal{M}_c$ is the cutoff Mach number. We take $\mathcal{M}_c^2 = 0.95$. In order to guarantee stability of the scheme the upwinding must be turned on at the foot of the shock where the flow is subsonic. To accomplish this, several iterations of "expanding μ" are applied. In each such iteration, we set $\mu_{i+\frac{1}{2}} = \mathrm{MAX}[\mu_{i-\frac{1}{2}}, \mu_{i+\frac{1}{2}}, \mu_{i+\frac{3}{2}}]$.

In the difference formula (19), stabilization is achieved by replacing $\rho_{i+\frac{1}{2}}$ by

$$\tilde{\rho}_{i+\frac{1}{2}} = \rho_{i+\frac{1}{2}} - \mu_{i+\frac{1}{2}}(\rho_{i+\frac{1}{2}} - \rho_{i-\frac{1}{2}})$$

Mass Flux Biasing It has been found that sonic lines are more accurately resolved by using upwind biasing of the mass flux to stabilize the finite difference scheme rather than density biasing. In this formulation, the switching function is not used. Instead we let ρ^* be the sonic density (in this case assumed to be 1.0). The sonic value of the velocity, u^* is also 1.0. In the difference formula (19), stabilization is achieved by replacing $\rho_{i+\frac{1}{2}}$ by $\tilde{\rho}_{i+\frac{1}{2}}$ given by

$$\tilde{\rho}_{i+\frac{1}{2}} = \rho_{i+\frac{1}{2}} - \frac{1}{u_{i+\frac{1}{2}}}\left[\overline{\rho_{i+\frac{1}{2}}u_{i+\frac{1}{2}}} - \overline{\rho_{i-\frac{1}{2}}u_{i-\frac{1}{2}}}\right] \tag{20}$$

where

$$\overline{\rho_{i+\frac{1}{2}}u_{i+\frac{1}{2}}} = \begin{cases} 0 & \text{if } \mathcal{M}_{i+\frac{1}{2}} < 1.0 \\ \rho_{i+\frac{1}{2}}u_{i+\frac{1}{2}} - \rho^* u^* & \text{if } \mathcal{M}_{i+\frac{1}{2}} \geq 1.0 \end{cases} \tag{21}$$

Computational Results All computations reported below were performed with first order mass flux biasing. Similar asymptotic results are found with first order density biasing except that the numbers of Newton steps are uniformly smaller. A complete Newton method was implemented with analytic computation of the Jacobian matrix. Figure 1 shows the geometry of the nozzle and the solution obtained on a very fine grid. The value of K used is 1.15, resulting in a shock wave of moderate strength.

Figure 2 shows the convergence of Newton's method for this problem on a series of uniform grids. Globalization was done using a non-derivative line search that minimized the $L2$ norm of the residual of (19). The line search routine has a tolerance that controls to what level of error the minimum of the norm of the residual is sought. A value of 10^{-6} was used in these calculations corresponding to a very accurate line search. As can be seen from the second panel of the figure, the step lengths were very close to zero except near convergence.

In figure 3, we show the results of applying the nonlinear elimination method to this problem. The eliminated equations and unknowns correspond to a subdomain containing roughly 1/8th of all the grid points in the problem centered on the ultimate location of the shock wave. As can be seen from the

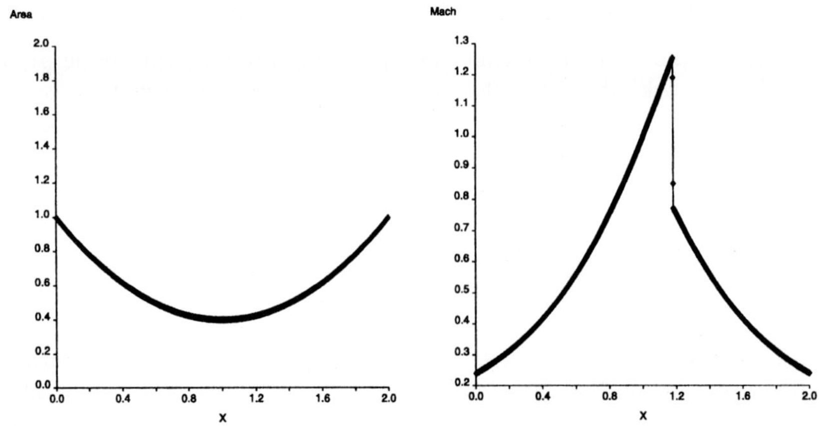

Fig. 1. Model converging diverging nozzle problem.

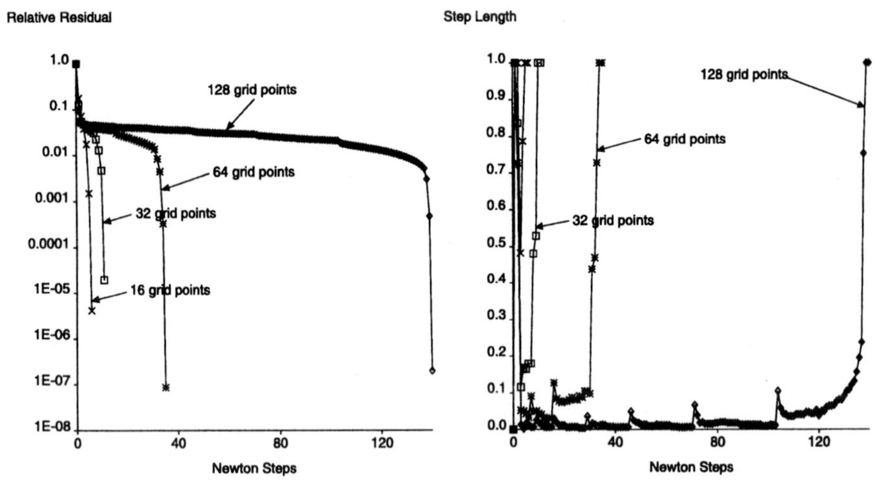

Fig. 2. Newton convergence for model problem with standard globalization.

figure, the convergence of Newton's method is dramatically improved and becomes almost independent of the density of the grid. For this problem, the nonlinearity is almost all concentrated near the shock wave. This is obvious from a consideration of the nature of equations (19), (20), and (21). With this setup, the size of the eliminated system grows linearly with the size of the original problem. The cost of the line search is increased in the nonlinear elimination method over this cost in the standard Newton method because the latter only involves computing the residual whereas the former involves solving nonlinear subproblems over the eliminated subdomain. However, total CPU time is dramatically lower with the elimination method. One effective way to control the computational cost for the nonlinear solves needed in the globalization is to use continuation in the step length ϵ. This strategy can reduce the computational cost by an order or magnitude.

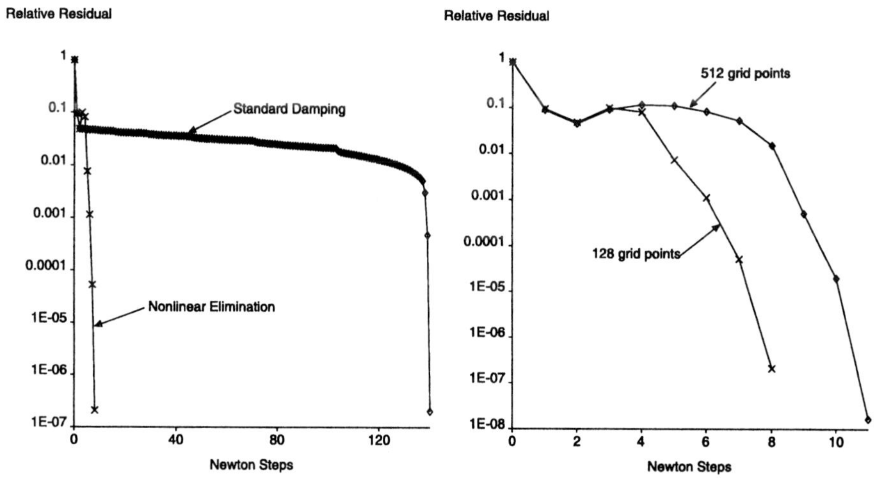

Fig. 3. Newton convergence with nonlinear elimination of the shock region.

We note that for the case of shock waves, there is usually sufficient regularity with respect to grid density to allow grid sequencing [4, 26] to work effectively. However, for second order dissipation methods, this regularity is not as reliable as would be desired. In the case of boundary layers near separation, grid regularity is poor and nonlinear elimination seems to be essential to achieving robust convergence [1, 8, 10].

As the preceding example demonstrates, it is not necessarily true that introducing additional variables will make a nonlinear system easier to solve using Newton's method. Further, the difference between Newton's method applied to the full system and Newton's method applied to the reduced system is purely a difference in globalization. An interesting extension of these ideas has been proposed by Cai and Keyes [5] in which Newton's method is nonlinearly preconditioned by a nonlinear Schwarz method. This method has also been successfully applied to the nozzle problem discussed here with similar acceleration of convergence.

5.3 Globalization Methods Implemented in TRANAIR

All Newton based methods either for analysis or optimization require some kind of globalization to prevent divergence in the case of a poor initial approximation to the solution. There is a plethora of such methods in the literature, many of which are classical such as the Levenberg Marquardt method or the use of an augmented Lagrangian merit function. In our optimization method, step size control is sometimes achieved through user specified constraints that bound the change in some aerodynamic quantity. In design as in analysis, changes in local Mach number have proven to be effective in some cases even though the bounds required are often problem dependent. There is some reason based on experience in analysis cases to believe that convergence would be enhanced by some type of more automatic step size control for the step in the design variables once the direction is determined by solving the optimization subproblem. The methods described below are rather obviously motivated by nonlinear elimination applied to the Lagrange Newton method [29]. To take into account the inequality constraints we introduce a merit function Φ_M [11, 12]. A popular choice is based on the augmented Lagrangian. We do not have a method to easily apply nonlinear elimination to these constraints with the exception of certain constraints on coefficients of lift and pitching moment which can be enforced exactly in the state equation solve.

Globalization Method #1

Choose ϵ^l to minimize $I(X^l + \epsilon^l Q\delta u^l, u^l + \epsilon^l \delta u^l) + \Phi_M$

Based on the observation that a poor initial guess for the state equations can result in stagnation of even a globalized Newton's method and that this situation was often encountered in step 2(b) of the solution adaptive algorithm discussed in section 4 when using Method #1, a globalization method that maintains the satisfaction of the state equations to some level of accuracy for each change in the design variables was implemented.

Globalization Method #2

Choose ϵ^l to minimize $I(\hat{X}^l, u^l+\epsilon^l\delta u^l)+\Phi_M$, where $\hat{X}^l$ is determined by solving to some level of accuracy the state equation $F(\hat{X}^l, u^l + \epsilon^l\delta u^l) = 0$. Set $X^{l+1} = \hat{X}^l$ and $u^{l+1} = u^l + \epsilon^l\delta u^l$.

An alternative is to partitioning the state equations into two subsets F_1 and F_2 and the state variables into corresponding subsets X_1 and X_2 such that given X_1 one can solve F_2 for X_2. Then one can apply nonlinear elimination to F_2. One obvious choice is to take F_2 to be the boundary layer equations and X_2 the boundary layer variables.

Globalization Method #3

Choose ϵ^l to minimize $I(\hat{X}_2^l, X_1^l + \epsilon^l\delta X_1^l, u^l + \epsilon^l\delta u^l) + \Phi_M$, where $\hat{X}_2^l$ is determining by solving to some level of accuracy the state equation $F_2(\hat{X}_2^l, X_1^l + \epsilon^l\delta X_1^l, u^l + \epsilon^l\delta u^l) = 0$ and δX_1^l is determined by restricting $Q\delta u^l$. Set $X_2^{l+1} = \hat{X}_2^l$, $X_1^{l+1} = X_1^l + \epsilon^l\delta X_1^l$, and $u^{l+1} = u^l + \epsilon^l\delta u^l$.

The computational cost of globalization method #2 can be dramatically reduced by using continuation in ϵ to generate initial guesses for the state equation solves. In many cases, the initial guess generated by this continuation strategy is good enough that only a single Newton step is required for adequate convergence for a new value of ϵ. We can usually use the same preconditioner for all values of ϵ. Using these strategies, the cost of the globalization can be reduced to that of one or two nonlinear state equation solves on a comparable grid.

Generally, one of the salient features of compressible viscous flows is the fact that the region of attraction for Newton's method is very small. In fact, for the case of strong shocks and also for boundary layer coupling, careful attention to globalization strategies is absolutely essential to achieving the kind of reliable convergence required in the industrial engineering environment. For shock waves [26], either grid continuation, continuation in artificial viscosity, or nonlinear elimination is effective. For boundary layer coupling [1, 8], nonlinear elimination has proven the most satisfactory strategy especially near or beyond separation. This is probably due to the bifurcations known to be present in nonlinear systems representing such flow fields. Another continuation method for viscous flows that has been unexplored by the authors is continuation in Reynolds number. It is also true that generally aerodynamic optimization problems are not well posed without regularization which adds another element of difficulty. Since there are an essentially infinite number of ways to pose a given engineering optimization problem, it not always easy to determine which way is most amenable to numerical optimization or how to regularize the problem if it is ill-posed.

5.4 Computational Examples in Aerodynamic Design

In this section, we discuss the results of applying these globalization methods to four nonlinear least squares problems. One of the motivations for the introduction of these methods was the possibility that a design step could result in a sufficiently large change in the geometry that on the next grid the analysis problem nonlinear solution process would not converge. In our experience, failure to converge on a given grid is rarely improved by going on to a new grid. This did in fact happen in several cases examined here when using globalization #1. In some cases, this can be mitigated by user specified constraints. Unfortunately, specification of these constraints is problem dependent.

We note that in this case of a least squares objective, the method employed in TRANAIR is the Gauss Newton method. In the case that the residuals are nearly zero at the solution, we should expect the rapid convergence characteristic of Newton's method from a properly formulated sensitivity method. Thus, this type of test case is in some sense less challenging than the problem of minimizing a scalar objective. Suppose the cost function I is given by

$$I = \frac{1}{2}\sum_{l=1}^{q} c_l^2.$$

In this case, the reduced gradient and reduced Hessian are given by

$$\frac{dI}{du_j} = \sum_{l=1}^{q} c_l \frac{dc_l}{du_j}, \qquad j = 1, 2, ..., m,$$

$$\frac{d^2 I}{du_j du_i} = \sum_{l=1}^{q} \frac{dc_l}{du_i}\frac{dc_l}{du_j} + \sum_{l=1}^{q} c_l \frac{d^2 c_l}{du_j du_i}, \qquad i, j = 1, 2, ..., m. \tag{22}$$

If the c_l are small, the second term in equation (22) for the Hessian can be neglected and a good approximation to the reduced Hessian is given only in terms of the sensitivities

$$R_{l,j} = \frac{dc_l}{du_j} = \frac{\partial c_l}{\partial u_j} + \sum_{k=1}^{n} \frac{\partial c_l}{\partial X_k}\frac{dX_k}{du_j}$$

We can now write equation (7) in the form $R^T R\delta u = -R^T c$. This is equivalent to the least squares problem MIN $||R\delta u + c||_2$ which can be solved by applying a constrained linear least squares solver.

The test cases discussed in this section are pressure matching cases involving a target pressure distribution obtained by analyzing an airfoil section similar to the ONERA M6 section. The initial airfoil was the NACA0012. The objective function is

$$I = \frac{1}{2}\sum_{l=1}^{q} W_k[(c_p)_l - (\hat{c}_p)_l]^2$$

with the weights W_l taken to be the surface panel length for the surface point in question. The pressure coefficients on the airfoil are $(c_p)_l$ and $(\hat{c}_p)_l$ are the target pressure coefficients. The first case used the flow conditions $M_\infty = 0.75$ and $\alpha = 1.0$. Figure 4 shows the initial geometry, the target geometry, and also the initial pressure distribution (on a fine grid) and the target pressure distribution.

In these cases a large number of grids was used to examine asymptotic convergence issues. For this case the grid was frozen after grid 15. The gridding is solution adaptive and the numbers of finite elements in the grids were 198, 397, 731, 1399, 2650, 5549, 6388, 7355, 8478, 9234, 10606, 12246, 13140, 14836, and 15288. This final grid is repeated a number of times. The case was run with globalization methods #1 and #2. In the inviscid case, there is little difference between the globalizations because the step size selected is almost always unity. There are some grids for which globalization #1 seems to suffer

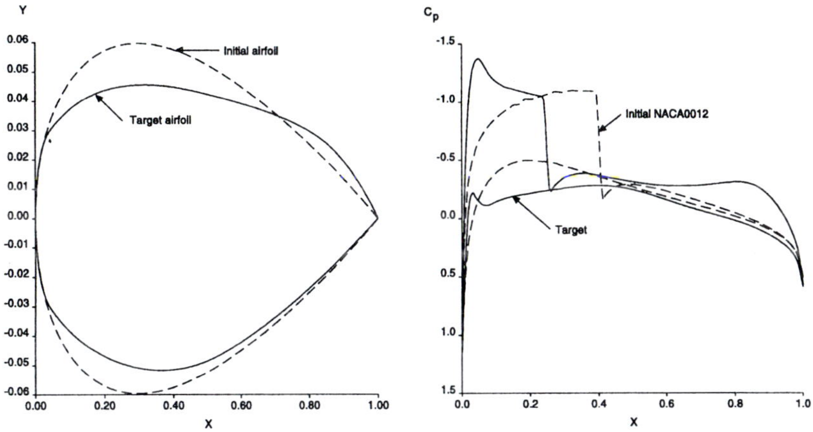

Fig. 4. Geometry and solution for Design Least Squares Case.

from a poor step where the agreement between the sensitivity prediction and the actual result is poor, but they are infrequent.

Figure 5 shows the convergence history for this case. Both the objective function and the estimated reduced gradient are shown as functions of the grid number. There are two values of the objective shown for each grid. The first results from the analysis of a fixed transpiration due to the initial values of the design variables on that grid. The second value is the best estimate of the objective after the updating of the design and flow variables on that grid. Severe oscillation indicates that the estimated values of the objective are seriously inaccurate. This can be due to a design step that is beyond the range of validity of the linearized model of the state variables used in the optimization subproblem. On early grids, it can result from grid to grid changes in the solution due to solution adaptive grid refinement. It should be noted that finite differences are used to linearize various nonlinear components of the objective and state equations. Because of this fact, it is not expected that the reduced gradient can be reduced by a factor greater than the square root of machine precision (roughly 10^{-7} in this case) and this is born out by the computational results. The vertical line at grid 15 indicates the point at which the grid is frozen. The fixed grid results are for convenience shown starting at this point.

In these cases, solution adaptive gridding introduces a possible difficulty that might place a strain on global convergence because the target pressure distribution was taken from an analysis on a very fine grid and thus matching this distribution exactly on a coarse grid would result in the "wrong" geometry. To eliminate this effect as well as to dispense with the effect of potentially large changes in the grid in the solution adaptive case, another test was run. First a good solution adaptive grid was determined for the target airfoil and pressure distribution in an initial TRANAIR run. Then the

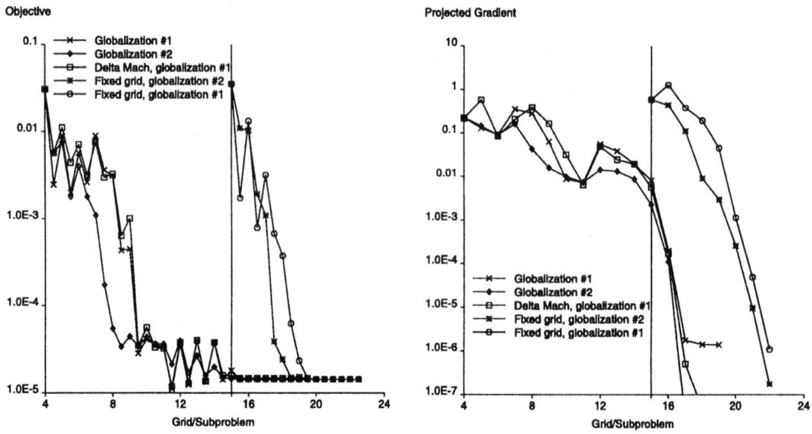

Fig. 5. Convergence of Inviscid Design Least Squares Case with Various Globalization Options.

design was run from the initial NACA0012 geometry with each grid being this solution adaptive grid. In this case, as can be seen in Figure 5, both globalization methods converged even though method #2 was a little faster.

A second test case illustrates the globalization methods as applied to a viscous flow case formulated using the same airfoils with a Reynolds number of 6.5 million. The C_l was specified to be the same as the inviscid case, namely, 0.3655. The pressure distributions for the initial airfoil and the target are shown in Figure 6. In this case, the effect of globalization method #2 is very significant. Using globalization method #1, convergence is achieved only by using constraints on the change in local Mach number in the optimization subproblems and line search. The value chosen was 0.2 based on experience in analysis cases. However, the final value of the objective is not quite as good as is achieved with globalization method #2. The convergence histories are given in figure 7. Using globalization method #1, the analysis portion of the code cannot converge on grid 5 with the current design variables. With globalization method #2, the optimization converges rather rapidly, the result being good even before the grid is frozen. Also in the fixed grid test, there is a significant advantage for globalization method #2. This is due to the better estimation of the true solution on the next grid taking into account the optimization step.

Another more challenging set of flow conditions was also run. The target pressure distribution was obtained by running the target airfoil at $M_\infty = 0.75$ and $\alpha = 2.0$. The resulting pressure distribution is shown in Figure 6 for both the inviscid and viscous cases. In this case, both globalizations perform well. The convergence histories are shown in figure 8.

This more challenging case was also run in viscous flow with a Reynolds number of 6.5 million. The C_l was once again set to match the inviscid case.

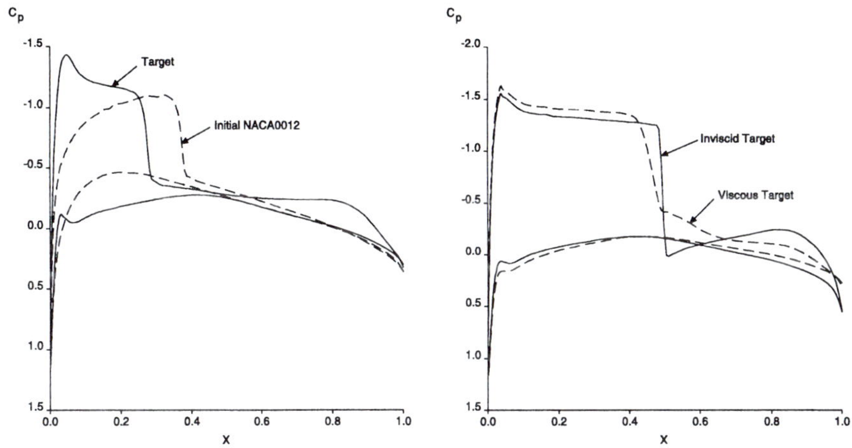

Fig. 6. Target Pressure Distributions for Viscous Case and for Challenging Case

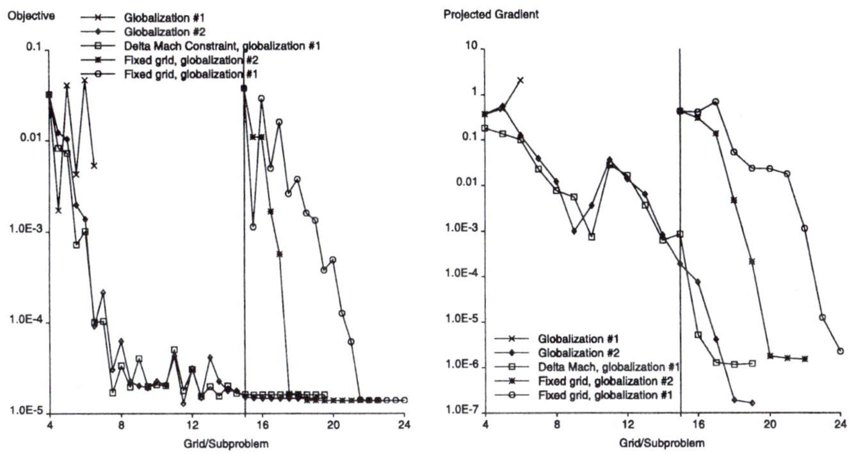

Fig. 7. Convergence of Viscous Design Least Squares Case with Various Globalization Options.

This case has shock induced separation of the boundary layer. The convergence histories are shown in figure 9. In this case, there are significant convergence difficulties and even method #2 has some difficulty matching the pressure distribution in the separated flow region. Globalization method #1 even with a constraint on the change in Mach number fails to sufficiently reduce the residuals of the analysis problem on grid #5. However, the results obtained with method #2 are acceptable from an engineering point of view.

All of the cases in this section were run with first order density biasing, the least nonlinear form of upwinding available in TRANAIR. With more

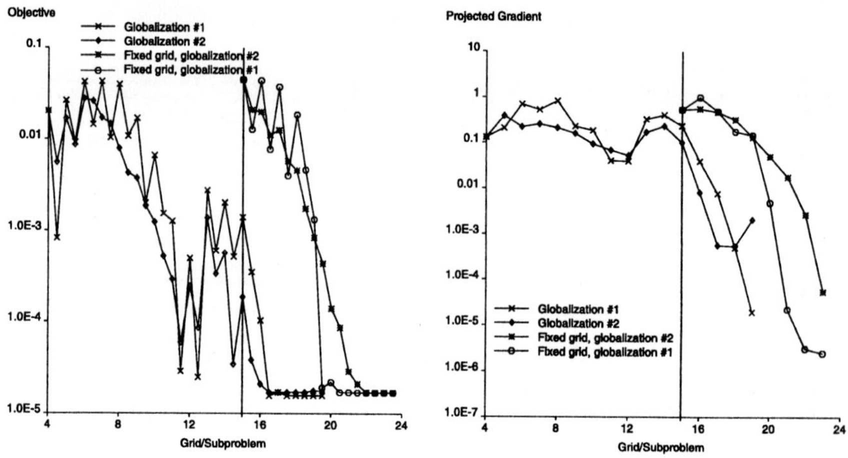

Fig. 8. Convergence of Challenging Design Least Squares Case with Various Globalization Options.

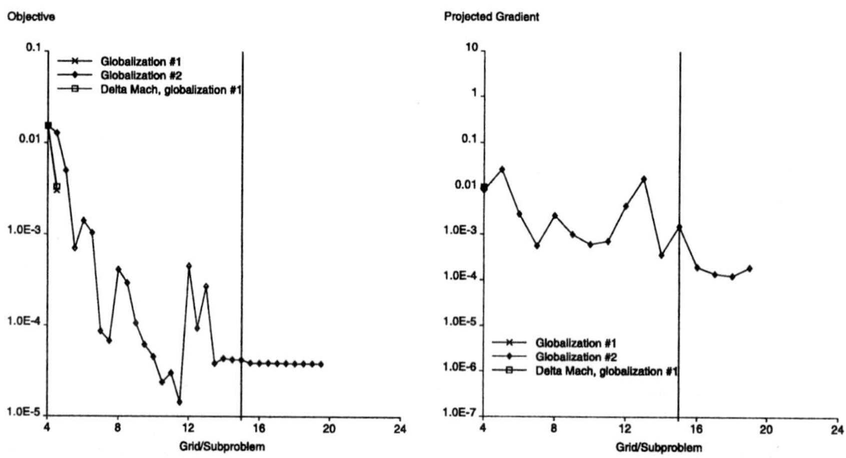

Fig. 9. Convergence of Separated Viscous Design Least Squares Case with Various Globalization Options.

"accurate" forms of upwinding, less reliable results can be expected. Finally, we have noted that often in cases of drag minimization such as discussed in section 8, globalization method #2 can result in designs for which the objective is somewhat lower and for which the inequality constraints are more nearly satisfied than globalization method #1. This may seem to be a minor matter, but when coupled with more robust convergence, this becomes important in multi-point design.

6 Multi-point Design

The TRANAIR design and optimization method described in section 4 has been extended to optimization at multiple flow conditions. Multi-point design is desirable for at least two reasons. First, it is obvious that a single point design has no control of the performance at other conditions (except through constraints that may be *ad hoc* or difficult to formulate) and in practice it often leads to poor off-design performance. Second, there is hope that multi-point design will be more stable than single point design; for example, in minimizing drag for an airfoil there are often many shock free single point airfoil designs none of which is viable from an engineering point of view.

The algorithm for multi-point design is very similar to that described in section 4. Step 1 is to give each flow condition solver the initial geometry (i.e., values for the design variables). Steps 2(a) and 2(b) can proceed in parallel and are completely independent for different flow conditions. Steps 2(c) and 2(d) require an optimization that must get input from each of the flow conditions. The objective function is assembled, and the constraint and sensitivity information combined to form one large optimization problem. The resulting updates to the design variables are passed back to each of the flow conditions and steps 2(e), 2(f), and 2(g) can again proceed independently for each flow condition. Details of the method and some simple computational examples can be found in [21].

With regard to globalization, in a multi-point design, lack of convergence of any analysis/sensitivity run on any grid for any flight condition brings the whole process to a halt. As is often the case in complex engineering processes, the overall probability of success is the manifold product of the probability of success on any analysis case. Thus, improvements in robustness tend to be given very high priority by the industrial engineering community. Similarly, many multi-point designs boil down to trading drag at different operation conditions against each other and against many inequality constraints. These trades are often sufficiently sensitive that a merit function based globalization process such as #1 results in less than satisfying results. Globalization method #2 can more accurately do the required trades.

7 Limitations and Remaining Difficulties

In this section, we will indicate the remaining limitations of aerodynamic design optimization by considering the problem of single point drag minimization for an airfoil in transonic flow. This problem is not well-posed in the classical sense because there may be multiple solutions. However, consideration of this problem illustrates the difficulties often encountered in practice in aerodynamic optimization.

In our test case, the drag to be minimized is the sum of the wave and profile drag. The design variables consist of 9 camber and 11 thickness modes.

The thickness knots are located at 0,10,20,30,40,50,60,70,80,90, and 100% of chord. The thickness modes at 0 and 100% allow changes in the slope of thickness so that the trailing edge closure angle and the leading edge radius of curvature can change. The camber knots are located at 10,20,30,40,50,60,70,80, and 90% of chord. Values of camber and thickness are determined by a special spline fit to the values at the knots. The angle of attack is also a design variable. In figure 10, we show the initial airfoil and the final airfoil achieved with globalization methods #1 and #2. In table 1, we give the final value of C_d which is the objective function. Globalization #2 results in a somewhat lower level of the objective function. It should be noted that in this case, NPSOL was almost never able to achieve convergence for the optimization subproblems. The typical return condition was that the current solution cannot be improved and the estimated condition number of the Hessian typically became very large, roughly the reciprocal of machine precision, about 10^{14}. The estimated reduced gradient was never reduced more than a single digit indicating a remaining undiagnosed difficulty.

Table 1. Aerodynamic Forces for Various Optimization Options.

Case	C_d	Wave C_d	Profile C_d	C_l
baseline NACA0012	246.98	125.35	121.63	0.550
Camber/thickness, globalization #1	91.57	0.75	90.82	0.550
Camber/thickness, globalization #2	90.30	0.00	90.30	0.550
Curvature variables, globalization #2	90.92	0.14	90.78	0.550
Monotone curvature, globalization #2	91.09	0.08	91.01	0.550

One form of regularization involves restricting the design geometry space to "smooth" airfoil shapes. We were able to test the effect on optimization performance by using design variables that parameterized directly the curvature of the airfoil. Once again there were 20 variables describing the curvature and angle of attack was also a design variable. The variables represented coefficients for a Legrendre polynomial expansion of the curvature. The square root of arc length from the leading edge of the airfoil was used as the independent variable for these polynomials and globalization method #2 was used. As can be seen from figure 10, the resulting airfoil is much smoother even though the drag is somewhat higher. In this case, NPSOL was able to find optimal solutions for most of the subproblems and the estimated Hessian condition numbers were between 10^3 and 10^6 indicating that the optimization problem is better posed. In Figure 11, the curvatures of the optimized airfoils are shown. The airfoils optimized using splines show very rough curvature distributions while those design with curvature variables show smoother distributions. Also shown is the effect of imposing monotonicity on the curvature on a large portion of the upper and lower surfaces. As can be seen from table 1, there is a small objective function increase associated with these additional inequality constraints.

As mentioned above, an outstanding question is whether adding more state equation curvature information to the optimization subproblems would produce more reliable convergence. If the last 4 terms in equation (10) were added to the optimization subproblem, or if the simple quadratic program given by equation (12) were formed and solved, the result would be a true Lagrange Newton method and global convergence might be accelerated. Certainly, this issue is one of the most potentially important algorithmic issues

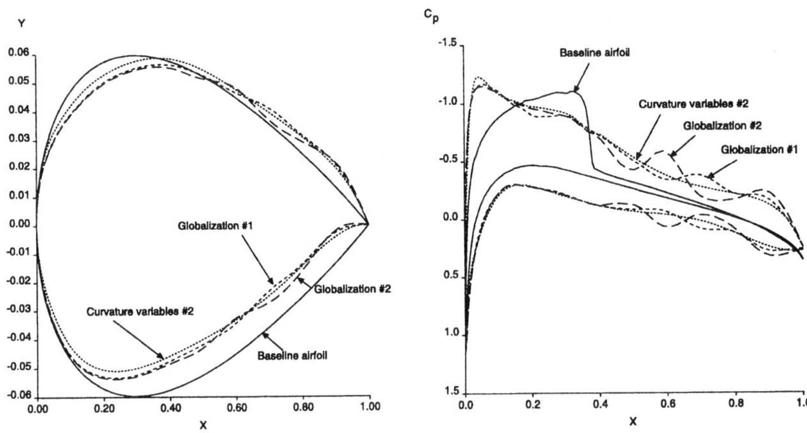

Fig. 10. Optimized Airfoils and Pressures Distributions with Various Optimization Methodologies

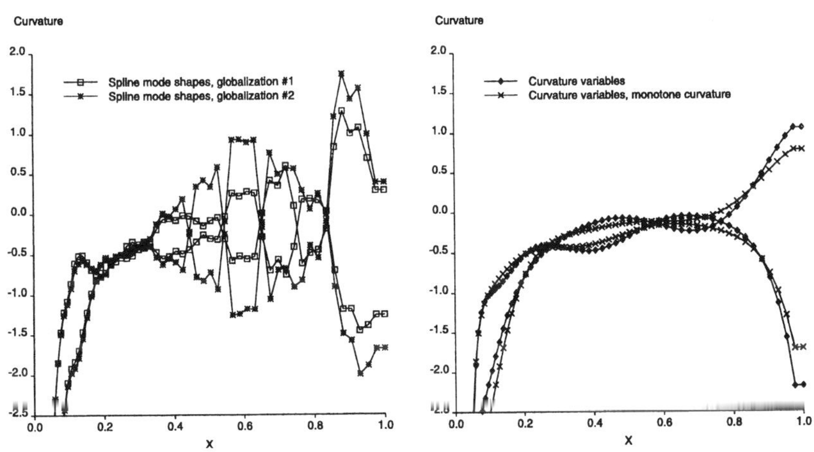

Fig. 11. Curvature Distributions of Optimized Airfoils and the Effect of Curvature Monotonicity Constraints

left unaddressed by our work so far even though the computational cost and programming issues involved in examining it systematically for the case of viscous compressible flow are formidable.

8 Summary

We have improved the single point design and optimization methodology in TRANAIR to include a range of globalization strategies and extended it to a multi-point design capability. Both enhancements improve the robustness and usefulness of the capability. We have offered a rationale for preferring the sensitivity method to the adjoint method in the context of a reduced gradient method by showing that some second order information is inexpensively available in the sensitivity method. We have discussed the effect in practice of staying relatively close to the manifold of solutions to the state equations which in this case are highly nonlinear.

References

1. M. B. Bieterman, R. G. Melvin, F. T. Johnson, J. E. Bussoletti, D. P. Young, W. P. Huffman, C. L. Hilmes, and M. Drela. Boundary Layer Coupling in a General Configuration Full Potential Code. Technical Report BCSTECH-94-032, Boeing Computer Services, 1994.
2. G. Biros and O. Ghattas. Parallel Lagrange-Newton-Krylov-Schur Methods for PDE-Constrained Optimization. Part I: The Krylov-Schur Solver. Technical Report, Laboratory for Mechanics, Algorithms, and Computing, Carnegie Mellon University, 2000.
3. A. E. Bryson and Y. C. Ho. *Applied Optimal Control: Optimization, Estimation, and Control.* Hemisphere, New York, 1975.
4. X.-C. Cai, W. D. Gropp, D. E. Keyes, R. G. Melvin, and D. P. Young. Parallel Newton-Krylov-Schwarz Algorithms for the Transonic Full Potential Equation. *SIAM J. Sci. Comput.*, **19**: 246–265, 1998.
5. X.-C. Cai, and D. E. Keyes. Nonlinearly Preconditioned Inexact Newton Algorithms. submitted to *SIAM J. Sci. Comput.*
6. E. J. Cramer, J. E. Dennis, P. D. Frank, R. M. Lewis, and G. R. Shubin. Problem Formulation for Multidisciplinary Optimization. *SIAM J. Optimization*, **4**: 754–776, 1994.
7. R. S. Dembo, S. C. Eisenstat, and T. Steihaug. Inexact Newton Methods. *SIAM Journal on Numerical Analysis* **19**: 400–408, 1982.
8. M. Drela. *Two-Dimensional Transonic Aerodynamic Design and Analysis Using The Euler Equations.* PhD Dissertation, MIT, 1985.
9. M. Drela. Pros and Cons of Airfoil Optimization. *Frontiers of Computational Fluid Dynamics 1998*, World Scientific, 1998.
10. M. Drela. Viscous and Inviscid Inverse Schemes Using Newton's Method. *Special Course on Inverse Methods for Airfoil Design for Aeronautical and Turbomachinery Applications*, AGARD Report No. 780, 1990.
11. R. Fletcher. *Practical Methods of Optimization.* 2nd edn., John Wiley and Sons, New York, 1987.

12. P. E. Gill, W. Murray, and M. H. Wright. *Practical Optimization.* 1st edn., Academic Press, Inc., San Diego, 1981.
13. P. E. Gill, S. J. Hammerling, W. Murray, M. A. Saunders, and M. A. Wright. *User's Guide for LSSOL (Version 1.0): A FORTRAN Package for Constrained Linear Least-Squares and Convex Quadratic Programming.* Stanford University Technical Report, Department of Operations Research, 1986.
14. P. E. Gill, W. Murray, M. A. Saunders, and M. A. Wright. *User's Guide for NPSOL (Version 4.0): A FORTRAN Package Nonlinear Programming.* Stanford University Technical Report SOL86-2, Department of Operations Research, 1986.
15. R. T. Haftka and Z. Gürdal. *Elements of Structural Optimization*, Kluwer Academic Publishers, Boston, 1992.
16. W. P. Huffman, R. G. Melvin, D. P. Young, F. T. Johnson, J. E. Bussoletti, M. B. Bieterman, and C. L. Hilmes. Practical Design and Optimization in Computational Fluid Dynamics. AIAA Paper 93-3111, 1993.
17. A. Jameson. Aerodynamic Design via Control Theory. *Journal of Scientific Computing*, **3**: 233–260, 1988.
18. A. Jameson. Airfoils Admitting Non-unique Solutions of the Euler Equations. AIAA Paper 91-1625, 1991.
19. W. H. Jou, W. P. Huffman, D. P. Young, R. G. Melvin, M. B. Bieterman, C. L. Hilmes, and F. T. Johnson. Practical Considerations in Aerodynamic Design Optimization. AIAA Paper 95-1730, 1995.
20. P. J. Lanzdron, D. J. Rose, and J. T. Wilkes. An Analysis of Approximate Nonlinear Elimination. *SIAM Journal on Scientific Computing*, **17**: 538–559, 1996.
21. R. G. Melvin, D. P. Young, W. P. Huffman, F. T. Johnson, C. L. Hilmes, and M. B. Bieterman, 'Recent Progress in Aerodynamic Design Optimization', International Journal for Numerical Methods in Fluids, **30**, 205–216 (1999).
22. C. S. Morawetz. *Comm. Pure Appl. Math.*, **10**: 400, 1957.
23. C. S. Morawetz *Comm. Pure Appl. Math.*, **11**: 129, 1958.
24. P. E. Rubbert. CFD and the Changing World of Airplane Design. AIAA Wright Brothers Lecture, Sept. 1994.
25. Y. Saad. *Iterative Methods for Sparse Linear Systems.* PWS Publishing Company, Boston, 1995.
26. D. P. Young, R. G. Melvin, M. B. Bieterman, F. T. Johnson and S. S. Samant. Global Convergence of Inexact Newton Methods for Transonic Flow. *Int. J. Num. Meth. Fluids*, **11**: 1075–1095, 1990.
27. D. P. Young, R. G. Melvin, M. B. Bieterman, F. T. Johnson, S. S. Samant, and J. E. Bussoletti. A Locally Refined Rectangular Grid Finite Element Method: Application to Computational Fluid Dynamics and Computational Physics. *J. Comp. Phys.* **92**: 1–66, 1991.
28. D. P. Young, W. P. Huffman, R. G. Melvin, M. B. Bieterman, C. L. Hilmes, and F. T. Johnson. Inexactness and Global Convergence in Design Optimization. AIAA Paper 94-4386, 1994.
29. D. P. Young and D. E. Keyes. Newton's Method and Design Optimization. ISSTECH-96-011, May 1996.

Optimization of Large-Scale Reacting Flows using MPSalsa and Sequential Quadratic Programming

A.G. Salinger[1], R.P. Pawlowski[1], J.N. Shadid[1], B. van Bloemen Waanders[2], R. Bartlett[23], G.C. Itle[3], and L. Biegler[3]

[1] Parallel Computational Sciences Department
[2] Optimization and Uncertainty Quantification Department
Sandia National Laboratories,[†] Albuquerque, NM 87185 USA
[3] Department of Chemical Engineering
Carnegie Mellon University, Pittsburgh, PA, USA

Abstract. An rSQP optimization algorithm has been linked to the MPSalsa parallel reacting flows code. The goal is to develop SAND (Simultaneous Analysis aNd Design) methods for use with large-scale PDE simulations. MPSalsa is a unstructured grid finite element code that uses a fully coupled Newton method to solve the PDEs governing fluid flow, heat transfer, and non-dilute mass transfer. In this paper, we present results for optimization of a Chemical Vapor Deposition reactor for growing thin films of Gallium Nitride. In particular, we address issues of inexactness in the Jacobian matrix and of solution multiplicity.

1 Introduction

In this paper we present our current work in applying rSQP (reduced Sequential Quadratic Programming) optimization algorithms to steady state reacting flow applications. This is a SAND (Simultaneous Analysis aNd Design) approach that has the potential to be more efficient than traditional black box (NAND) approaches. The advantage of the SAND approach is that the nonlinear PDE problem (or constraints) is converged simultaneously to the optimization problem, while in the NAND approach, the PDE problem is converged every time the design parameters are changed. The drawback of the SAND approach is that the interface of the optimizer to the application code is more intrusive, requiring the ability to request linear solves of the Jacobian matrix for the PDE problem, the residuals of the discretized PDEs for a given state vector, and gradients of the objective function. Another outstanding issue, which we will touch on in this paper, is the level of accuracy in the Jacobian matrix needed for the optimization method to converge to the right solution.

[†] Sandia is a multiprogram laboratory operated by Sandia Corporation, a Lockheed-Martin Company, for the United States Department of Energy under Contract DE-AC04-94AL85000.

Reacting flow applications are modeled by coupled sets of PDEs, and can have nonlinearities due to convection, reaction rates, and dependency of physical properties on the local state (e.g. calculating density from the ideal gas law). Some examples of reacting flow systems of engineering interest are combustion systems, catalytic reactors, and Chemical Vapor Deposition (CVD) reactors. In this paper we will study a CVD reactor design problem of interest to researchers in the Chemical Processing Science department at Sandia.

In Section 2 we present the CVD reactor optimization problem. In Section 3.1 the governing PDEs, the solution methods, and the stability analysis algorithms implemented in MPSalsa are presented. In Section 3.2 we briefly present the optimization methods and interface to MPSalsa. In Section 4 we present the outcomes of our inital runs, including some interesting results concerning the effects of inexactness in the Jacobian, the effects of using second order corrections to keep constraint residuals small, and the effects of solution stability and multiplicity.

2 CVD Reactor Optimization Problem

The rotating disk reactor is a common configuration for performing Chemical Vapor Deposition (CVD) of thin films, including many important semiconducting materials. The optimization problem formulated in this paper is generated from the work of Sandia researchers attempting to improve the design of the inlet of a rotating disk CVD reactor for use in growing thin films of Gallium Nitride (GaN). GaN is used in blue light emitting diodes and other photonic devices. The quality of the electronic device is highly dependent on the uniformity of the growth rate at different positions in the reactor. We are attempting to use simulations and optimization algorithms to determine if a new reactor, designed with a restricted inlet for reducing the costs of reactant gases, can achieve highly uniform GaN film growth.

The finite element mesh for the base shape of the reactor is shown in Figure 1(a). This is an axisymmetric (2D) model, where the left side is the axis of symmetry. A mixture of trimethylgallium, ammonia, and hydrogen gases ($Ga(CH_3)_3$, NH_3, and H_2) enter the top of the reactor, flow over the disk, which is heated, and then flow down the annular region out the bottom of the mesh. At the heated disk, the $Ga(CH_3)_3$ and NH_3 react to deposit a GaN film and release three molecules of methane (CH_4). This simplified mechanism has been shown to work well in modeling GaN film uniformities since the growth rate is predominantly transport limited [7]. This mesh depicts a restricted inlet design, where the top of the reactor has a smaller radius than the lower part of the reactor.

The main parameter used in this paper is the inlet velocity of the gases, V. Two additional parameters in this model define the shape of the inlet, namely the Shoulder Radius and Shoulder Height, which define the position

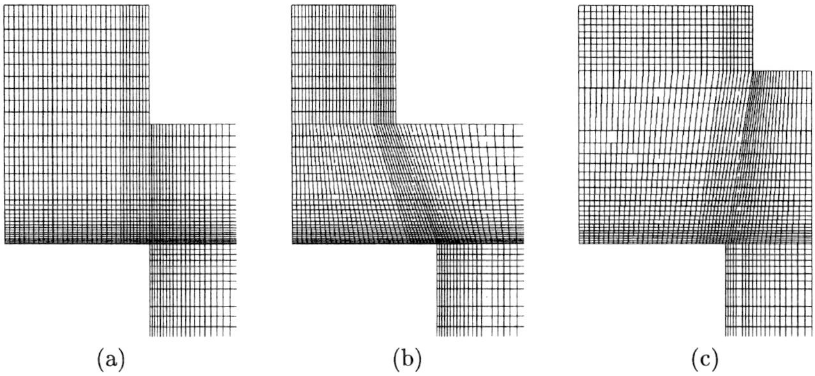

Fig. 1. Three different meshes for the restricted inlet design of the rotating disk reactor are shown: (a) the baseline case mesh where the shoulder radius is above the edge of the disk and the height is half of the inlet height; (b) a mesh when the shoulder radius parameter is decreased; (c) a mesh where the shoulder radius and height are both increased above the base case.

where the mesh transitions from the inlet radius to the larger reactor radius. The mesh is moved algebraically and continuously as a function of these geometric design parameters. Figure 1(b) shows how the mesh changes for a decreased shoulder radius, and Figure 1(c) shows how the mesh deforms continuously for larger values of the shoulder radius and shoulder height. If the optimum occurs too far away from where the initial mesh is generated, it would be appropriate to remesh the new geometry from scratch.

The objective function measures the uniformity of the growth rate of GaN over the disk. We chose an L^2 norm over an $L^{\inf}$ norm so that the objective is continuous and has a continuous derivative. Since the L^2 norm had very small values over a range of parameters, the log was taken. The final form of the objective function is

$$\text{Objective Function} = F = log(SD + 10^{-10}) \tag{1}$$

where SD is the standard deviation squared and is defined as

$$SD = \frac{1}{N_n} \sum_{i=1}^{N_n} (\frac{g_i - g_{ave}}{g_{ave}})^2, \tag{2}$$

Here N_n is the number of nodes on the surface, g_i is the growth rate of GaN at node i, and g_{ave} is the average growth rate.

3 Numerical Methods

3.1 Reacting Flow Simulation

The governing equations and numerical methods summarized in this section have been implemented in the MPSalsa computer code, developed at Sandia National Laboratories. More complete descriptions of the code and capabilities can be found in the following references [4, 7, 9, 10, 12, 13]. The fundamental conservation equations for momentum, heat, and mass transfer are presented for a reacting flow application. The equations for fluid flow consist of the incompressible Navier-Stokes equations for a variable-density fluid and the continuity equation, which express conservation of momentum and total mass. The steady-state momentum equation takes the form:

$$\rho(\mathbf{u} \bullet \nabla)\mathbf{u} - \nabla \bullet \mathbf{T} - \rho\mathbf{g} = 0, \tag{3}$$

where $\mathbf{u}$ is the velocity vector, ρ is the mixture density, and $\mathbf{g}$ is gravity vector. $\mathbf{T}$ is the stress tensor for a Newtonian fluid:

$$\mathbf{T} = -P\mathbf{I} - \frac{2}{3}\mu(\nabla \bullet \mathbf{u})\mathbf{I} + \mu[\nabla\mathbf{u} + \nabla\mathbf{u}^T] \tag{4}$$

Here P is the isotropic hydrodynamic pressure, μ is the mixture viscosity, and $\mathbf{I}$ is the unity tensor. The total mass balance is given by:

$$\nabla \bullet (\rho\mathbf{u}) = 0 \tag{5}$$

The density depends on the local temperature and composition via the ideal gas law. For nondilute systems, the multicomponent formulation is used:

$$\rho = \frac{P_o \sum_{j=1}^{N_g} W_j X_j}{RT}, \tag{6}$$

where P_o is the thermodynamic pressure, R is the gas constant, T is the temperature, X_j is the mole fraction of the jth species, W_j is the molecular weight of the jth species, and N_g is the number of gas-phase species (which is 4 for the model in this paper).

The steady-state energy conservation equation is given as:

$$\rho\hat{C}_p(\mathbf{u} \bullet \nabla)T = \nabla \bullet (\lambda\nabla T) - S, \tag{7}$$

where $\hat{C}_p$ is the mixture heat capacity and λ is the mixture thermal conductivity. The last term on the right hand side S is the source term due to the heat of reaction, which is negligible under the process conditions in this example problem.

The species mass balance equation is solved for N_g-1 species:

$$\rho(\mathbf{u} \bullet \nabla)Y_k = \nabla \bullet \mathbf{j}_k + W_k\dot{\omega}_k \quad \text{for } k = 1, \ldots, N_g\text{-}1, \tag{8}$$

where Y_j is the mass fraction of the jth species, $\mathbf{j}_k$ is the flux of species k relative to the mass averaged velocity $\mathbf{u}$ and $\dot{\omega}_k$ is the molar rate of production of species k from gas-phase reactions. A special species equation, which enforces the sum of the mass fractions to equal one, replaces one of the species balances (usually the species with the largest mass fraction):

$$\sum_{k=1}^{N_g} Y_k = 1 \quad \text{for } k = N_g \tag{9}$$

The diffusive flux term (Multicomponent Dixon-Lewis Formulation) includes transport due to both concentration gradients and thermal diffusion (Soret effect):

$$\mathbf{j}_k = \rho Y_k \left(\frac{1}{X_k \overline{W}} \sum_{j \neq k}^{N_g} W_j D_{kj} \nabla X_j - \frac{D_k^T}{\rho Y_k} \frac{\nabla T}{T} \right) \tag{10}$$

Where X_j is the mole fraction of species j, D_{kj} is the ordinary multicomponent diffusion coefficient, and D_k^T is the thermal diffusion coefficient. $\overline{W}$ is the mean molecular weight of the mixture given by:

$$\overline{W} = \sum_{k=1}^{N_g} X_k W_k = \frac{1}{\sum_{k=1}^{N_g} \frac{Y_k}{W_k}} \tag{11}$$

The conversion between mass (Y_k)and mole (X_k) fractions is:

$$Y_k = \frac{W_k}{\overline{W}} X_k \tag{12}$$

At the disk surface, surface chemical reactions take place. In general these can be very complicated, but for this model problem the reaction has been shown to be approximated very well by a transport limited model. In this case, the growth rate of GaN on the surface (as well as the consumption of $Ga(CH_3)_3$ and NH_3, and the production of CH_4) is proportional to the concentration of trimethylgallium ($Ga(CH_3)_3$) at the surface.

In general, the numerous physical properties in the above equations are dependent on the local temperature and composition. In the MPSalsa code, we use the Chemkin library and database format to obtain these physical properties. These terms add considerable nonlinearity to the problem.

The above system of 9 coupled PDEs (for unknowns u_r, u_z, u_θ, P, T, $Y_{Ga(CH_3)_3}$, Y_{CH_4}, Y_{NH_3} and Y_{H_2}) are solved with the MPSalsa code. MPSalsa uses a Galerkin/least-squares finite element method [13] to discretize these equations over the spatial domain. While this code is designed for general unstructured meshes in 2D and 3D, and runs on massively parallel computers, this application is 2D, uses the mesh shown in Figure 1(a), and was run on a single processor workstation. The discretized system contains 22000 unknowns.

A fully coupled Newton's method is used to robustly calculate steady-state solutions. While analytic Jacobian entries are supplied for derivatives with respect to the solution variables and the density, derivatives of the other physical properties are only calculated with the numerical Jacobian option. This option uses first order finite differencing on the element level. The resulting linear system at each iteration is solved using the Aztec package of parallel, preconditioned iterative solvers. In this paper, we exclusively used an ILU preconditioner and the GMRES solver with no restarts. On a single processor SGI workstation, a typical matrix formulation required 9 seconds for the inexact analytic Jacobian and 96 seconds to calculate the (nearly) exact finite difference numerical Jacobian. A typical linear solve required 40 seconds.

Parameter continuation methods have been implemented in MPSalsa via the LOCA library [8, 11]. LOCA includes an arclength continuation algorithm for tracking solution branches even when they go around turning points (folds). As will be seen in Section 4, this is a powerful tool for uncovering solution multiplicity. In addition, a turning point tracking algorithm has been implemented to directly delineate the region of multiplicity as a function of a second parameter. A complementary tool for performing linearized stability analysis by approximating the few rightmost eigenvalues of the linearized time dependent problem has also been successfully implemented [3, 6, 11].

3.2 rSQP optimization

The optimization problem was solved by interfacing MPSalsa with the rSQP code from Carnegie Mellon university. Future work will use the updated rSQP++ code [1]. The optimization problem is formulated as follows:

$$
\begin{aligned}
\min \quad & f(y,z) \\
\text{s.t.} \quad & c(y,z) = 0 \\
& x^L \le x \le x^U \\
& \mathbf{x} = \begin{bmatrix} y \\ z \end{bmatrix} \\
& x \in R^n, y \in R^m, m \le O(10^6) \\
& z \in R^{n-m}, n-m = O(1-50)
\end{aligned}
\tag{13}
$$

where f is the objective function, c are the constraint equations (i.e., the residual vector for the discretized PDE problem), y is the vector of length m of state variables corresponding to the velocity, temperature, pressure, and mass fraction unknowns, and z is a vector of length $n-m$ of decision or optimization variables.

We use the reduced SQP optimization algorithm, as it generally requires the fewest number of function and gradient evaluations and is considered more robust than other optimization methods. The SQP method can be derived from the application of a Newton method to the KKT conditions of (13). At each iteration k, SQP forms and solves a quadratic programming subproblem written as:

$$\begin{array}{ll} \mathrm{Min}_{d\in R^n} & g_k^T d + \frac{1}{2} d^T W_k d \\ \text{s.t.} & A_k^T d + c_k = 0 \\ & d^L \le x_k + d \le d^U \end{array} \tag{14}$$

where g_k is the gradient of the objective function, W_k is the Hessian of the Lagrangian or its approximation, A_k is the Jacobian of the constraints, and d is the step size calculated from the optimization problem. The subscript k indicates that these quantities are calculated at x_k. (For convenience we will suppress this subscript for the remainder of this section.)

If second derivative information is not available then the Hessian matrix is often approximated with quasi-Newton updates. However, under these conditions (14) can be prohibitive to solve because either a dense matrix W_k or its quasi-Newton update factors of dimension n must be stored. Instead, we modify the formulation of (14) through a reduced space decomposition. Here the search direction d is represented by a range space step (Yp_Y) and a null space step (Zp_Z). This decomposition is summarized as follows:

$$\begin{array}{l} A^T = [N|C] \\ d = Yp_Y + Zp_Z \\ \mathbf{Y} = \begin{bmatrix} N^T C^{-T} \\ I \end{bmatrix} \\ \mathbf{Z} = \begin{bmatrix} I \\ -C^{-1}N \end{bmatrix} \end{array} \tag{15}$$

where C is the Jacobian of constraints with respect to the state variable (which is the Jacobian matrix from the PDE model) and N is the matrix of constraint gradients with respect to the design variables. Note that $A^T Z = 0$ and $Y^T Z = 0$. Also, an alternate choice for the range space basis, $Y^T = [I \mid 0]$ leads to a cheaper (but sometimes less robust) computation if $n - m$ is large.

The vectors p_Y and p_Z are obtained by substituting (15) into (14) to yield a reduced space quadratic programming problem [2] that we solve here using a variation of the Goldfarb-Idnani algorithm (called QPKWIK).

$$\begin{array}{ll} \mathrm{Min} & (Z^T g + w)^T p_Z + \frac{1}{2} p_Z^T B p_Z \\ \text{s.t.} & a - x_k \le Zp_Z + Yp_Y \le b - x_k \\ & p_Y = -(A^T Y)^{-1} c \\ & B \approx Z^T W Z \qquad (BFGS \text{ approximation}) \\ & x^{k+1} = x^k + d \end{array} \tag{16}$$

The following steps are performed for each iteration of the optimizer.

1. An initial guess of the variables must be supplied to start the algorithm.
2. Calculate $f, \nabla f, c, C, N$.
3. Calculate $p_Y = -(A^T Y)^{-1} c$ using the linear solver.
4. Calculate $C^{-1}N$, the sensitivities of the state variable with respect to the design variable, using $n - m$ calls to the linear solver.
5. Solve the reduced QP using QPKWIK to generate p_Z.
6. Calculate d=$Y p_Y + Z p_Z$.
7. Are the Karush Kuhn Tucker conditions satisfied? If Yes, then stop.
8. Apply a line search to find a stepsize α that satisfies the Armijo conditions.
9. Set $x_{k+1} = x_k + \alpha d$ and $k = k + 1$. Goto step 2.

Logic has been added so that multiple solves of the same matrix with different right hand sides can reuse the same preconditioner.

One addition to the algorithm is the option to take second order correction steps between Steps 7 and 8. Here we apply the following procedure:

- Set $x_{k,0} = x_k + d$ and $l = 0$.
- For $l \leq L_{max}$
  - If Armijo conditions are satisfied, set $\alpha = 1.0$ and go to Step 9.
  - Solve $p_{Y,l} = -(A(x_k)^T Y(x_k))^{-1} c(x_{k,l})$
  - Set $x_{k,l+1} = x_{k,l} + Y_k p_{Y,l}$ and evaluate $f(x_{k,l+1})$ and $c(x_{k,l+1})$.
- endfor

For nonlinear constraints, the second order correction allows limited constraint following in order to improve the line search. In our implementation we set $L_{max} = 5$. A more detailed description of the rSQP algorithm is presented in [5].

4 Results

4.1 One Parameter Model

The first results are shown in Figure 2 for the one parameter system. Here the inlet velocity V is the design parameter while the Shoulder Radius and Shoulder Height parameters are held fixed at 6.35 and 5.08 as in Figure 1(a). Starting at a velocity of $V = 20$ (cm/sec), a simple continuation run down to a velocity of $V = 7$ showed a clear minimum near $V = 11.7$ and Objective Function $F = -6.9$.

Two runs of this problem using the rSQP optimizer were performed. For this run, the exact numerical Jacobian was used, and up to 5 second order correction steps per iteration were allowed. The linear solver tolerance was set at a relative residual reduction of 10^{-8}. When starting at $V = 20$ and converged PDE constraints, the optimizer converged in 15 iterations to a point

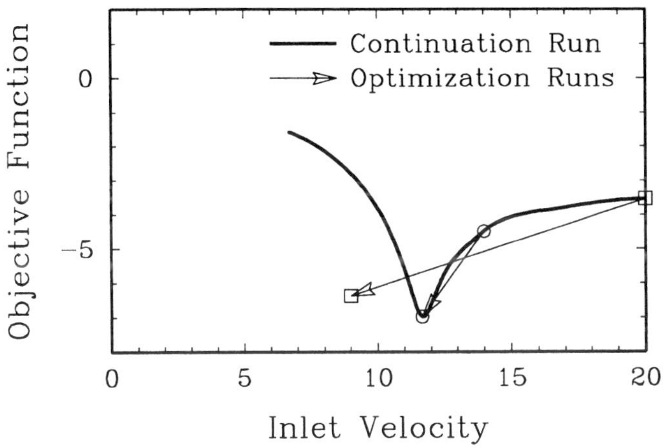

Fig. 2. Results for a 1 parameter continuation run (bold line), showing the Objective Function as a function of the inlet velocity of the reactant gases. Two results for the rSQP optimizer are shown, where the run starting at $V = 14$ (circle symbols with connecting arrow) converged to the expected local minimum while the run starting at $V = 20$ (square symbols with connecting arrow) converged to a point not seen on the continuation run.

at $V = 9.00$ and $F = -6.36$ (in about 3 hours compute time). However, when starting at $V = 14$ and with a converged steady-state solution, the optimizer reached the minimum at $V = 11.67$ and $F = -6.967$ in 14 iterations. As can be seen in Figure 2, the first run does not appear to be even on the solution branch of converged PDE constraints.

Three deposition profiles as a function of radial position are shown in Figure 3. The profile at the initial conditions of $V = 20$ has a minimum growth rate at the center and has a 8.5% nonuniformity. The solution found by the optimizer at $V = 11.67$, that also appears to be the minimum from the continuation run, shows a much flatter profile with an internal maximum, and an overall nonuniformity of 1.2%. The other solution found by the optimizer at $V = 9.00$ has a very similar shape, a smaller overall growth rate, and a 1.8% nonuniformity. Growth rate nonuniformities in the neighborhood of 1.0% are desirable.

Subsequent parameter continuation and linearized stability analysis calculations revealed that this solution is indeed a solution to the PDE constraints, yet a solution that is linearly unstable. The results of an arclength parameter continuation run with linear stability determinations are shown in Figure 4. The dashed line indicated physically unstable solutions while the solid lines are locally stable. One can see that there are three local minima in the objective function, only one of which is linearly stable. Over a large range of inlet velocities, $6.11 < V < 15.86$, there are three solutions that exist at the same parameter values. The rSQP optimizer, when started at $V = 20$, jumped into

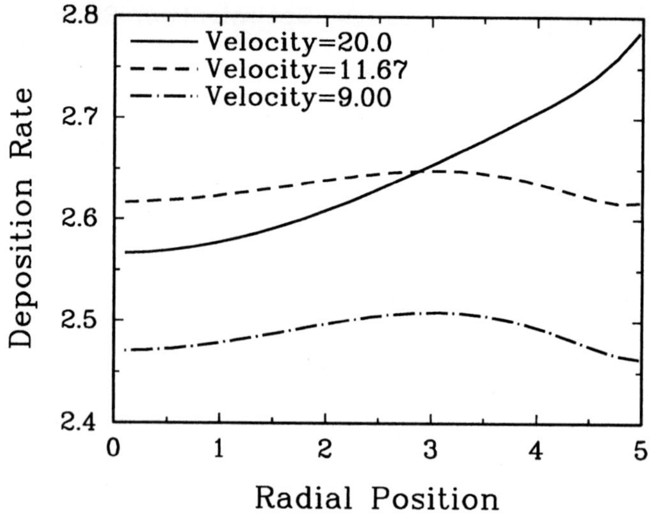

Fig. 3. Radial profiles of the surface deposition rate at three different solutions: the inital guess at $V = 20$, and the final solutions from the two optimization runs at $V = 11.67$ and $V = 9.00$.

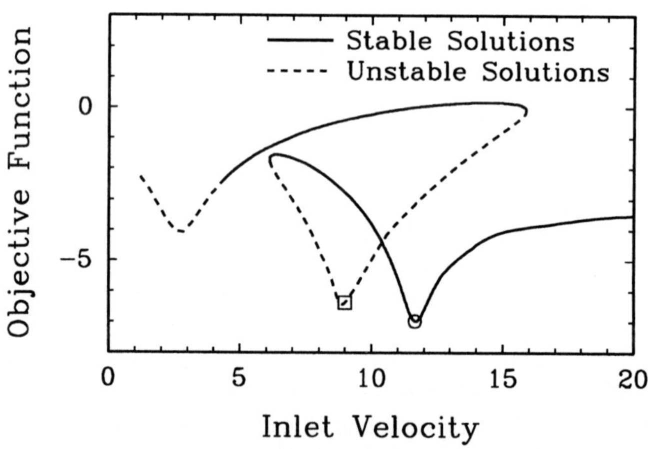

Fig. 4. Results for a 1 parameter continuation run with arclength continuation and linearized stability analysis are shown. The dashed lines represent unstable solution branches. The symbols show the results of the two optimization runs from Figure 2.

the basin of attraction for a local minimum at $V = 9.00$. The physical basis for the multiplicity is well understood. Recirculation flow cells can develop as a result of the buoyancy force of the heated reactor surface.

4.2 Three Parameter Model

The one parameter model showed that it is imperative to be aware of solution multiplicity and unstable solution branches. Continuation runs on the turning points defining the boundaries of multiplicity were performed to see how the region of multiplicity changes as a function of the additional geometric parameters. The effect of Shoulder Radius on the multiplicity region is shown in Figure 5, and the effect of Shoulder Height on the region of multiplicity is shown in Figure 6. The results show that the maximum velocity where multiplicity occurs has a direct dependence on the Shoulder Radius and is relatively insensitive to the Shoulder Height. The minimum velocity where multiplicity occurs is insensitive to the Shoulder Radius but has an inverse dependence on the Shoulder Height.

A single three-parameter optimization run was performed, starting at the same conditions where the one-parameter run that converged to the stable minimum was started: Velocity $= 14.0$, Shoulder Radius $= 6.35$, and the Shoulder Height $= 5.08$. The run was performed with up to 5 second order correction steps per optimization iteration. After 60 iterations, the objective function had been driven down to $F = -6.32$, which is not as low as the

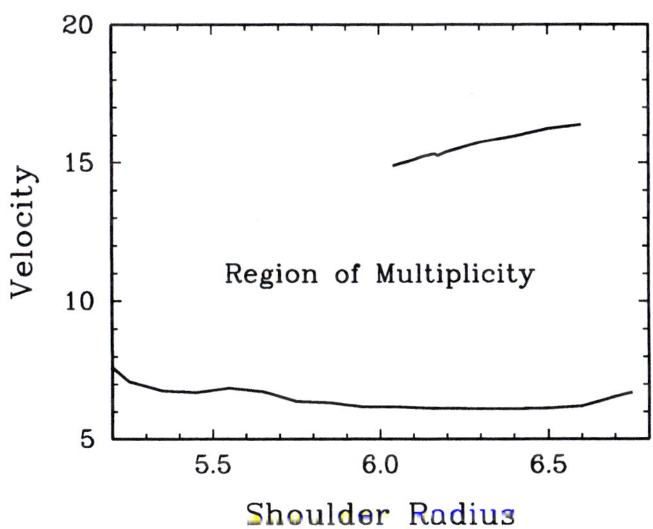

Fig. 5. Results of turning point continuation runs showing how the region of multiplicity identified in Figure 4 changes as a function the geometric Shoulder Radius parameter.

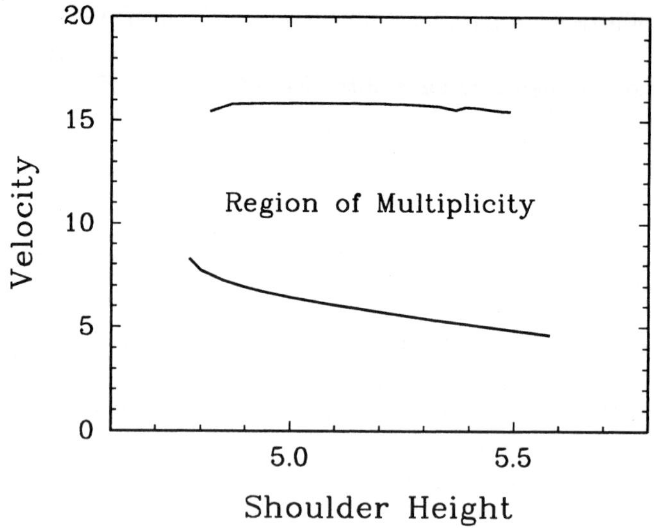

Fig. 6. Results of turning point continuation runs showing how the region of multiplicity identified in Figure 4 changes as a function the geometric Shoulder Height parameter.

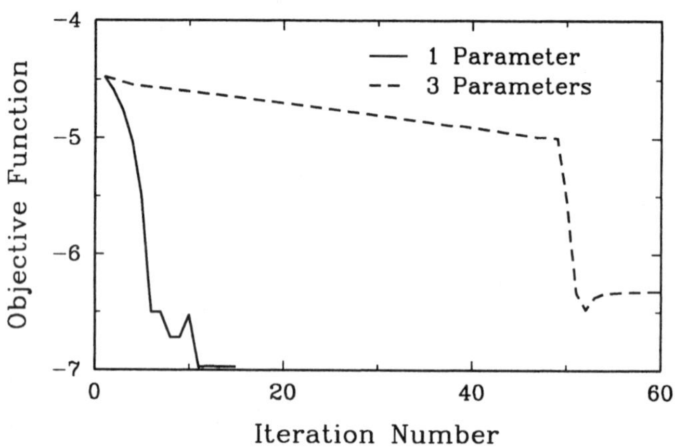

Fig. 7. A comparison of the 3-parameter optimization run after 60 iterations and the 1-parameter run, started at the same conditions, which converged after 14 iterations.

$F = -6.967$ achieved in the 1 parameter optimization. Possible reasons for this are that the three-parameter model is converging to a local minimum or that the singularities in the region are causing convergence problems. Future runs will need to be made to fully understand this preliminary result. The result of the three-parameter run is compared to the one-parameter run in Figure 7.

4.3 Effects of Jacobian Inexactness and Second Order Corrections

To test the effects of inexactness in the Jacobian and Second Order Correction Steps on the convergence of the optimization algorithm, three more runs of the 1-parameter model were performed. These all started at $V = 14$ for comparison with the successful optimization run, which was computed with a full numerical Jacobian and up to 5 second order correction steps per iteration. The results are shown in Figure 8. In the first additional run, the analytic (inexact) Jacobian was used, and the second order corrections were retained. This Jacobian leaves out the derivatives of all the physical properties with respect to the local state (temperature and composition), only including the correct density dependence. The Figure shows that this run converges visibly to the same optimum as the original case, both in iteration 11, though the original case reached the optimum in 14 iterations and the inexact case failed to meet the convergence criterion after 40 iterations. Two more runs were performed where no second order correction steps were allowed. The run with the inexact Jacobian converged visibly to the optimum after 86 iterations though had not converged within the tolerance after 100 iterations. The run with the exact numerical Jacobian without second order corrections had not yet converged to the optimum and was prematurely stopped after 120 iterations, surprisingly performing worse than the run with the inexact Jacobian.

For this problem, MPSalsa required 96 seconds to fill the full numerical Jacobian as compared to only 9 seconds for the analytic Jacobian, while an iterative linear solve required approximately 40 seconds. The runs with sec-

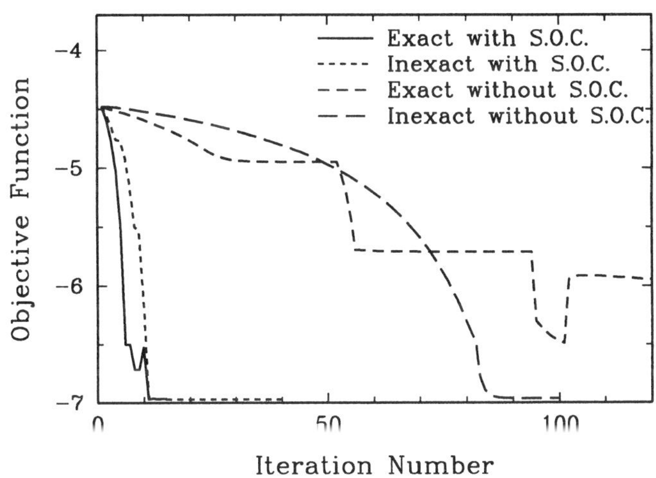

Fig. 8. A comparison of 4 runs for the 1-parameter model, comparing exact and inexact Jacobians, and with and without second order correction steps (S.O.C.).

ond order corrections required, on average, 5 linear solves per iteration, while the runs without second order corrections required exactly 2 linear solves per iteration. Therefore for this problem, the quickest numerical approach for visibly reaching the optimum was using the inexact analytic Jacobian and with the second order correction steps. The runs with the inexact Jacobian did not trigger the convergence tolerance set in the algorithm, and therefore performed many wasted iterations after visibly reaching the optimum. Since there are numerous approximations in the model, particularly with the chemistry mechanisms, the optimum needs only be converged to two digits of accuracy.

5 Summary and Conclusions

We have successfully coupled an rSQP code with MPSalsa, a large-scale reacting flow code. We used the algorithms to study the restricted inlet design of the rotating disk chemical vapor deposition reactor for growth of thin GaN films. We have verified the results obtained from the optimization algorithms by comparison with a parameter continuation run for a 1-parameter model. We found that solution multiplicity can lead to problems, since the optimizer converged to a local minimum on an unstable solution branch on the very first run. Continuation of the turning points with respect to the additional geometric parameters, using algorithms in the LOCA library, delineated the region of solution multiplicity. The optimization of the full three-parameter model was run for 60 iterations, but had not yet reached convergence.

From this preliminary experience, we can draw an important conclusion: solution multiplicity of nonlinear steady-state problems must be recognized and can be diagnosed using stability analysis tools. The technique in this paper of tracking the region of multiplicity is not scalable to larger numbers of design parameters, and is more expensive than the optimization calculations. At a minimum, the stability of the candidate optimum must be checked with a linear stability analysis tool. Concerning inexactness in the Jacobian matrix, and the effect of second order correction steps, we have gathered some evidence. For this run, it appears that inexactness in the Jacobian does not seriously hinder convergence, particularly if second order correction steps are used.

Acknowledgements

We would like to thank William Hart, Omar Ghattas, David Womble, and Sudip Dosanjh for their contributions and support. This work was primarily funded by the US DOE through the Sandia LDRD program and Sandia's Computational Science Research Institute.

References

1. Bartlett, R. *Object Oriented Approaches to Large-Scale Nonlinear Programming for Process Systems Engineering.* PhD thesis, Department of Chemical Engineering, Carnegie Mellon University, Pittsburgh, PA, 2001.
2. L.T. Biegler, J. Nocedal, and C. Schmid. A reduced hessian method for large-scale constrained optimization. *SIAM J. Opt.*, 5:314, 1995.
3. E. A. Burroughs, L. A. Romero, R. B. Lehoucq, and A. G. Salinger. Large scale eigenvalue calculations for computing the stability of buoyancy driven flows. *Sandia Technical Report*, SAND2001-0113, 2001.
4. M.S. Eldred, W.E. Hart, W.J. Bohnhoff, V.J. Romero, S.A. Hutchinson, and A.G. Salinger. Utilizing object-oriented design to build advanced optimization strategies with generic implementation. *Proceedings of the 6th AIAA/NASA/ISSMO Symposium on Multidisciplinary Analysis and Optimization, AIAA-96-4164-CP, Bellevue, WA*, pages 1568–1582, 1996.
5. G.C. Itle, A.G. Salinger, R.P. Pawlowski, J.N. Shadid, and L.T. Biegler. A tailored optimization strategy for pde-based design: Application to a cvd reactor. *Computers in Chem. Eng.*, 2002. submitted.
6. R. B. Lehoucq and A. G. Salinger. Large-scale eigenvalue calculations for stability analysis of steady flows on massively parallel computers. *International Journal for Numerical Methods in Fluids*, 36:309–327, 2001.
7. R. P. Pawlowski, C. Theodoropoulos, A. G. Salinger, T. J. Mountziaris, H. K. Moffat, J. N. Shadid, and E. J. Thrush. Fundamental models of the metalorganic vapor-phase epitaxy of galluim nitride and their use in reactor design. *Journal of Crystal Growth*, 221:622–628, 2000.
8. A. G. Salinger, N.M. Bou-Rabee, E.A. Burroughs, R.B. Lehoucq, R.P. Pawlowski, L.A. Romero, and E.D. Wilkes. LOCA: A library of continuation algorithms - Theroy manual and user's guide. Technical report, Sandia National Laboratories, Albuquerque, New Mexico 87185, 2002. SAND2002-0396.
9. A. G. Salinger, K. D. Devine, G. L. Hennigan, H. K. Moffat, S. A. Hutchinson, and J. N. Shadid. MPSalsa: A finite element computer program for reacting flow problems - part II user's guide. Technical report, Sandia National Laboratories, Albuquerque, New Mexico 87185, 1996. SAND96-2331.
10. A. G. Salinger, J. N. Shadid, S. A. Hutchinson, G. L. Hennigan, K. D. Devine, and H. K. Moffat. Analysis of gallium arsenide deposition in a horizontal chemical vapor deposition reactor using massively parallel computations. *Journal of Crystal Growth*, 203:516–533, 1999.
11. A.G. Salinger, R.B. Lehoucq, and L.A Romero. Stability analysis of large-scale incompressible flow calculations on massively parallel computers. *CFD Journal*, 9(1):529–533, 2001.
12. J. N. Shadid, H. K. Moffat, S. A. Hutchinson, G. L. Hennigan, K. D. Devine, and A. G. Salinger. MPSalsa: A finite element computer program for reacting flow problems - Part I theoretical development. Technical report, Sandia National Laboratories, Albuquerque, New Mexico 87185, 1996. SAND95-2752.
13. J.N. Shadid. A fully-coupled Newton-Krylov solution method for parallel unstructured finite element fluid flow, heat and mass transport. *IJCFD*, 12:199–211, 1999.

Part III

Multifidelity Models and Inexactness

First-Order Approximation and Model Management in Optimization

Natalia M. Alexandrov[1] and Robert Michael Lewis[2]

[1] NASA Langley Research Center, Hampton VA, n.alexandrov@larc.nasa.gov
[2] College of William & Mary, Williamsburg VA, buckaroo@math.wm.edu

Abstract. We discuss first-order approximation and model management optimization, an approach to the optimization of systems governed by differential equations. Our approach tries to alleviate the expense of relying exclusively on high-fidelity simulations, e.g., the solution of the governing differential equations on very fine meshes or the use of very detailed physics, while still guaranteeing global convergence of the overall optimization process to a solution of the high-fidelity problem. We focus here on several model management methods and experience with their performance.

1 Introduction

We discuss an approach to the optimization of systems governed by differential equations. Such optimization problems can be formulated as a nonlinear program (NLP) of the form

$$\begin{array}{ll} \underset{x}{\text{minimize}} & f(x, u(x)) \\ \text{subject to} & c_E(x, u(x)) = 0 \\ & c_I(x, u(x)) \geq 0 \\ & x_L \leq x \leq x_U, \end{array} \tag{1}$$

with the distinctive feature that the quantities $u(x)$, which describe the physical state of the system, are obtained via a computational simulation. Given a set of design variables x, the simulation computes the quantities $u(x)$ of engineering or scientific interest, by solving a system of differential equations:

$$A(x, u(x)) = 0. \tag{2}$$

For instance, this may represent a coupled set of PDE, with each equation a particular aspect of the physical system (e.g., aerodynamics or structures in aerospace design) or the behavior of a subsystem (e.g., panel structural analysis in structural design of a wing).

The simulation-based NLP (1) is a "large-scale" problem due to the large number of state variables $u(x)$ and the associated computational cost of (2), even if the number of variables x explicitly manipulated in the course of optimization is relatively small. The simulation (2) leads to a frequently prohibitive cost of function and derivative evaluations in solving (1). The cost derives from a fine level of discretization in the numerical solution of (2), perhaps compounded by detailed models of physical phenomena (such as viscous effects in fluid dynamics).

One approach to addressing this difficulty is to apply optimization procedures that use variable-fidelity approximations and models of the phenomena described by the governing equations. Engineers have used models of different fidelity in optimization for a long time. See, for instance, [9] for a review of approximations in structural optimization and [23] for a review of models in aerodynamic optimization. One can be satisfied with intermediate models of lower fidelity, provided they lead to progress in the solution of the optimization problem (see, e.g., [10] and the references therein). However, the use of models in optimization had been largely heuristic. Purely heuristic approaches to employing low-fidelity models in optimization are not guaranteed to converge to high-fidelity answers.

This paper discusses first-order approximation and model management optimization (AMMO). The development of AMMO from unconstrained to constrained problems is discussed in [1, 2, 5, 6]. AMMO seeks to alleviate the expense of relying exclusively on high-fidelity simulations, while still guaranteeing global convergence of the overall optimization process to a solution of the high-fidelity problem. We describe AMMO frameworks based on three representative optimization algorithms and some accumulated numerical experience.

We have focused on models that insensitive to the number of design variables. Such models fall into three broad categories:

- Variable-accuracy models arise when a particular simulation can be solved to some user-specified tolerance.
- Variable-resolution models arise when using a single physical model on meshes of varying degrees of refinement.
- Variable-fidelity physics models arise when certain physical aspects of the system are relaxed. For instance, in aerodynamics, the physical models range from linear potential models for inviscid, irrotational, incompressible flow to the Navier-Stokes equations of nonlinear viscous flow.

Related approaches based on based on fitting surfaces to samples of high-fidelity responses are discussed in [18, 26]. However, data-fitting models are, in general, suitable only for problems with a relatively small number of design variables.

In the NLP (1), $u(x)$ is computed by solving (2) at each optimization iteration. It can thus be viewed as a variable reduction method insofar as it treats u as a function of the design variables x, removing the state variables u from the optimization problem. Alternative problem formulations may treat both the design and state variables as optimization variables, while (2) is treated as explicit equality constraints, thus giving rise to optimization problems with large numbers of variables directly manipulated by the optimization algorithm. Simultaneous Analysis and Design (SAD or SAND) (e.g., [20]) is an example of such a formulation.

The approach we discuss is applicable to SAND and other formulations that maintain varying degrees of feasibility with respect to the simulation constraints (see [3, 4] for details on the relationship among various problem formulations). We focus here on (1), since this formulation allows us to use existing simulation software in the most straightforward way. Treatment of (2) as explicit constraints may entail a significant effort in restructuring existing codes for the solution of the governing equations. Altering production simulation codes is rarely feasible. Thus, we assume throughout the discussion that evaluating the functions and constraints of (1) involves the solution of the governing equations (i.e., a simulation), although we will omit the dependence on $u(x)$ from (1), without loss of generality, to simplify notation.

In order to show the wide applicability of the AMMO idea, we survey a number of earlier developments, as well as introduce a new AMMO approach, the $S\ell_1QP$ algorithm in Section 4.2. For brevity, we have omitted any complete proofs of convergence. However, we outline how some of the proofs would proceed.

2 The Basic Idea of First-Order Model Management

AMMO combines well-established engineering approximation concepts with the trust-region idea from nonlinear programming to ensure robust behavior. Derivative-based optimization techniques, including trust-region methods, rely on local models. The tacit assumption is that the model is a local first or second-order Taylor series, with various approximations to the first and second derivatives of the contributing functions.

In AMMO, we take the idea of local models a step further by replacing the Taylor series in the optimization subproblems with general models that have local *trends* that are similar to those obtained with high-fidelity analyses. The step toward general models is motivated by the availability of a wide range of models in applications, as noted in the introduction. AMMO constructs models using high-fidelity and low-fidelity information and ascertains that the model has local trends similar to the high-fidelity counterpart by requiring that low-fidelity responses satisfy first-order consistency conditions with respect to the high-fidelity model, as follows.

Let $\tilde{f}$, $\tilde{c}_E$, and $\tilde{c}_I$ be lower-fidelity models of the objective, equality and inequality constraints, respectively. At each major iteration x_k of an AMMO algorithm, the models are required to satisfy

$$\tilde{f}(x_k) = f(x_k), \quad \tilde{c}_E(x_k) = c_E(x_k), \quad \tilde{c}_I(x_k) = c_I(x_k), \tag{3}$$

$$\nabla\tilde{f}(x_k) = \nabla f(x_k), \quad \nabla\tilde{c}_E(x_k) = \nabla c_E(x_k), \quad \nabla\tilde{c}_I(x_k) = \nabla c_I(x_k). \tag{4}$$

One can also impose higher-order consistency conditions, though for the problems discussed in this paper higher-order derivatives are not available.

Conditions (3)–(4) ensure that $\tilde{f}$, $\tilde{c}_E$, and $\tilde{c}_I$ mimic the local behavior of Taylor series models around our current best design x_k. First-order consistency is easily ensured in practice. The work reported here uses a technique we call the β-correction, due to Haftka [11, 19]. Given a high-fidelity function ϕ_{hi} (say, f) and any low-fidelity model ϕ_{lo} of ϕ_{hi}, we correct ϕ_{lo} as follows. Define

$$\beta(x) = \frac{\phi_{hi}(x)}{\phi_{lo}(x)}$$

and construct the linear approximation

$$\beta_k(x) = \beta(x_k) + \nabla\beta(x_k)^T(x - x_k).$$

Then

$$\tilde{\phi}(x) = \beta_k(x)\phi_{lo}(x)$$

satisfies the consistency conditions (3)–(4).

An alternative, additive correction scheme arises in the context of variable-resolution modeling that attempts to mimic the full approximation scheme in multigrid [22]:

$$\tilde{\phi}(x) = \phi_{lo}(x) + [\phi_{hi}(x_k) - \phi_{lo}(x_k)] + [\nabla\phi_{hi}(x_k) - \nabla\phi_{lo}(x_k)]^T(x - x_k).$$

The AMMO approach can be used in conjunction with any gradient-based optimization algorithm and any suite of variable-fidelity models. However, practical efficiency of any particular AMMO scheme depends on the ability to transfer the computational load to the lower-fidelity, cheaper computations. This will depend on the predictive qualities of the corrected low-fidelity models for the purposes of optimization.

3 AMMO for Unconstrained Optimization

An AMMO trust-region algorithm for the unconstrained minimization problem, minimize $f(x)$, is summarized as follows:

Algorithm 1: AMMO for unconstrained minimization

Choose the initial iterate $x_0 \in I\!R^n$, the initial trust radius $\Delta_0 > 0$, the maximum allowable trust radius $\Delta_{\max} > 0$, and scaling factors $0 < \kappa_1 \leq \kappa_2 < 1 < \kappa_3$.

Do $k = 0, 1, \ldots$ until convergence:

Choose $\tilde{f}_k$ that satisfies $\tilde{f}_k(x_k) = f(x_k)$ and $\nabla\tilde{f}_k(x_k) = \nabla f(x_k)$.

Find an approximate solution s_k to the subproblem:

$$\begin{aligned} &\underset{s}{\text{minimize}} \quad \tilde{f}_k(x_k + s) \\ &\text{subject to } \| s \| \leq \Delta_k. \end{aligned}$$

Update x_k according to

$$x_{k+1} = \begin{cases} x_k + s_k & \text{if } f(x_k + s_k) < f(x_k) \\ x_k & \text{otherwise} \end{cases}$$

and Δ_k according to

$$\Delta_{k+1} \in \begin{cases} [\| s_k \|, \ \min(\kappa_3 \| s_k \|, \Delta_{\max})] & \text{if } f(x_k + s_k) < f(x_k) \\ [\kappa_1 \| s_k \|, \ \kappa_2 \| s_k \|] & \text{otherwise} \end{cases}$$

End do

Conditions (3) and (4) guarantee that sufficiently close to x_k, the model $\tilde{f}_k$ is a suitable approximation of f for the purposes of optimization. The trust-region approach gives us a mechanism to regulate the use of the model $\tilde{f}_k$ in the course of an optimization iteration: if the model $\tilde{f}_k$ fails to find steps that improve f, we decrease Δ_k and fall back on a region in which $\tilde{f}_k$ is increasingly an accurate approximation of f. On the other hand, if $\tilde{f}_k$ is doing a good job of approximating the behavior of f, then we do not decrease the length of the steps we take, and thereby avoid unnecessarily restricting the progress of the optimization.

The model can change from iteration to iteration, so we denote it by $\tilde{f}_k$. One has the freedom to change the model to reflect the current region of the design space. Moreover, while only a single model is specified in Algorithm 1, nothing precludes recursive levels of subproblems and models of varying fidelity and computational cost in solving the subproblem, as in [22].

The trust-region subproblem in Algorithm 1,

$$\begin{array}{ll} \underset{s}{\text{minimize}} & \tilde{f}_k(x_k + s) \\ \text{subject to} & \| s \| \leq \Delta_k, \end{array} \tag{5}$$

involves a general first-order model $\tilde{f}_k$ instead of a conventional quadratic model. Algorithm 1 requires only that this subproblem be solved approximately, as is typical of trust-region methods [13]. The solution of each iteration, s_k, is acceptable as long as it satisfies a fraction of Cauchy decrease (FCD) condition on the change in the model $\tilde{f}_k$ from the point x_k to the point $x_k + s_k$. We use the following form of the FCD condition: there exist $\beta > 0$ and $C > 0$, independent of k, for which the step s_k satisfies

$$\tilde{f}_k(x_k) - \tilde{f}_k(x_k + s_k) \geq \beta \| \tilde{g}_k(x_k) \| \min\left(\Delta_k, \frac{\| \tilde{g}_k(x_k) \|}{C}\right), \tag{6}$$

where $\tilde{g}_k(x) = \nabla \tilde{f}_k(x)$.

Satisfaction of (6) has not proven to be an issue, since, the steps generated by most algorithms applied to (5) satisfy the FCD condition by design. Solving (5) is itself an iterative process. During this process, the first step typically yields FCD. Since any steps after the first only produce further decrease in $\tilde{f}_k$, the solution of the subproblem satisfies (6).

Given (6), the convergence analysis of our algorithm follows that in [24]. We make the additional hypothesis that the Hessians $\nabla^2 \tilde{f}_k(x_k)$ (or any quasi-

Newton approximations of $\nabla^2 \tilde{f}_k(x_k)$ used in the solution of (5)) remain bounded uniformly in k.

Theorem 1. *Suppose f is bounded below and uniformly continuously differentiable. Then the sequence of iterates generated by Algorithm 1 satisfies* $\liminf_{k\to\infty} \| \nabla f(x_k) \| = 0$.

By changing the criterion for accepting a step s_k slightly, one can ensure $\lim_{k\to\infty} \| \nabla f(x_k) \| = 0$, as in conventional trust-region methods [13].

Unlike the trust-region approach, line-search strategies [17] do not generalize in such a straightforward way to AMMO. Suppose the current iterate is point A, and we apply a line-search method to minimize the model $\tilde{f}_k$ to find a candidate step. If, in this latter process, the line-search method first visits the point B and then the point C, then we know that the direction from A to B is a descent direction for f, and that the direction from B to C is a descent direction for $\tilde{f}_k$. However, there is no guarantee that the direction from A to C is a descent direction for f. Consequently a backtracking line-search is not guaranteed to yield a step that decreases the objective as one takes shorter steps from A towards C. One would need to store the intermediate iterates in the minimization of $\tilde{f}_k$ and perform a search along the piecewise linear arc defined by those points.

4 AMMO for Constrained Optimization

Any derivative-based optimization method can be adapted for AMMO. We give the outlines of several algorithms in this section. The global convergence analysis for each AMMO adaptation is essentially that of the underlying algorithm because of the matching conditions (3)–(4). However, each proof depends on the computation of the step in each subproblem. In particular, this step must yield sufficient decrease for an associated merit function [13]. This was noted previously for unconstrained optimization in connection with the condition (6). These decrease conditions have not proven to be an issue in practice; the subproblem are solved in ways that guarantee sufficient decrease.

The pros and cons of the various AMMO methods also reflects the pros and cons of the underlying algorithms. For instance, while the augmented Lagrangian AMMO method of Section 4.1 is straightforward to implement, it can be slow to approach a solution. Both comments can be made about augmented Lagrangian algorithms.

4.1 AMMO Based on an Augmented Lagrangian Algorithm

The AMMO method described next is based on the augmented Lagrangian algorithm of [14]. We assume that all inequalities have been turned into equalities through the use of non-negative slacks, and the NLP has the form

$$\begin{aligned} &\text{minimize } f(x) \\ &\text{subject to } c(x) = 0 \\ &\qquad\qquad x_L \le x \le x_U . \end{aligned} \tag{7}$$

The associated augmented Lagrangian is

$$\Phi(x; \lambda, \mu) = f(x) + \lambda^T c(x) + \frac{1}{2\mu} \| c(x) \|^2 . \tag{8}$$

The vector λ is the Lagrange multiplier estimate for the constraints, and μ is a penalty parameter. The bounds on the variables are left explicit. Let P be the projection onto the set $B = \{ x \mid x_L \le x \le x_U \}$. Given $x \in B$ and v, define $P(x, v) = x - P[x - v]$.

Algorithm 2: Augmented Lagrangian AMMO

Initialization: Set $k = 0$. Select $\bar{x}_0$, the initial penalty parameter $\mu_0 < 1$, and the initial convergence criteria ω_0 and η_0. Specify the least allowable decrease, τ, in the penalty parameter.

1. Inner Iteration:
Approximately solve
$\underset{x}{\text{minimize}} \quad \Phi(x; \lambda_k, \mu_k)$
subject to $x_L \le x \le x_U$
as described below to find x_k satisfying
$\| P(x_k, \nabla\Phi(x_k; \lambda_k, \mu_k)) \| \le \omega_k$.
If $\| c(x_k) \| \le \eta_k$, go to 2. Otherwise go to 3.

2. Lagrange multiplier update: (Set $k = k + 1$)
Update Lagrange multipliers with any standard update formula, e.g., Hestenes-Powell update. Choose $\mu_{k+1} < \mu_k$, $\omega_{k+1} > 0$, $\eta_{k+1} > 0$, so that $\lim_{k\to\infty} \omega_k = 0$ and $\lim_{k\to\infty} \eta_k = 0$. Go to 1.

2. Reduce the penalty parameter: (Set $k = k + 1$)
Set $\lambda_{k+1} = \lambda_k$, $\mu_{k+1} \le \tau\mu_k$. Choose ω_{k+1}, η_{k+1} (as in 2). Go to 1.

The resulting algorithm closely resembles the basic algorithm in [14]. The subproblem

$$\begin{aligned} &\underset{x}{\text{minimize}} \quad \Phi(x; \lambda_k, \mu_k) \\ &\text{subject to } x_L \le x \le x_U , \end{aligned} \tag{9}$$

we solve by a sequence of bound constrained trust region steps using approximations of $\Phi(x; \lambda, \mu_k)$ constructed of models of f and c satisfying (3) and (4). Because the behavior of the augmented Lagrangian algorithm is independent of the method used to solve (9), the AMMO adaptation inherits the analytical properties reported in [14]. In particular, we are guaranteed convergence of subsequences of the x_k to constrained stationary points. Choices of the constants in Algorithm 2 are discussed in [12, 14].

The AMMO adaptation of the augmented Lagrangian algorithm was tested on aerodynamic optimization of a 3D wing with variable-resolution models [5]. It produced significant savings in function and derivative evaluation compared to the original augmented Lagrangian algorithm. However, the underlying method is intrinsically slower for our problems of interest than other methods we have attempted. A version of the augmented Lagrangian AMMO may still prove use in some multidisciplinary applications.

4.2 AMMO Based on SQP-Related Methods

We turned to SQP-like methods in search of a more efficient basis for AMMO algorithms. A straightforward SQP-AMMO proved to be a promising candidate [5]. Implementing a robust SQP code is a time-consuming, nontrivial undertaking, so using well-tested, commercially or publicly available codes would be preferable, with the AMMO mechanism superimposed. Unfortunately, none of the software we examined proved suitable for simulation-based optimization because the software for solving the explicitly constrained trust region subproblem could not easily recover from trial steps that caused the analysis to fail. As a result of this difficulty, we are developing nonlinear programming software suited for simulation-driven optimization. In the meantime, nonsmooth exact penalty functions present themselves as a potential alternative to SQP that is attractive in the simplicity of its merit functions, while maintaining convergence properties similar to those of conventional SQP. In this section, we describe AMMO based on the Sℓ_1QP method of Fletcher [15].

Consider a composite penalty function

$$\mathcal{P}(x; h) \equiv f(x) + h(c(x)), \tag{10}$$

where f and c are smooth and h is a convex but perhaps only continuous functional. Fletcher's choice of $\mathcal{P}$ [15] is the penalty function

$$\mathcal{P}(x; \sigma) = f(x) + \sigma \sum_{i \in E} |c_i(x)| + \sigma \sum_{i \in I} \max\{0, c_i(x)\}. \tag{11}$$

This is an exact penalty function associated with (1) if the penalty parameter σ satisfies

$$\sigma > \min_{i \in L} |\lambda_i|, \tag{12}$$

where L is the set of all multipliers for (1). The model of $\mathcal{P}$ is chosen to be

$$m(x_k, s; \sigma) \equiv q(x_k, s) + \sigma \sum_{i \in E} |l_i(x_k, s)| + \sigma \sum_{i \in I} \max\{0, l_i(x_k, s)\}, \tag{13}$$

where

$$q(x_k, s) = f(x_k) + \nabla f(x_k)^T s + \frac{1}{2} s^T H_k(x_k) s \text{ and} \tag{14}$$

$$l_i(x_k, s) = c_i(x_k) + \nabla c_i(x_k)^T s \text{ for } i \in E \text{ or } i \in I. \tag{15}$$

H_k approximates $\nabla^2 f(x_k)$ and the value of σ is usually fixed, although it can be adjusted at the initial steps of solution. The prototype Sℓ_1QP algorithm solves for s_k subproblems of the form

$$\begin{array}{ll} \underset{s}{\text{minimize}} & m(x_k, s; \sigma) \\ \text{subject to} & \| s \|_\infty \le \Delta_k. \end{array} \tag{16}$$

The step s_k is assumed to be a global solution of (16). A relaxation of this condition is given in Section 4.3. The step is evaluated by comparing the actual and predicted reduction in $\mathcal{P}$ defined by

$$\rho_k = \frac{\mathcal{P}(x_k; \sigma_k) - \mathcal{P}(x_k + s_k; \sigma_k)}{m(x_k, 0; \sigma_k) - m(x_k, s_k; \sigma_k)}. \tag{17}$$

The step evaluation and update stages may be summarized as follows.

Algorithm 3: Updates for the Sℓ_1QP Algorithm

Select $0 < r_1 < r_2 \le 1$ and $0 < \kappa_1 < 1 < \kappa_2$.
Typical values are $r_1 = 0.25$, $r_2 = 0.75$, $\kappa_1 = 0.25$, $\kappa_2 = 2$.

$$\text{Set } x_{k+1} = \begin{cases} x_k & \text{if } \rho_k \le 0 \\ x_k + s_k & \text{otherwise.} \end{cases}$$

$$\text{Set } \Delta_k = \begin{cases} \kappa_1 \| s_k \| & \text{if } \rho_k < r_1 \\ \kappa_2 \Delta_k & \text{if } \rho_k > r_2 \text{ and } \| s_k \| = \Delta_k \\ \Delta_k & \text{otherwise.} \end{cases}$$

The Sℓ_1QP algorithm shares the global convergence properties of the SQP method. Variants that improve the local convergence properties may be found in [15].

Sℓ_1QP-AMMO also chooses $\mathcal{P}$ as a merit function. Now, instead of the model (13) we choose

$$m(k, x_k, s; \sigma) \equiv \tilde{f}(k, x_k, s) + \sigma \sum_{i \in E} |\tilde{c}_{E,i}(k, x_k, s)| + \sigma \sum_{i \in I} \max\{0, \tilde{c}_{I,i}(k, x_k, s)\} \tag{18}$$

whose components satisfy the consistency conditions (3)–(4). Note that the model m depends on k, as in Algorithm 1. The Sℓ_1QP-AMMO is stated as follows.

Algorithm 4: $S\ell_1$QP-AMMO

Initialization: Choose x_0, Δ_0, and constants as in Algorithm 3.
Do $k = 0, 1, \ldots$ until convergence:
  Model construction:
    Construct a model $m(k, x_k, s; \sigma_k)$ of $\mathcal{P}$
    whose components satisfy (3)–(4).
  Step computation:

$$\text{Solve for } s_k \begin{cases} \underset{s}{\text{minimize}} & m(k, x_k, s; \sigma_k) \\ \text{subject to} & \| s \| \le \Delta_k \end{cases}$$

  Step evaluation: Compute ρ_k as in (17)
    Accept or reject the step based on ρ_k as in Algorithm 3.
  Updates:
    Update x_k, Δ_k based on ρ_k as in Algorithm 3.
End do

After each iteration, σ_k is updated according to the rule in [25].

The global convergence result is a minor modification of Theorem 14.5.1 in [15].

Theorem 2. *Let $f, c_E, c_I \in C^2(\Omega)$ have bounded second derivatives on a bounded $\Omega \subset I\!R^n$. Let $\tilde{f}$, $\tilde{c}_E$, $\tilde{c}_I \in C^2(\Omega)$ be any models of f c_E, and c_I, respectively, that satisfy the first order consistency conditions (3)–(4) and have uniformly bounded second derivatives on Ω. Let $\{x_k\} \in \Omega$ be the sequence of iterates generated by $S\ell_1$QP–AMMO. The there exists an accumulation point x_* at which the first-order optimality conditions for minimizing $\mathcal{P}$ hold, that is,*

$$\underset{\lambda \in \partial h_*}{\mathit{maximize}}\ (g_* + \nabla c_* \lambda)^T s \ge 0 \ \mathit{for\ all}\ s, \tag{19}$$

where ∂h_* is the generalized derivative of h.

The proof of the theorem is nearly identical to that in [15] and similar to all proofs of this type. Namely, one assumes that there are descent directions from x_* and then derives a contradiction that arises from the iterate and trust-region updating rules. The only simple distinction in the proof of $S\ell_1$QP–AMMO convergence is that the bound on the error in approximating the objective and constraints by general models relies not only on direct Taylor series representation (as in $S\ell_1$QP), but also on the first-order consistency conditions.

4.3 An Alternative $S\ell_1$QP-AMMO.

The analysis of the variant of $S\ell_1$QP-AMMO discussed above assumes that the subproblem (16) is solved exactly. An alternative approach is to require the trial step s_k to satisfy a sufficient decrease condition. In this case, we do not specify how the subproblem is solved, but place the following conditions on the model m and a sufficient decrease condition on the step s_k:

- **Smoothness:** The model m is locally Lipschitz continuous and regular with respect to s for all (x, σ) and continuous in (x, σ) for all s.
- **Zero-order consistency:** The values of the function and model coincide when $s = 0$.
- **First-order consistency:** The generalized directional derivatives of the function and model coincide when $s = 0$.
- **Bounded parameters:** The set of problem parameters is closed and bounded.
- **Sufficient decrease:** For any x_*, there exist constants $\delta, \epsilon, \kappa \in (0,1)$ such that s_k satisfies

$$m(k, x_k, 0, \sigma_k) - m(k, x_k, s_k, \sigma_k) \geq \kappa \parallel g(x_k) \parallel \min\{\delta, \Delta_k\}, \tag{20}$$

where $g = \arg\min_{g \in \partial f} \parallel g \parallel$. These conditions are summarized in [13].

With regard to Sℓ_1QP-AMMO, the smoothness and boundedness conditions are satisfied by assumption, and the zero- and first-order matching is guaranteed by the first-order consistency. To assure convergence, the following specifications must be made in the general Sℓ_1QP-AMMO framework.

Algorithm 5: Updates for Sℓ_1QP-AMMO with sufficient decrease

Select $\Delta_{max} > 0$, $0 < r_1 \leq r_2 \leq 1$ and $0 < 1/\kappa_3 \leq \kappa_1 \leq \kappa_2 < 1 < \kappa_3$.

$$\text{Set } (x_{k+1}) = \begin{cases} x_k + s_k & \text{if } \rho_k \geq r_1 \\ x_k & \text{otherwise.} \end{cases}$$

$$\text{Set } \Delta_{k+1} \in \begin{cases} [\kappa_1 \Delta_k, \kappa_2 \Delta_k] & \text{if } \rho_k < r_1 \\ [\kappa_2 \Delta_k, \Delta_k] & \text{if } \rho_k \in [r_1, r_2) \\ [\kappa_3 \Delta_k, \kappa_2 \Delta_{max}] & \text{if } \rho_k \geq r_2. \end{cases}$$

Convergence of Sℓ_1QP-AMMO to a first-order critical point is immediate under these conditions (see, e.g., Theorem 11.2.5 in [13]).

4.4 Example: AMMO with Variable-Fidelity Physics

To test the applicability and limitations of AMMO, we construct computational experiments that exercise a wide range of models with various descent characteristics, so that the relationship between models ranges from favorable to potentially the worst-case scenario. In qualitative terms, the relationship is favorable if the lower-fidelity model provides a long sequence of descent steps for the higher-fidelity merit function before the lower-fidelity model has to be re-calibrated. The worst-case scenario occurs when the lower-fidelity model does not capture the trends in the high-fidelity objective at all.

We evaluate the performance of AMMO as follows. The problems are solved with high-fidelity function and derivative evaluations, using well-known commercial optimization software[1], such as NPSOL [16] and PORT [21], in order to obtain a baseline for the amount of work needed to find an optimum. The problems are then solved in the AMMO framework under investigation.

Results of experiments with variable-resolution models can be found in [5]. The example in this section is abstracted from [6]. It was constructed with variable-fidelity models in anticipation of an adverse situation. In fact, it appeared to be the worst-case scenario which serves well to demonstrate the principles of AMMO.

We have considered aerodynamic optimization of a multi-element airfoil designed to operate in transonic conditions [28]. The transonic free-stream Mach number and the multi-element nature of the airfoil make inclusion of the viscous effects very important for obtaining physically correct results. This is confirmed in Figure 1, which depicts the Mach number in the flow for the high-fidelity and low-fidelity models, where the boundary and shear layers are clearly visible in the viscous case. To capture the viscous effects,

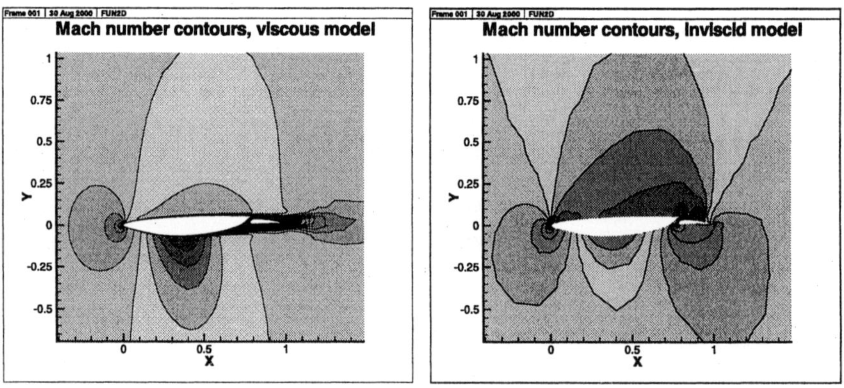

Fig. 1. Mach number contours for viscous vs. inviscid model. In the viscous model on the right, the viscous effects are clearly visible around the flap.

the governing equations of the high-fidelity model are the Reynolds-averaged Navier-Stokes (RANS) equations. Detailed descriptions can be found in [7, 8]. The Euler equations represent the low-fidelity model. The analysis in both the RANS and Euler modes is provided by FUN2D, an unstructured mesh flow solver [7]. The sensitivity derivatives are computed via a hand-coded adjoint approach [8].

The mesh for the viscous model consists of 10449 nodes and 20900 triangles. The mesh for the inviscid model comprises 1947 nodes and 3896 trian-

[1] The use of names of commercial software in this paper is for accurate reporting and does not constitute an official endorsement, either expressed or implied, of such products by the National Aeronautics and Space Administration or ICASE.

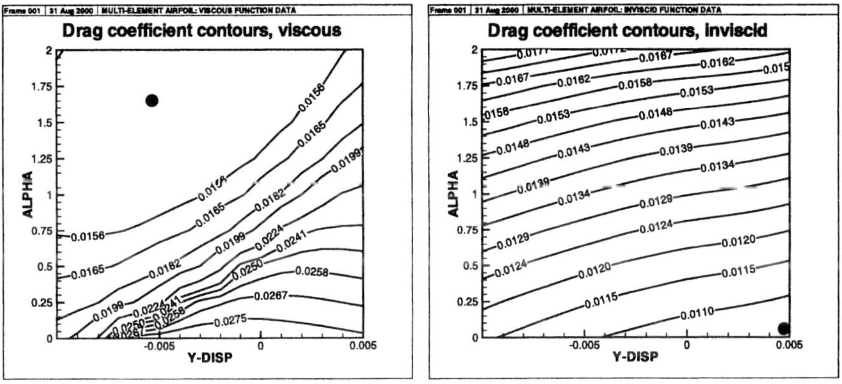

Fig. 2. Drag coefficient contours of the viscous and inviscid models. In each figure the circle represents the optimal solution. Note the opposite trends in the models.

gles. The free-stream Mach number is $M_\infty = 0.75$, the Reynolds number is $Re = 9 \times 10^6$, and the global angle of attack is $\alpha = 1°$.

The objective of this problem is to minimize the drag coefficient by adjusting the global angle of attack and the y-displacement of the flap. The first test case is restricted to two design variables to enable visualization. The baseline case for both models was constructed at $\alpha = 1°$ and zero y-displacement of the flap. Figure 2 depicts the level sets of the drag coefficient for the viscous and inviscid models. The solution for each problem is marked with a circle. The problem manifests the most adverse situation: not only is the low-fidelity model not a good approximation of the high-fidelity model but, in fact, the descent trends in the two models are reversed.

We conducted our experiments on an SGI™ Origin™ 2000 workstation with four MIPS RISC R10000 processors. One low-fidelity analysis took approximately 23 seconds and one low-fidelity sensitivity analysis took between 70 and 100 seconds. In contrast, one high-fidelity analysis took approximately 21 minutes and one high-fidelity sensitivity analysis took approximately 42 minutes to compute. The measures were taken in CPU time.

The problem was first solved with single-fidelity models alone using PORT, then with AMMO. As the time for low-fidelity computations was negligible

Test	hi-fi eval	lo-fi eval	total t	factor
PORT with hi-fi analyses, 2 var	14/13		$\approx$ 12 hrs	
AMMO, 2 var	3/3	19/9	$\approx$ 2.41hrs	$\approx$ 5
PORT with hi-fi analyses, 84 var	19/19		$\approx$ 35 hrs	
AMMO, 84 var	4/4	23/8	$\approx$ 7.2hrs	$\approx$ 5

Table 1. AMMO performance vs PORT

in comparison to the high-fidelity computations, we estimated the savings strictly in terms of high-fidelity evaluations.

Because the ability to handle high-dimensional problems is an important attribute of first-order AMMO frameworks and because some numerical effects could be an artifact of low dimensionality, the experiments were repeated for the problem with 84 variables. The additional variables were represented by parameters describing the shape of the airfoil. The time per analysis for the 84-variable case was approximately the same as for the two-variable case. However, the time per sensitivity increased to nearly two hours per sensitivity calculation. Table 1 summarizes the number of function (first number) and derivative (second number) computations expended in PORT and in AMMO for both problem formulations. PORT and AMMO produced the same final designs.

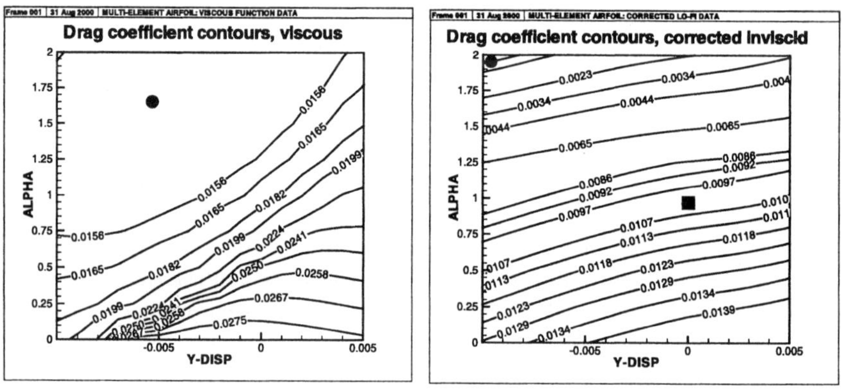

Fig. 3. Drag coefficient level sets of the viscous and corrected inviscid models. The square represents the initial iterate around which the first correction is built. The circle on the left is the optimum of the high-fidelity problem. The circle on the right is the optimum of the corrected high-fidelity model.

Despite the dissimilarity between the high-fidelity and low-fidelity models, the β-correction worked extremely well. This is illustrated in Figure 3 for the two-variable case. The plot on the left shows the level sets of the high-fidelity model with the solution. The plot on the right depicts the level sets of the low-fidelity model β-corrected at the initial point. The initial point is marked by a square. Here we have applied the correction to the entire region to visualize the effect of the correction on the low-fidelity function. The figure clearly shows that the correction, using the function and derivative information at the initial design, reversed the trend of the low-fidelity model, allowing the optimizer to find the next iterate in the northwest corner of the plot, marked by a circle. Similar analysis can be conducted for all iterations. In fact, AMMO located the solution ($\alpha = 1.6305°$, flap y-displacement $= -0.0048$) of the high-fidelity problem at the next iteration. The high-fidelity drag coefficient at the initial point was $C_D^{\text{initial}} = 0.0171$, the high-fidelity drag coefficient at the solution was $C_D^{\text{final}} = 0.0148$, a decrease of approximately 13.45%. The drag reduction in the 84-variable case was approximately 25%.

5 Concluding remarks

The proposed AMMO techniques can be used in conjunction with any method based on derivatives. Integration with simulation codes is straightforward in that it does not require alterations to simulation software. Convergence analysis is a direct consequence of the analysis for underlying optimization algorithms.

Sensitivity information is becoming increasingly available as part of simulation codes. Automatic differentiation techniques generate sensitivity information with relative simplicity, and appreciation of the utility of sensitivities in design and optimization has also led to a more widespread implementation of efficient adjoint techniques. Because sensitivity information is so useful in constructing models for the purposes of optimization, we believe the first-order framework we have described will prove very useful.

In our studies of AMMO and variable-fidelity models we have observed significant computational savings over conventional optimization. Despite the sometimes significant dissimilarity between high- and low-fidelity models, we were able to capture the descent behavior of the high-fidelity model with the assistance of first-order corrections. When the models are greatly dissimilar, first-order information appears indispensable in obtaining descent directions. The first-order corrected model leads to efficient optimization. Given these results, we are cautiously hopeful about several much larger single-discipline tests, multidisciplinary optimization cases, and a variety of alternative modeling options that we are currently studying.

Our experience so far with AMMO suggests several intriguing directions of inquiry concerning the nature of models for optimization. As noted, even though the low-fidelity model in the example of Section 4.4 was a lousy predictor for the high-fidelity model, the corrected low-fidelity model served us very well. In the case of this particular example, the cause is clear. The nearly linear uncorrected low-fidelity function is easily manipulated with first-order corrections, and, in fact, the corrected low-fidelity model is a good approximator for the nearly convex and quadratic high-fidelity function. The generality of this situation needs to be investigated. Colleagues at the University of Montreal have communicated to us an independent substantiation of this phenomenon for other CFD optimization problems [27]. We are also considering other correction schemes and appropriate termination criteria for the subproblem solutions.

Acknowledgments

The second author's research was supported by the National Aeronautics and Space Administration under NASA Contract No. NAS1-97046 while the author was in residence at ICASE, NASA Langley Research Center, Hampton, VA 23681.

References

1. N. M. ALEXANDROV, *On managing the use of surrogates in general nonlinear optimization*, September 1998. AIAA Paper 98-4798.
2. N. M. ALEXANDROV AND R. M. LEWIS, *A trust region framework for managing approximation models in engineering optimization*, September 1996. AIAA Papers 96-4101 and 96-4102.
3. ——, *Algorithmic perspectives on problem formulations in MDO.* AIAA Paper 2000-4719, 2000.
4. ——, *Analytical and computational properties of distributed approaches to MDO.* AIAA Paper 2000-4718, September 2000.
5. N. M. ALEXANDROV, R. M. LEWIS, C. R. GUMBERT, L. L. GREEN, AND P. A. NEWMAN, *Optimization with variable-fidelity models applied to wing design.* AIAA Paper 2000-0841, January 2000. Accepted for publication in the AIAA Journal of Aircraft.
6. N. M. ALEXANDROV, E. J. NIELSEN, R. M. LEWIS, AND W. K. ANDERSON, *First-order model management with variable-fidelity physics applied to multi-element airfoil optimization.* AIAA Paper 2000-4886, September 2000.
7. W. K. ANDERSON AND D. L.BONHAUS, *An implicit upwind algorithm for computing turbulent flows on unstructured grids*, Computers and Fluids, 23 (1994), pp. 1–21.
8. ——, *Aerodynamic design on unstructured grids for turbulent flows*, Tech. Rep. NASA TM 112867, NASA Langley Research Center, June 1997.
9. J.-F. M. BARTHELEMY AND R. T. HAFTKA, *Approximation concepts for optimum structural design—a review*, Structural Optimization, 5 (1993), pp. 129–144.
10. S. BURGEE, A. GIUNTA, V. BALABANOV, B. GROSSMAN, W. MASON, R. NARDUCCI, R. HAFTKA, AND L. WATSON, *A coarse-grained parallel variable-complexity multidisciplinary optimization problem*, The International Journal of Supercomputing Applications and High Performance Computing, 10 (1996), pp. 269–299.
11. K. J. CHANG, R. T. HAFTKA, G. L. GILES, AND P.-J. KAO, *Sensitivity-based scaling for approximating structural response*, Journal of Aircraft, 30 (1993), pp. 283–288.
12. A. R. CONN, N. GOULD, AND P. L. TOINT, *LANCELOT: a Fortran package for large-scale nonlinear optimization (release A)*, vol. 17 of Springer series in computational mathematics, Springer-Verlag, New York, 1992.
13. ——, *Trust-Region Methods*, MPS-SIAM Series on Optimization, SIAM-MPS, Philadelphia, 2000.
14. A. R. CONN, N. I. M. GOULD, AND P. L. TOINT, *A globally convergent augmented Lagrangian algorithm for optimization with general constraints and simple bounds*, SIAM Journal on Numerical Analysis, 28 (1991), pp. 545–572.

15. R. FLETCHER, *Practical Methods of Optimization*, John Wiley & Sons, Chichester, 1989. Second Edition.
16. P. E. GILL, G. H. GOLUB, W. MURRAY, AND M. H. WRIGHT, *User's guide for SOL/NPSOL: a Fortran package for nonlinear programming*, Department of Operations Research, Stanford University, Stanford, CA, 1983.
17. P. E. GILL, W. MURRAY, AND M. H. WRIGHT, *Practical Optimization*, Academic Press, London, 1981.
18. A. GIUNTA, *Aircraft Multidisciplinary Optimization Using Design of Experiments Theory and Response Surface Modeling Methods*, PhD thesis, Department of Aerospace and Ocean Engineering, Virginia Tech, Blacksburg, VA, 1997.
19. R. T. HAFTKA, *Combining global and local approximations*, AIAA Journal, 29 (1991), pp. 1523–1525.
20. R. T. HAFTKA, Z. GÜRDAL, AND M. P. KAMAT, *Elements of Structural Optimization*, Kluwer Academic Publishers, Dordrecht, 1990.
21. L. KAUFMAN AND D. GAY, *PORT Library: Optimization and Mathematical Programming*, Bell Laboratories, May 1997.
22. R. M. LEWIS AND S. G. NASH, *A multigrid approach to the optimization of systems governed by differential equations.* AIAA Paper 2000-4890, 2000.
23. J. C. NEWMAN, III, A. C. TAYLOR, III, R. W. BARNWELL, P. A. NEWMAN, AND G. J.-W. HOU, *Overview of sensitivity analysis and shape optimization for complex aerodynamic configurations*, Journal of Aircraft, 36 (1999), pp. 87–96.
24. M. POWELL, *Convergence properties of a class of minimization algorithms*, in Nonlinear Programming 2, O. Mangasarian, R. Meyer, and S. Robinson, eds., Academic Press, New York, NY, 1975, pp. 1–27.
25. M. J. D. POWELL, *Variable metric methods for constrained optimization*, in Mathematical Programming, the State of the Art, A. Bachem, M. Grotschel, and B. Korte, eds., Springer-Verlag, 1983, pp. 288–311.
26. J. RODRÍGUEZ, J. RENAUD, AND L. WATSON, *Trust region augmented Lagrangian methods for sequential response surface approximation and optimization*, in Proceedings of DETC '97, September 1997. ASME paper DETC97DAC3773, presented at the 1997 ASME Design Engineering Technical Conferences, September 14–17, Sacramento, California.
27. J.-Y. TREPANIER. Personal communication, 2001.
28. R. T. WHITCOMB AND L. R. CLARK, *An airfoil shape for efficient flight at supercritical mach numbers*, Tech. Rep. NASA TM-X-1109, NASA Langley Research Center, 1965.

Multifidelity Global Optimization Using DIRECT

Steven E. Cox and Raphael T. Haftka

University of Florida, Gainesville, FL

Abstract. Multifidelity global optimization methods have the potential to allow the optimization of problems which are computationally expensive and have high dimensionality. The DIRECT algorithm has been modified to incorporate multi-fidelity information. Two correction strategies have been employed with problems with up to ten dimensions and thousands of local optima. DIRECT has demonstrated the ability to reach the same global optimum using high fidelity analyses or corrected low fidelity function values. Initial work with two correction strategies have shown a minor reduction in the number of high-fidelity analyses required but refinements to the stopping criteria have the potential for greatly reducing the cost of the optimization.

1 Introduction

The high cost of computer simulations has driven interest in optimization based on multiple levels of fidelity–combining a large number of inexpensive less accurate analyses with a small number of expensive and accurate ones. For local optimization, one approach has been to approximate the ratio between a high-fidelity and a low-fidelity analysis as a linear polynomial based on derivatives, the so-called Global Local Approximation (GLA) [1]. GLA has been used successfully on a variety of engineering design problems (e.g., [2]. It was also used as a component for a rigorous trust region algorithm [3]. Another approach that has a more global flavor is to fit a polynomial response surface to the ratio (e.g., [4]). The objective of the present work is to extend the multi-fidelity approach to global optimization.

Many classes of algorithms handle different types of global optimization problems [5]. Genetic Algorithms, Simulated Annealing and other stochastic methods have been used extensively for problems with discrete or nondifferentiable design spaces [6–8] while Lipschitzian optimizers and multistart methods are popular for continuous design spaces [9, 10]. Approximations have been used extensively in global optimization to reduce the cost and locate the region of the global optimum before using expensive high-fidelity analyses for the final convergence [4, 11, 12]. This can be useful in locating the global optimum, but for some problems the global optimum is very different than the global optimum of the approximation. For this reason we seek a multifidelity global optimization algorithm that combines a low-fidelity approximation with high-fidelity analyses to improve the likelihood of locating the global optimum.

The DIRECT global optimization algorithm has been found to be an efficient optimization method for moderate numbers of dimensions [13]. We have used a modified form of the DIRECT algorithm to optimize the conceptual design of a High Speed Civil Transport (HSCT) configuration with up to 26 design variables as well as several algebraic test problems. DIRECT is known for being quick to get close to the global optimum but the final convergence is very slow. By using DIRECT to get close to the global optimum and switching to a local optimizer for the final convergence from a few of the most promising points, it is possible to combine the fast global search of DIRECT with the rapid convergence of a local optimizer. This method has compared favorably with two different multistart local optimization strategies, especially for problems with many widely scattered local optima.

Although DIRECT is relatively efficient and robust, it may still require tens of thousands of evaluations to locate the global optimum for more than about 10 design variables for nonconvex problems. To reduce the cost, we explore a multifidelity approach based on the GLA technique.

2 DIRECT Algorithm

The DIRECT algorithm, [8], is a variation of Lipschitzian optimization that uses all values for the Lipschitz constant. Lipschitzian optimization requires the user to specify the Lipschitz constant, K, which is used as a prediction of the maximum possible slope of the objective function. Lipschitzian optimization uses the value of the objective function at the corners of each box and K to find the box with potentially the lowest objective function value. The function is evaluated at the predicted minimum possible value and the process is repeated for a set number of iterations. (See Figure 1.)

DIRECT does not require an estimate of the Lipschitz constant. It uses the function value at the center of each box and the box size to find the boxes which potentially contain the optimum. A box is selected if using some Lipschitz constant K, that box could contain the lowest function value. (See Figure 2.) It can be shown that this requires the box to lie on the bottom part of the convex hull of the set of boxes in a graph such as Figure 2. In Figure 2, each of the boxes is one of four sizes. The best design from the smallest box and largest and next to largest boxes are potentially optimal because, for each point, there is some value of K where that box could contain a better design than any other box.

The DIRECT algorithm is as follows:

1. Normalize the search space to the unit box. Let c_1 be the center point of the box and evaluate $f(c_i)$.
2. Identify the set S of potentially optimal boxes.
3. For each box $j \in S$:
   (a) Identify the set of dimensions I with the longest side length. Let δ equal one-third of this length.

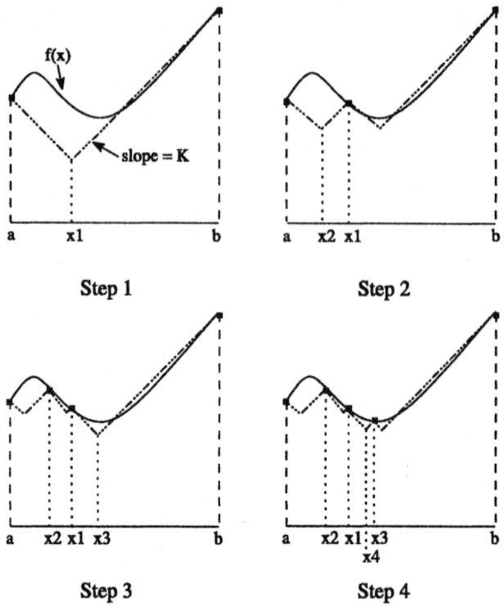

Figure 1. Lipschitzian Optimization

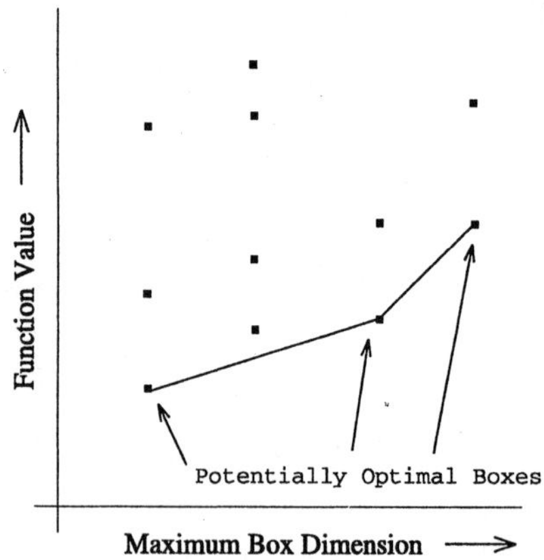

Figure 2. DIRECT Point Selection

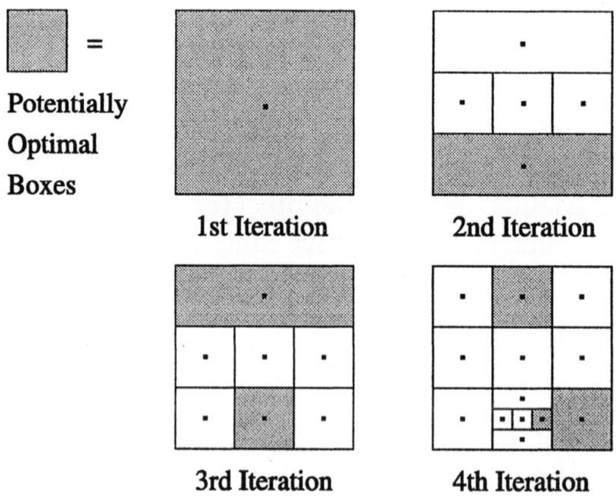

Figure 3. DIRECT box division example

(b) Sample the function at the points $(c \pm \delta e_i)$ for all $i \in I$, where c is the center of the box and e_i is the ith unit vector.
(c) Divide the box containing c into thirds along the dimensions in I, starting with the dimension with the lowest value of $f(c \pm \delta e_i)$ and continuing to the dimension with the highest $f(c \pm \delta e_i)$.

4. Repeat 2. - 3. until stopping criterion is met.

Graham's Scan routine is used to identify the potentially optimal boxes, [8]. Boxes that were not previously potentially optimal can become potentially optimal in later iterations as the boxes are divided. (See Figure 3.) DIRECT can also select more than one box to divide at each iteration. This can allow for a wider search than Lipschitzian optimization and makes DIRECT ideal for parallelization.

DIRECT was found to be quick to locate regions of local optima but slow to converge. To speed up the convergence, DIRECT is stopped once the smallest box reaches a specified percentage of the original box size and a local optimizer is used for the final optimization. For our experiments, the optimization was stopped when the smallest box reached 0.01% of the original box size. DOT was then used starting from up to 15 of the best points analyzed by DIRECT, which were at least 5% of the width of the design space away from the other starting points selected for local optimization.

3 Multifidelity DIRECT

In order to incorporate multifidelity data in the DIRECT algorithm, a correction ratio is used. The goal of the correction scheme was to improve the accuracy of the low-fidelity analysis over the entire design space while putting the most emphasis on points near the optima found by DIRECT. The DIRECT algorithm provides a ready-made method for selecting points to perform high-fidelity analyses without substantially adding complexity to the algorithm. DIRECT selects boxes for division that are either large, which puts their centers far away from other analyzed points, or have low function values, which implies that they are near an optimum. This is also the desired distribution of high-fidelity points. For the multifidelity implementation of the DIRECT algorithm, whenever DIRECT selects a box for division, a high-fidelity analysis is performed at the center of the box before it is divided. The correction factor at this point would then be used to calculate a correction for the low-fidelity analyses generated in that box in one of two ways.

3.1 Constant Correction Ratio

The first constant correction factor (CCF), multifidelity version of DIRECT uses a constant correction factor for each box. Figure 4 shows how this correction is applied. The correction factor is based on the center point of the last box that the low-fidelity point was in before the box was divided. The correction factor for the box to be divided is calculated immediately before the new points are analyzed and the correction is applied before the box is split into separate boxes. This avoids the need for response surfaces or other surrogates. If one of the corrected boxes is divided at a later iteration, a new high-fidelity value is calculated at the center of that box to correct the new low fidelity points that are generated and the center point.

Equation 1 shows how the low-fidelity values are corrected in the CCF multifidelity DIRECT. α is the ratio of $f_{high_{center}}$ and $f_{low_{center}}$, the high and low-fidelity function values at the center of the box being divided, while

f_{low} and f_{corr} are the low-fidelity and corrected function values at the new points that are generated in the box.

$$f_{corr} = \alpha f_{low} \tag{1}$$

The CCF multifidelity DIRECT algorithm is outlined as follows. The components added for the multifidelity version are in boldface.

1. Normalize the search space to the unit box. Let c_1 be the center point of the box and evaluate $f_l\,(c_i)$
2. Identify the set S of potentially optimal boxes.
3. For each box $j \in S$:
   (a) **Perform a high-fidelity analysis at the center point of the box and calculate f_h**
   (b) **Calculate the ratio $\alpha = f_h/f_l$ at the center**
   (c) Identify the set of dimensions I with the longest side length. Let δ equal one-third of this length.
   (d) Sample the function at the points $c \pm \delta e_i$ for all $i \in I$, where c is the center of the box and e_i is the ith unit vector.
   (e) **Apply the correction ratio to the low-fidelity values within the box $f_{corr} = f_l\, X\, \alpha$**
   (f) Divide the box containing c into thirds along the dimensions in I, starting with the dimension with the lowest value of $f_{cor}\left(c \pm \delta e_i\right)$ and continuing to the dimension with the highest $f_{cor}\left(c \pm \delta e_i\right)$.
4. Repeat 2. - 3. until stopping criterion is met.

3.2 Linear Correction Response Surface

The second correction method that we have studied uses a linear correction response surface (LCRS) to predict the correction factor for the low-fidelity points. A linear fit to the correction factor at two points per dimension is used to predict the slope of the correction factor in part of the design space as described next.

For each potentially optimal box, a high-fidelity analysis is performed at the center, the correction factor at that point is calculated, and this correction factor is then compared with the correction factor used originally at that point. If the correction factor was accurate enough, then the slopes used in that box will not be updated. The correction factor at that point will be combined with the slopes of the correction factor response surface for that region and used to calculate the correction factor for the low fidelity points generated in that box. If the correction factor was not accurate enough, then a high and a low-fidelity analysis are performed at each new point that is generated to divide the potentially optimal box. The corrected value of each new point will then be set to the high-fidelity value. The correction factors at these points will generate new slopes to be used in the boxes that result from dividing the potentially optimal box. For any dimensions that were not divided at this time, the slope of the response surface used to calculate the corrected value at the center of the potentially optimal box is copied over to the new response surface. For our initial work the allowable error in the correction factor was set to 20%.

This correction method refines the correction factor used within each box by incorporating a first order approximation to the correction factor. By checking the accuracy of the correction when each box is divided, the algorithm can identify where the correction response surface is still accurate and where it is in need of refinement. As each correction response surface is generated, it is numbered and the slopes are stored in an array. For each box in the design space a variable is used to indicate which response surface it should use to calculate the correction factor for the points used when it is divided.

This version of the multifidelity DIRECT took much longer to converge than the original form of DIRECT. The convergence criterion that was used is based on the size of the smallest box in the design space just as in the original version of DIRECT. For this version, the size of the smallest box size shrunk rapidly at the start of the optimization but took a large number of iterations to get any smaller. DIRECT considers the points on the lower, right side of the convex hull of points in Figure 2 to be potentially optimal. Points on the left side of the convex hull are not potentially optimal because one can find a competing point that is both lower and is in a larger box. For the LCRS version of DIRECT, after a few iterations, the smallest box generally had a corrected function value that was slightly higher than the corrected function value of a slightly larger box. This continued for a large number of iterations before a box with the smallest size would get divided. The difference between the function values of the smallest box and the smallest one selected for division was generally very small and comparable to the size of the error in the corrected value, but it resulted in an order of magnitude increase in the number of function evaluations used by DIRECT. For this reason, we chose to include points on the lower, left side of the convex hull provided their value was less than 0.1% higher than the next largest box on the convex hull. This was an arbitrary value based on limited experimentation, but it worked well for our test problem.

Equation 2 shows how the low-fidelity values are corrected in the LCRS multifidelity DIRECT for a box using the $j^{(th)}$ response surface. $f_{high_{center}}$ and $f_{low_{center}}$ are the high and low-fidelity function values at the center of the box being divided while f_{low} and f_{corr} are the low-fidelity and corrected function values at the new points that are generated. $RS_j(i)$ is the coefficient for the slope in the e_i direction of response surface j and $(x_i - x_{i_{center}})$ is the distance of the new point from the center of the box being divided.

$$f_{corr} = \left(\frac{f_{high_{center}}}{f_{low_{center}}} \sum_{i=1}^{n} (RS_j(i)(x_i - x_{i_{center}})) \right) f_{low} \tag{2}$$

The LCRS version of the multifidelity DIRECT algorithm is described as follows. The components added for this version are in boldface.

1. Normalize the search space to the unit box.
2. Divide Box 1
   (a) Perform **high** and low fidelity analysis at the points $candc \pm \delta e_i$ for all$i$ ** I, where c is the center of the design space and e_i is the ith unit vector.
   (b) $f_{cor} = f_{high}$ for each point analyzed
   (c) **Create correction response surface (RS) from initial points**
   (d) **Store coefficients in RS array and set 'RS to use' to 1 for this set of boxes**
   (e) Divide into thirds, starting with the dimension with the lowest value of $f_{cor}(c \pm \delta e_i)$ and continuing to the dimension with the highest $f_{cor}(c \pm \delta e_i)$.
3. Identify the set S of potentially optimal boxes.
4. For each box $j \in S$:
   (a) If no high fidelity analysis at center
      i. **Perform high fidelity analysis at center**
      ii. **Compare actual correction factor (CF) to CF used previously in that box**
      iii. **If correction is inaccurate**
         A. Identify the set of dimensions I with the longest side length. Let δ equal one-third of this length.
         B. Perform **high** and low fidelity analysis at the points $c \pm \delta e_i$ for all $i \in I$, where c is the center of the potentially optimal box and e_i is the ith unit vector.
         C. $f_{cor} = f_{high}$
         D. Create correction RS from initial points. Use coefficients from previous RS for undivided dimensions
         E. Store RS in next row of RS array
         F. Update 'RS to use' in new boxes and divided box to indicate new RS
      iv. **If correction is OK**
         – **Get 'RS to use' from potentially optimal box**
         – **Get RS from RS array**
         – Identify the set of dimensions I with the longest side length. Let δ equal one-third of this length.
         – Perform low-fidelity analysis at the points $c \pm \delta e_i$ for all $i \in I$, where c is the center of the box and e_i is the ith unit vector.

- **Apply RS and CF at the center of the box to low fidelity values**

(b) If previous high fidelity analysis at center

i. Identify the set of dimensions I with the longest side length. Let δ equal one-third of this length.

ii. Perform low-fidelity analysis at the points $c^+_- \delta c_i$ for all $i \in I$, where c is the center of the box and e_i is the ith unit vector.

iii. **Get 'RS to use' from potentially optimal box and the appropriate RS from the RS array**

iv. **Use RS and CF at the center of the box to calculate the CF for the low fidelity values**

(c) Divide into thirds, starting with the dimension with the lowest value of $f_{cor}\left(c^+_- \delta e_i\right)$ and continuing to the dimension with the highest $f_{cor}\left(c^+_- \delta e_i\right)$.

5. Repeat 3. - 4. until stopping criterion is met.

4 Test Problem

The test problem used on the multifidelity DIRECT was designed to provide a large number of local optima scattered throughout the design space. The high-fidelity test function is a 6th order function confined to the box $[-2, 2]^n$ defined as:

$$y = 25 + \sum_{i=1}^{n} \{2(x_i + e_i)^2 - 1.5(x_i + e_i)^4 + 0.2(x_i + e_i)^6\} \tag{3}$$

The one-dimensional graph of this function is given in Figure 5.

The low-fidelity function is a truncated Chebyshev expansion of the high-fidelity function. It drops the 6th order term of the Chebyshev expansion of the high-fidelity equation. This represents a problem where the simple model fails to capture the higher order behavior of the true function. The low-fidelity function is given by:

$$\begin{aligned} y = 25 + \sum_{i=1}^{n} \quad & \left[-1 - 2\left(2\left(\tfrac{x_i+e_i}{2}\right)^2 - 1\right)\right. \\ & \left.-0.6\left(8\left(\tfrac{x_i+e_i}{2}\right)^4 - 8\left(\tfrac{x_i+e_i}{2}\right)^2 + 1\right)\right] \end{aligned} \tag{4}$$

The one-dimensional graph of the function is given in Figure 6.

Test Results for the two versions of the multifidelity DIRECT were compared against the original version of DIRECT for optimizing the high-fidelity function. For each version, the optimization was stopped once the smallest box diameter reached 0.01% of the original box side length. At that point DOT was started from up to 15 starting points separated by at least 5% of the width of the design space. Each optimizer was run 30 times for each

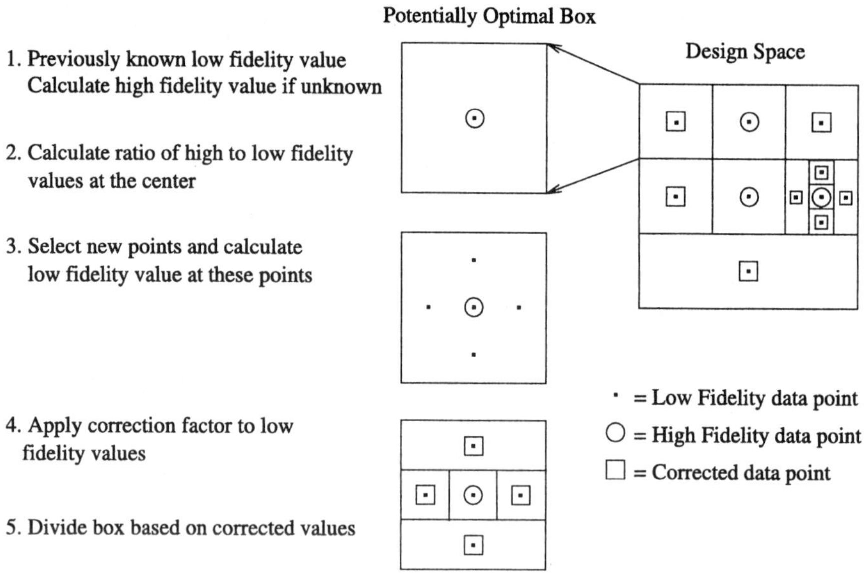

Figure 4. Constant Correction Factor application for multifidelity DIRECT

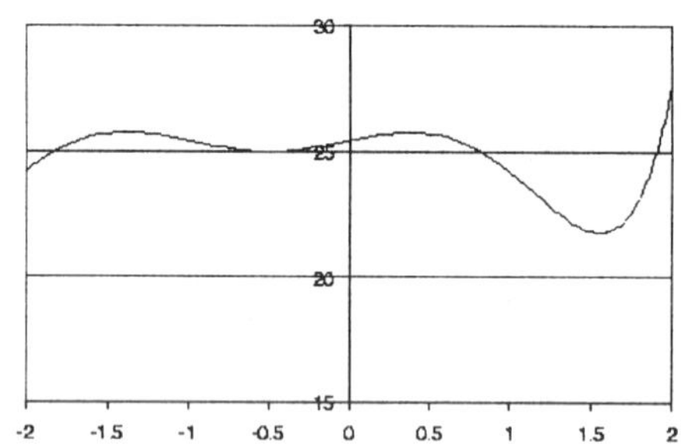

Figure 5. High-fidelity Function in One Dimension (e=.3)

case with random values for e and the results were averaged. Table I gives the results for the constant correction factor and response surface versions of DIRECT.

For the CCF version of DIRECT, the 5 and 10 dimensional case resulted in a slight improvement in the number of high-fidelity analyses used by the optimization but at the cost of a large number of low-fidelity analyses. The results from the two dimensional case indicate that this correction scheme can cause problems for DIRECT. As the function values at the center of each box change from corrected values to high-fidelity values, the points that were near the convex hull can change. This can cause the optimizer to select boxes for division that would normally not be selected by the original version of DIRECT. An animation of the point selection for the two dimensional case

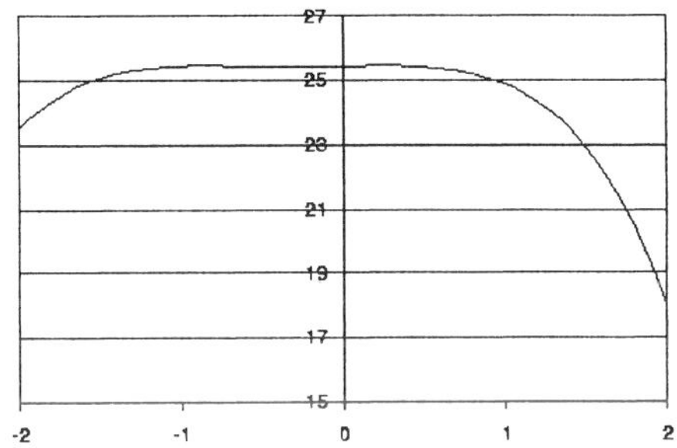

Figure 6. Low-fidelity Function in One Dimension (e=.3)

Table 1. Multifidelity versions of DIRECT vs. Single Fidelity DIRECT

	Original DIRECT	CCF DIRECT		LRS DIRECT	
	High	High	Low	High	Low
2 DV	688	1562	4170	588	439
5 DV	2202	1650	2500	2687	6710
10 DV	8345	6673	34642	10026*)	57555*

showed the optimizer jumping to boxes where the corrected low-fidelity value is lower than the high-fidelity value. For this case the optimizer performed a wider search of the design space than the single fidelity version and required more evaluations to reach the optimum.

For the LCRS version of DIRECT, the opposite trend was the case. As the number of dimensions increased, the optimizer became much less efficient at locating the global optimum. For the 10D case, 6 out of the 30 runs failed to locate the global optimum before converging normally and for 12 of the runs which did locate the global optimum, the optimizer was stopped by the upper limit on the number of points to be analyzed, which was set to 100,000 low-fidelity evaluations.

One difference between the two dimensional case and the other cases is that the low-fidelity function becomes negative in some areas for more than 2 dimensions. This can cause problems in areas where the objective function is changing sign. In order to test what effect this had on the optimization the high and low-fidelity functions were changed so that the low-fidelity function remained positive for the entire design space. The high-fidelity function was changed to:

$$y = \sum_{i=1}^{n} \{10 + 2(x_i + e_i)^2 - 1.5(x_i + e_i)^4 + 0.2(x_i + e_i)^6\} \tag{5}$$

The low-fidelity function was set to:

$$y = \sum_{i=1}^{n} \left[10 - 1 - 2\left(2\left(\frac{x_i+e_i}{2}\right)^2 - 1\right) - 0.6\left(8\left(\frac{x_i+e_i}{2}\right)^4 - 8\left(\frac{x_i+e_i}{2}\right)^2 + 1\right)\right] \tag{6}$$

This prevents both functions from becoming negative at any point in the design space for any number of dimensions. The results for this version of the test problems are given in Table II.

Table 2. Multifidelity DIRECT Comparison for Modified Problem

	Original DIRECT	CCF DIRECT		LRS DIRECT	
	High	High	Low	High	Low
2 DV	688	1553	4069	612	541
5 DV	2181	26842	150000*	2205	6288
10 DV	8333	17885	150000*	8237*	48884*

For the two DV problem, both versions performed approximately the same as for the original problem. This was expected since the problems were basically the same. However, for the other two cases the LCRS version of DIRECT was slightly cheaper than for the original problem. This seems to indicate that it is more efficient when the high and low-fidelity functions remain the same sign during the optimization. For the 10 DV case, the LCRS version failed to locate the global optimum for seven out of the 30 runs but none of them reached the maximum allowable number of function evaluations. For the CCF version of DIRECT, the optimizer ran until it reached the maximum number of allowable function evaluations for the five and 10 DV cases which was 150,000. For these problems the optimizer had the same problem that LCRS version had originally. After a small number of iterations, the smallest boxes were not divided until most of the boxes that were slightly larger were divided. In order to reduce the number of function evaluations that were required the same modification to the potentially optimal box division that was used in the LCRS version was used for the CCF version. Boxes on the convex hull that were smaller than the one containing the best function value but whose function value was less than 0.1% worse were selected as potentially optimal. The results from this version of the CCF DIRECT are given in Table III.

By allowing DIRECT to divide boxes that are smaller than the best function value but only slightly worse, the CCF version converged at approximately the same number of function evaluations as the LCRS version but none or the CCF runs resulted in a sub-optimal solution. By using a constant correction factor, the CCF version avoids spurious corrected function values

Table 3. Multifidelity Constant Correction Factor DIRECT with Modified Potentially Optimal Box Selection

	CCF DIRECT (*ModifiedVersion*)	
	High	Low
2 DV	610	512
5 DV	2350	7477
10 DV	8557	49557

caused by large differences in the predicted slope of the correction factor vs. the actual slope.

For both versions of the Multifidelity DIRECT, the actual number of boxes was much higher than for the single fidelity version of DIRECT. Due to the errors in the corrected values, the boxes with the lowest corrected function values tended to have a larger true function value when a high-fidelity analysis was performed. This resulted in a much slower reduction to the minimum box size. However, when the optimization was stopped after a smaller number of iterations than it would require using the box size criterion, in most cases it could still locate the global optimum. This is an indication that a different stopping criterion may be useful for the Multifidelity DIRECT.

After the boxes become small, the difference between the corrected value at a point and the high-fidelity value can be relatively small. However, the method that we are currently using performs a high fidelity analysis at each potentially optimal box. By reducing the number of high-fidelity analyses in the congested regions we could greatly reduce the cost of the optimization. Similarly, by increasing the order of the correction response surface, the correction can be applied to a much larger region. Further research will explore the use of a single correction response surface over the entire design space.

5 Conclusions

None of the multifidelity methods used in this study resulted in a substantial reduction in the number of high fidelity analyses used in the optimization. However, the results have shown that it is possible to correct the low fidelity function and use the corrected value in a DIRECT optimization. Some of the main issues that remain to be explored for the correction factor based multifidelity direct are: the proper convergence criteria to use, the possibility of using a correction factor for multiple divisions instead of using a high-fidelity analysis for each potentially optimal box, and the use of a global correction response surface instead of the localized surfaces used so far.

6 Acknowledgment

This research was supported in part by NASA grant NAG2-1179.

References

1. Haftka, R. T. "Combining global and local approximations." *AIAA Journal*, Vol. 29, No. 9, 1991, pp. 1523-1525.
2. McQuade, P.D., Eberhardt, S. and Livne, E., "CFD-Based Aerodynamic Approximation Concepts Optimization of a Two-Dimensional Scramjet Vehicle", *Journal of Aircraft*, Vol. 32, NO. 2, March-April 1995, pp. 262-269
3. Alexandrov, N. M., Dennis, J. E., Lewis, R. M., and Torczon, V., "A Trust-region Framework for Managing the use of Approximation Models in Optimization", *Structural Optimization*, Vol 15, 1998, pp. 16-23
4. Floudas C.A. and Pardalos P.M. (1996), State of the Art in Global Optimization, Kluwer Academic Publishers, Dordrecht, The Netherlands
5. Ramrez-Rosado, I. J. and Bernal-Augustn, "Genetic Algorithms Applied to the Design of Large Power Distribution Systems", *IEEE Transactions on Power Systems*, Vol. 13, NO. 2, May 1998, pp. 696-703
6. Lamberti, L., Venkataraman, S., Haftka, R. T. and Johnson, T. F., "Preliminary Design Optimization of Stiffened Panels Using Approximate Analysis Models", *Submitted For Review to the International Journal for Numerical Methods in Engineering*, 2001
7. Swift, R. A. and Batill, S. M., "Simulated Annealing Utilizing Neural Networks for Discrete Variable Optimization Problems in Structural Design", *33rd AIAA/ASME/ASCE/AHS/ASC Structures, Structural Dynamics and Materials Conference*, Dallas, TX, April 1992, pp. 2536-2544
8. Jones, D.R., Perttunen, C.D., Stuckman, B.E. (1993), "Lipschitzian Optimization Without the Lipschitz Constant", *Journal of Optimization Theory and Application*, Vol.79, pp. 157-181.
9. Baker, C. A., Grossman, B., Haftka, R. T., Mason, W. H. and Watson, L. T., "HSCT Configuration Design Space Exploration using Aerodynamic Response Surface Approximations", *7th AIAA/USAF/NASA/ISSMO Symposium on Multidisciplinary Analysis and Optimization*, St Louis, Missouri, Sept. 2-4, 1998, pp. 769-777, AIAA Paper-98-4803
10. Ishikawa, T., Tsukui, Y. and Matsunami, M., "A Combined Method for the Global Optimization Using Radial Basis Function and Deterministic Approach", *IEEE Transactions on Magnetics*, Vol. 35, NO. 3, May 1999, pp. 1730-1733
11. Jones, D. R., Schonlau, M. and Welch, W. J., " Efficient Global Optimization of Expensive Black-Box Functions", *Journal of Global Optimization*, Vol.13, 1998, pp. 455-492
12. Knill, D. L., .Giunta, A. A., Baker, C. A., Grossman, B., Mason, W. H., Haftka, R. T. and Watson, L. T. "Response surface model combining linear and euler aerodynamics for super sonic transport design." *Journal of Aircraft*, Vol. 36, No. 1, 1999, pp. 75-86.
13. Cox, S., Haftka, R.T., Baker, C., Grossman, B., Mason, W. and Watson, L. T., "Global Multidisciplinary Optimization of a High Speed Civil Transport", *Proc. Aerospace Numerical Simulation Symposium '99*, Tokyo, Japan, June, 1999, pp. 23-28

Inexactness Issues in the Lagrange-Newton-Krylov-Schur Method for PDE-constrained Optimization

George Biros[1] and Omar Ghattas[2]

[1] Courant Institute of Mathematical Sciences, New York University, New York, NY 10012, USA (biros@cs.nyu.edu).

[2] Laboratory for Mechanics, Algorithms, and Computing, Departments of Biomedical Engineering and Civil & Environmental Engineering, Carnegie Mellon University, Pittsburgh, Pennsylvania, 15213, USA (oghattas@cs.cmu.edu).

Abstract. We present an overview of the Lagrange-Newton-Krylov-Schur (LNKS) method for solution of optimization problems that are governed by systems of partial differential equations. We discuss how to improve LNKS's work efficiency by carrying out certain computations inexactly, without compromising convergence. LNKS solves the Karush-Kuhn-Tucker optimality conditions using a Newton-Krylov algorithm. Its key component is a preconditioner based on variants of quasi-Newton reduced space Sequential Quadratic Programming (QN-RSQP) methods. LNKS combines the fast convergence properties of a Newton method with the ability of preconditioned Krylov methods to solve very large linear systems. Nevertheless, even with good preconditioners, solution of the optimization problem is several times more expensive than solution of the underlying PDE problem. To accelerate LNKS, its computational components are carried out inexactly: premature termination of iterative algorithms, inexact evaluation of gradients and Jacobians, and approximate line searches. We discuss several issues that arise with respect to these inexact computations, and the resulting trade-offs between speed and robustness.

1 Introduction

We discuss algorithmic and implementation aspects of the Lagrange-Newton-Krylov-Schur method for solution of large-scale nonlinearly-constrained optimization problems. The proposed techniques have applications to a broad category of optimization problems, including optimal control, optimal design, and parameter identification for systems governed by partial differential equations (PDEs). These are often known as *PDE-constrained optimization* problems.

In this article we will refer to the unknown PDE field quantities as the *state variables*; the PDE constraints as the *state equations*; solution of the PDE constraints as the *forward problem*; the inverse, design, or control variables as the *decision variables*; and the problem of determining the optimal values of the inverse, design, or control variables as the *optimization problem*.

Perhaps the most popular technique for large-scale nonlinearly-constrained optimization is the reduced space Sequential Quadratic Programming (RSQP) method [2], [17]. In its elaboration for PDE-constrained optimization, the linearized state equations and state variables are eliminated at each iteration, and a quadratic optimization problem is solved in the reduced space of decision variables. LNKS, which was introduced in [3, 6, 7], employs a family of methods that use Newton-Krylov algorithms to solve the Karush-Kuhn-Tucker optimality conditions in the full space of state and decision variables, and invokes a preconditioner motivated by reduced space ideas. We refer to the (nonlinear) Newton iterations as *outer* iterations, and use the term *inner* to denote the (linear) Krylov iterations for the Karush-Kuhn-Tucker (KKT) system that arises at each Newton iteration.

The inner iterative solver and the preconditioner that accelerate the computations of a Newton step for the KKT optimality conditions are essential ingredients of LNKS. We introduced and analyzed several variants of the preconditioner in [4, 6]. The parallelizability and scalability of the LNKS algorithm were also examined in those papers. The basic form of the outer Newton solver is analyzed in [7].

Like adjoint-based quasi-Newton RSQP methods (see e.g. [13]), this approach requires just two linearized forward solves per iteration, and in addition exhibits the fast convergence associated with Newton methods. Moreover, the two forward solves can be approximate (since they are used within the LNKS preconditioner); for example we replace them by an appropriate PDE preconditioner. LNKS builds on existing parallel PDE preconditioners and generally parallelizes and scales as well as the forward solver itself. In this paper, we extend our earlier work with a discussion of the inexact components of the LNKS method—in particular the inexact solves within the QN-RSQP steps used for globalization.

The paper is organized as follows. In Section 2 we overview the LNKS method. In Section 3 we discuss globalization approaches for the outer iteration and briefly present line-search based QN-RSQP methods. Section 4 discusses several inexact computations within LNKS. We conclude in Section 5 with numerical results from the application of the method to the optimal boundary control of the steady incompressible Navier-Stokes equations.

Some notational conventions follow. We use boldface characters to denote vector-valued functions and vector-valued function spaces. We use Roman characters to denote discretized quantities and italics for their continuous counterparts. For example $\boldsymbol{u}$ will be the continuous velocity field and $\mathbf{u}$ will be its discretization. Greek letters are overloaded and whether we refer to the discretization or the continuous fields should be clear from context. We also use $(+)$ as a subscript or superscript to denote variable updates within an iterative algorithm. Finally, quantities with a tilde on the top indicate that they are results of inexact computations.

2 LNKS Method

Let us consider the constrained optimization problem,

$$\min_{\mathbf{x}\in\mathbb{R}^N} f(\mathbf{x}) \quad \text{subject to} \quad \mathbf{c}(\mathbf{x}) = \mathbf{0}, \tag{1}$$

where $\mathbf{x} \in \mathbb{R}^N$ are the optimization variables, $f : \mathbb{R}^N \to \mathbb{R}$ is the objective function and $\mathbf{c} : \mathbb{R}^N \to \mathbb{R}^n$ are the constraints, which we assume consist of only the discretized state equations.

In order to exploit the structure of the problem we partition $\mathbf{x}$ (which lies in the full space of optimization variables) into state variables $\mathbf{x}_s \in \mathbb{R}^n$, and decision variables $\mathbf{x}_d \in \mathbb{R}^m$,

$$\mathbf{x} = \begin{Bmatrix} \mathbf{x}_s \\ \mathbf{x}_d \end{Bmatrix}, \tag{2}$$

where $m + n = N$. The Lagrangian,

$$\mathcal{L}(\mathbf{x}, \boldsymbol{\lambda}) := f(\mathbf{x}) + \boldsymbol{\lambda}^T \mathbf{c}(\mathbf{x}), \tag{3}$$

is used to convert the constrained optimization problem into a system of nonlinear equations. For convenience we introduce the following notation:

$$\begin{array}{lll}
\mathbf{A}(\mathbf{x}) := \partial_x \mathbf{c}(\mathbf{x}) & \in \mathbb{R}^{n\times N} & \text{Jacobian matrix of the constraints,} \\
\mathbf{W}(\mathbf{x}, \boldsymbol{\lambda}) := \partial_{xx} f(\mathbf{x}) + \sum_i \lambda_i \partial_{xx} \mathbf{c}_i(\mathbf{x}) & \in \mathbb{R}^{N\times N} & \text{Hessian matrix of the Lagrangian,} \\
\mathbf{g}(\mathbf{x}) := \partial_x f(\mathbf{x}) & \in \mathbb{R}^{N} & \text{gradient vector of the objective.}
\end{array}$$

The first order optimality conditions state that at a local minimum the gradient of the Lagrangian must vanish:

$$\begin{Bmatrix} \partial_x \mathcal{L} \\ \partial_\lambda \mathcal{L} \end{Bmatrix} (\mathbf{x}, \boldsymbol{\lambda}) = \begin{Bmatrix} \mathbf{g}(\mathbf{x}) + \mathbf{A}(\mathbf{x})^T \boldsymbol{\lambda} \\ \mathbf{c}(\mathbf{x}) \end{Bmatrix} = \mathbf{0} \quad (\text{or } \mathbf{h}(\mathbf{q}) = \mathbf{0}). \tag{4}$$

Customarily, these equations are called the Karush-Kuhn-Tucker or KKT optimality conditions, and points at which the gradient of the Lagrangian vanishes are called *KKT points.*

A Newton step on the optimality conditions is given by

$$\begin{bmatrix} \mathbf{W} & \mathbf{A}^T \\ \mathbf{A} & \mathbf{0} \end{bmatrix} \begin{Bmatrix} \mathbf{p}_x \\ \mathbf{p}_\lambda \end{Bmatrix} = - \begin{Bmatrix} \mathbf{g} + \mathbf{A}^T \boldsymbol{\lambda} \\ \mathbf{c} \end{Bmatrix} \quad (\text{ or } \mathbf{K}\mathbf{p} = -\mathbf{h}), \tag{5}$$

where $\mathbf{p}_x$ and $\mathbf{p}_\lambda$ are used to update $\mathbf{x}$ and $\boldsymbol{\lambda}$ from current to next iterations. The KKT optimality conditions (4) define a system of nonlinear equations. The Jacobian $\mathbf{K}$ of this system is termed the *KKT matrix.* Assuming sufficient smoothness, and that the initial guess is sufficiently close to a solution, updates obtained by (5) will converge quadratically to the solution [11]. Thus, the forward solves required for reduced methods can be avoided by remaining in the full space of state and decision variables, since it is the reduction onto the decision space that necessitates the forward solves [13]. Nevertheless, the full space approach also presents difficulties: a descent direction is not guaranteed, second derivatives are required, and the KKT system itself is very difficult to solve. The size of the KKT matrix is more than twice that of the forward problem, and it is typically very ill-conditioned. Ill-conditioning results not only from the forward problem but also from the different scales between first and second derivative submatrices. Moreover, the system is indefinite; mixing negative and positive eigenvalues is known to slow down Krylov solvers. Therefore, a good preconditioner is essential for making the method efficient.

In LNKS we use a Newton method to solve the KKT optimality conditions. To compute the Newton step we solve the KKT system using an appropriate Krylov method. At the core of the algorithm lies the preconditioner $\mathbf{P}$ for the Krylov method, which is an inexact version of the QN-RSQP algorithm. An outline of the basic LNKS method is given by Algorithm 1.

```
Algorithm 1
Choose x, λ
loop
   Check for convergence
   Compute c, g, A, W
   Solve P^{-1}Kp = P^{-1}h                              (Newton Step)
   Update x = x + p_x
   Update λ = λ + p_λ
end loop
```

To derive the preconditioner we rewrite the KKT system (5) in a block-partitioned form:

$$\begin{bmatrix} \mathbf{W}_{ss} & \mathbf{W}_{sd} & \mathbf{A}_s^T \\ \mathbf{W}_{ds} & \mathbf{W}_{dd} & \mathbf{A}_d^T \\ \mathbf{A}_s & \mathbf{A}_d & \mathbf{0} \end{bmatrix} \begin{Bmatrix} \mathbf{p}_s \\ \mathbf{p}_d \\ \mathbf{p}_\lambda \end{Bmatrix} = - \begin{Bmatrix} \mathbf{g}_s + \mathbf{A}_s^T \boldsymbol{\lambda} \\ \mathbf{g}_d + \mathbf{A}_d^T \boldsymbol{\lambda} \\ \mathbf{c} \end{Bmatrix}. \tag{6}$$

RSQP is equivalent to a block-row elimination: given $\mathbf{p}_d$, solve the last block of equations for $\mathbf{p}_s$, then solve the first to find $\mathbf{p}_\lambda$, and finally solve the middle one for the decision variable update $\mathbf{p}_d$. Therefore RSQP can be written as a particular block-LU factorization of the KKT matrix:

$$\mathbf{K} = \begin{bmatrix} \mathbf{W}_{ss}\mathbf{A}_s^{-1} & \mathbf{0} & \mathbf{I} \\ \mathbf{W}_{ds}\mathbf{A}_s^{-1} & \mathbf{I} & \mathbf{A}_d^T\mathbf{A}_s^{-T} \\ \mathbf{I} & \mathbf{0} & \mathbf{0} \end{bmatrix} \begin{bmatrix} \mathbf{A}_s & \mathbf{A}_d & \mathbf{0} \\ \mathbf{0} & \mathbf{W}_z & \mathbf{0} \\ \mathbf{0} & \mathbf{W}_{sd} - \mathbf{W}_{ss}\mathbf{A}_s^{-1}\mathbf{A}_d & \mathbf{A}_s^T \end{bmatrix}. \tag{7}$$

Note that these factors are permutable to block triangular form (this is why we refer to the factorization as block-LU) and that $\mathbf{W}_z$ is the Schur-complement for $\mathbf{p}_d$ and is given by

$$\mathbf{W}_z = \mathbf{W}_{ss} + \mathbf{A}_d^T \mathbf{A}_s^{-T} \mathbf{W}_{ss} \mathbf{A}_s^{-1} \mathbf{A}_d - \mathbf{A}_d^T \mathbf{A}_s^{-T} \mathbf{W}_{sd} - \mathbf{W}_{ds} \mathbf{A}_s^{-1} \mathbf{A}_d. \tag{8}$$

Based on the Schur-type factorization we use the following preconditioner for the KKT system:

$$\mathbf{P} = \begin{bmatrix} \mathbf{0} & \mathbf{0} & \mathbf{I} \\ \mathbf{0} & \mathbf{I} & \mathbf{A}_d^T \tilde{\mathbf{A}}_s^{-T} \\ \mathbf{I} & \mathbf{0} & \mathbf{0} \end{bmatrix} \begin{bmatrix} \tilde{\mathbf{A}}_s & \mathbf{A}_d & \mathbf{0} \\ \mathbf{0} & \tilde{\mathbf{W}}_z & \mathbf{0} \\ \mathbf{0} & \mathbf{0} & \tilde{\mathbf{A}}_s^T \end{bmatrix}. \tag{9}$$

The key components of the preconditioner are $\tilde{\mathbf{A}}_s^{-1}$ and $\tilde{\mathbf{W}}_z^{-1}$, the preconditioners for the forward problem and the reduced space (i.e. decision space) equations respectively. A natural choice for $\tilde{\mathbf{W}}_z^{-1}$ is a BFGS-like method which is commonly used to approximate the reduced Hessian in QN-RSQP methods. For an analysis of the RSQP-based preconditioner and more details on the derivations, see [6]. In [4–7] we give theoretical and numerical evidence that these preconditioners work well.

3 Globalization

Algorithm 1 is only locally convergent. Popular methodologies to globalize Newton's method—that is, allow convergence to a local minimum from any initial guess—include line search, trust region, and filter algorithms. Details can be found in [23]. Trust region methods have been successfully applied to PDE-constrained optimization [17], [19], [20]. Global convergence proofs for these methods can be found in [8]. Trust region methods are often based on the Steihaug modification [24] of the Conjugate Gradient (CG) algorithm. However, CG requires a positive (semi)definite system. For reduced space approaches, this is not a problem, since the reduced Hessian is assumed to be positive definite. To employ a trust region method within the full space, one could minimize the l_2 norm of the KKT conditions, effectively applying CG to the normal equations, but this would entail a squaring of the already very large condition number of the KKT system. Instead, we have opted for a line search algorithm. Since the first order necessary conditions (4) we are solving admit non-optimal stationary points as solutions, the line search procedure may fail to find a point that sufficiently reduces the merit function. In this case, we revert from a full-space Newton step back to a quasi-Newton-based RSQP step, which under appropriate conditions guarantees finding such a point. Our final strategy for globalization is the use of continuation on a nonlinear parameter. Such techniques are very popular and effective for solution of nonlinear PDEs that are otherwise difficult to solve, and it makes sense to exploit them for PDE optimization. In the remainder of this section, we discuss our line search, combined QN-RSQP–LNKS, and continuation strategies.

3.1 The line search

The basic component of a line search algorithm is the choice of a merit function, a scalar function (of $\mathbf{x}$ and λ) that monitors the progress of the algorithm. In contrast with unconstrained optimization, the choice of a merit function is not straightforward, since we are trying to balance optimality with feasibility. The two most popular choices are the l_1 and the augmented Lagrangian exact merit functions. The l_1 merit function is given by

$$\phi(\mathbf{x}) := f + \rho_\phi \|\mathbf{c}\|_1, \tag{10}$$

and the augmented Lagrangian by

$$\phi(\mathbf{x}, \boldsymbol{\lambda}) := f + \mathbf{c}^T \boldsymbol{\lambda} + \frac{\rho_\phi}{2} \mathbf{c}^T \mathbf{c}. \tag{11}$$

The scalar ρ_ϕ is the *penalty parameter*—a weight chosen to balance minimization of the objective function and the minimization of the residuals of the constraints. Both merit functions are exact provided the penalty parameter is large enough. By exact we mean that if $(\mathbf{x}_*, \boldsymbol{\lambda}_*)$ is a minimizer for (1), then it is also a minimizer for the merit function. A crucial property of a merit function is that it should accept unit step lengths close to a solution, and therefore permit Newton quadratic convergence to be observed. The l_1 merit function often suffers from the "Maratos" effect, that is, it sometimes rejects good steps and slows down the algorithm. The augmented Lagrangian merit function does not exhibit such behavior, but its drawback is that it requires accurate estimates of the Lagrange multipliers to perform well. (This is not a problem in LNKS since second-order accurate Lagrange multipliers estimates are computed.) Both algorithms are sensitive to the penalty parameter, which must be chosen judiciously to avoid an unbounded merit function, or very slow convergence.

In LNKS we use an Armijo-type line search algorithm. A safeguarded backtracking procedure is used to search for a scalar $\alpha \in [\alpha_{min}, 1]$ so that the so-called Armijo criterion

$$\phi(\alpha) \leq \phi(0) + \alpha \delta_A \mathbf{p}^T \nabla \phi(0), \tag{12}$$

of sufficient descent is satisfied. The algorithm used to compute the search direction $\mathbf{p}$ is left intentionally unspecified. All that matters to ensure global convergence is the properties of the merit function and the properties of $\mathbf{p}$. If ϕ is bounded and takes its minimum at a finite point, and if $\mathbf{p}$ is bounded, the safeguarded Armijo search is guaranteed to converge to a local minimum [22]. The line search algorithm we use is simple backtracking.

3.2 Continuation

One of the standard assumptions in global convergence proofs is that the Jacobian of the constraints is nonsingular for all optimization iterations. For highly nonlinear PDEs this may be an unrealistic assumption. Even if the Jacobian is nonsingular, severe ill-conditioning will cause both QN-RSQP and LNKS algorithms to stall. Indeed, in our numerical experiments (for iterates far from the solution), difficulties in the line search algorithm are correlated with difficulties converging the $\mathbf{A}_s$ and $\mathbf{K}$ linear solves.

Continuation is a popular method for addressing such convergence difficulties. Many nonlinear problems are characterized by a scalar parameter whose increase tracks the shrinking diameter of the attraction basin of Newton's method. In its simplest form, one begins with a value of this parameter for which Newton's method converges easily, and then solves a sequence of problems with larger values of the parameter, until the target value is reached. Each problem is initialized with the converged state field of the previous iterate. Examples of such parameters are the Reynolds number in viscous flow, the Mach number in compressible flow, the Hartman number in magnetohydrodynamics, the load parameter for nonlinear solid mechanics problems, and more generally the grid size for many problems that exhibit increasing nonlinearity with finer numerical resolution. In problems where such a parameter is not natural, one can construct a surrogate problem that is easy to solve in the parameter limit; an example is the pseudo-transient continuation method [18].

The popularity of continuation for solving PDEs suggests their use in PDE optimization methods, which after all must converge the PDEs as part of the optimality conditions. The extension is natural; the continuation steps are promoted up to the level of the optimization iteration. Continuation allows uphill steps (unlike monotone line search methods) to be taken and generates good initial guesses, not only for the optimization variables, but also for the penalty parameter in the merit function. The most important feature of the continuation algorithm is that it globalizes trivially (when the initial optimum can be computed reliably, and when all iterates on the continuation path are far from turning and bifurcation points). If the continuation step places the next iterate outside the attraction basin of the Newton method then we simply reduce the continuation step size. In principle, the method can be made to work without incorporating any other globalization strategy. Nevertheless, taking a large number of continuation steps can significantly slow down the algorithm. Therefore additional globalization strategies are necessary.

3.3 Combining QN-RSQP with LNKS

Quasi-Newton methods are well known for their robustness. Since LNKS already uses the reduced space structure for preconditioning, it is natural to ask whether QN-RSQP can be utilized to enhance the robustness of the overall algorithm. Global convergence proofs require the reduced Hessian, $\mathbf{W}_z$, to be strictly positive definite. If $\mathbf{W}_z$ is positive definite (and assuming the system (5) is solved exactly), then the resulting step $\mathbf{p}$ satisfies the descent criterion.

This is where quasi-Newton methods have an advantage over Newton methods. For example, by using a BFGS approximation, $\tilde{\mathbf{W}}_z$, positive definiteness can be guaranteed. LNKS does maintain a BFGS approximation—not for driving the outer iteration but for preconditioning purposes. Therefore, the remedy for an indefinite reduced Hessian is simple: if a computed search direction fails to satisfy the line search conditions, we discard it, and fall back on a search direction computed by QN-RSQP. For this reason, even when using a different preconditioner for the reduced Hessian, we insist on maintaining a BFGS approximation of $\mathbf{W}_z$ for globalization purposes.

As the factorization (7) illustrates, reduced space methods can be derived formally by a linear elimination of the state space step $\mathbf{p}_s$. The resulting unconstrained optimization problem has a gradient that includes second derivatives. These second derivative terms are dropped from the right hand sides of the decision and adjoint steps, at the expense of a reduction from one-step to two-step superlinear convergence [2]. The resulting QN-RSQP method is defined by Algorithm 2. An important advantage of this quasi-Newton method

Algorithm 2 Quasi-Newton RSQP

Choose $\mathbf{x}_s, \mathbf{x}_d, \tilde{\mathbf{W}}_z$
loop
  Evaluate $\mathbf{c}$, $\mathbf{g}$, $\mathbf{A}$
  $\mathbf{A}_s^T \boldsymbol{\lambda} = -\mathbf{g}_s$ — solve for $\boldsymbol{\lambda}$ **(Adjoint Step)**
  $\mathbf{g}_z = \mathbf{g}_d + \mathbf{A}_d^T \boldsymbol{\lambda}$
  Update $\tilde{\mathbf{W}}_z$ — **(Quasi-Newton approximation)**
  if $\|\mathbf{g}_z\| \leq tol$ and $\|\mathbf{c}\| \leq tol$ then
    Converged
  end if
  $\tilde{\mathbf{W}}_z \mathbf{p}_d = -\mathbf{g}_z$ — solve for $\mathbf{p}_d$ **(Decision step)**
  $\mathbf{A}_s \mathbf{p}_s = -(\mathbf{A}_d \mathbf{p}_d + \mathbf{c})$ — solve for $\mathbf{p}_s$ **(State step)**
  $\mathbf{x}_+ = \mathbf{x} + \mathbf{p}_x$
end loop

is that only two linearized forward problems need to be solved at each iteration, as opposed to the m needed by N-RSQP for constructing $\mathbf{A}_s^{-1}\mathbf{A}_d$ in $\mathbf{W}_z$, e.g. [13].

Let us review how QN-RSQP is used in combination with the l_1 and augmented Lagrangian merit functions. For the l_1 merit function we have

$$\nabla\phi^T \mathbf{p}_x = \mathbf{g}^T \mathbf{p}_x - \rho_\phi \|\mathbf{c}\|_1 = -\mathbf{g}_z^T \tilde{\mathbf{W}}_z^{-1} \mathbf{g}_z - \boldsymbol{\lambda}^T \mathbf{c} - \rho_\phi \|\mathbf{c}\|_1.$$

If $\tilde{\mathbf{W}}_z^{-1}$ is positive definite then the first term is always positive and we can choose ρ_ϕ by making the remaining terms positive. By setting

$$\rho_\phi = \|\boldsymbol{\lambda}\|_\infty + \delta, \quad \delta > 0, \tag{13}$$

we obtain a descent direction.

Similarly for the augmented Lagrangian we get

$$\begin{aligned}\nabla\phi^T\mathbf{p} &= (\mathbf{g}+\mathbf{A}^T\boldsymbol{\lambda}+\rho_\phi\mathbf{A}^T\mathbf{c})^T\mathbf{p}_x+\mathbf{p}_\lambda,\\ &= -\mathbf{g}_z^T\tilde{\mathbf{W}}_z^{-1}\mathbf{g}_z-\boldsymbol{\lambda}^T(\mathbf{c}+\mathbf{A}^T\mathbf{p}_x)+\mathbf{c}^T\mathbf{A}\mathbf{p}_x+\mathbf{c}^T\mathbf{p}_\lambda,\\ &= -\mathbf{g}_z^T\tilde{\mathbf{W}}_z^{-1}\mathbf{g}_z-\rho_\phi\mathbf{c}^T\mathbf{c}+\mathbf{c}^T\mathbf{p}_\lambda.\end{aligned}$$

If we assume that $\tilde{\mathbf{W}}_z^{-1}$ is strictly positive definite and choose

$$\rho_\phi \geq \frac{\mathbf{c}^T\mathbf{p}_\lambda}{\mathbf{c}^T\mathbf{c}}+\delta, \quad \delta>0, \tag{14}$$

then a descent direction is guaranteed. In our quasi-Newton formulations we assume that second derivatives are available. In this case,

$$\mathbf{p}_\lambda = (\partial_x\boldsymbol{\lambda})\mathbf{p}_x = -\mathbf{A}_s^{-T}\begin{bmatrix}\mathbf{W}_{ss} & \mathbf{W}_{sd}\end{bmatrix}\mathbf{p}_x \approx -\tilde{\mathbf{A}}_s^{-T}\begin{bmatrix}\mathbf{W}_{ss} & \mathbf{W}_{sd}\end{bmatrix}\mathbf{p}_x. \tag{15}$$

QN-RSQP can be parallelized very efficiently for moderate numbers of decision variables [21]. But when QN-RSQP is used to globalize LNKS, the need for exact forward solves greatly increases the cost of the LNKS method. By carrying out the state and adjoint solves that make up QN-RSQP inexactly, we can make the globalization much more efficient. How to do this will be discussed in the next section.

4 Inexact computations within LNKS

In large-scale computations inexactness is a powerful way to accelerate computations. In addition, it is often the case that inexactness robustifies algorithms (e.g. by damping Newton steps). In LNKS both outer and inner iterations are performed inexactly. We use inexact Lagrange-Newton solves within continuation loops, inexact Krylov-Schur solves to compute the Newton direction, and an inexact reduced Hessian in the KKT preconditioner and the QN-RSQP globalization. The various types of inexactness influence the algorithm in two basic ways: global convergence to a KKT point and local convergence rates. Analysis is required not only to provide theoretical guarantees for the robustness of the proposed algorithms, but also to suggest choices for truncation tolerances.

4.1 Inexact Newton's method

Before we discuss inexact Newton's method in the context of LNKS, we briefly summarize a few results for a general nonlinear system of equations. Assume we want to solve $\mathbf{h}(\mathbf{q}) = \mathbf{0}$. Further assume the following: (1) $\mathbf{h}$ and

$\mathbf{K} := \partial_q \mathbf{h}$ are sufficiently smooth in a neighborhood of a solution $\mathbf{q}_*$; (2) at each iteration an inexact Newton method computes a step $\mathbf{p}$ that satisfies

$$\|\mathbf{K}\mathbf{p} + \mathbf{h}\| \leq \eta_N \|\mathbf{h}\|, \tag{16}$$

where η_N is often called the *forcing term.* It can be shown that if $\eta_N < 1$ then $\mathbf{q} \to \mathbf{q}^*$ linearly; if $\eta_N \to 0$ then $\mathbf{q} \to \mathbf{q}^*$ superlinearly; and if $\eta_N = \mathcal{O}(\|\mathbf{h}\|)$ then we recover the quadratic convergence rate of a Newton method. The forcing term is usually given by

$$\eta_N = \frac{\|\mathbf{h}_{(+)} - \mathbf{h} - \mathbf{K}\mathbf{p}\|}{\|\mathbf{h}\|}. \tag{17}$$

For other choices of η_N and details on inexact Newton methods, see [10] and the references therein.

The extension of inexact methods to optimization is immediate, especially for unconstrained optimization. In [17] a global analysis is provided for a trust region RSQP-based algorithm. Close to a KKT point the theory for Newton's method applies and one can use the analysis presented in [9] to show that the inexact version of the LNKS algorithm converges. However, the line search we are using is not based on the residual of the KKT equations but instead on the merit function discussed in the previous session. That means that an inexact step that simply reduces $\|\mathbf{h}\|$ may not necessarily satisfy the merit function criteria. In [7] we show that for points that are close enough to the solution inexactness does not interfere with the line search.

Here we extend our discussion to the inexact QN-RSQP algorithm since, especially in the absence of a continuation scheme, QN-RSQP globalization is a crucial component of the LNKS method. In the analysis that follows we compare the inexact computation with an exact one. We assume that we have chosen a penalty parameter that gives sufficient decrease (based on the exact steps), and we establish conditions for the inexact computation so that this sufficient decrease is not compromised.

In the presence of inexactness the QN-RSQP steps (Algorithm 2) become

$$\begin{aligned}
\mathbf{A}_s^T \tilde{\boldsymbol{\lambda}} + \mathbf{g}_s &= \mathbf{r}_z, \\
\tilde{\mathbf{g}}_z &= \mathbf{g}_d + \mathbf{A}_d^T \tilde{\boldsymbol{\lambda}}, \\
\tilde{\mathbf{W}}_z \tilde{\mathbf{p}}_z &= -\tilde{\mathbf{g}}_z, \\
\mathbf{A}_s \tilde{\mathbf{p}}_s &= -(\mathbf{A}_d \tilde{\mathbf{p}}_z + \mathbf{c}) + \mathbf{r}_c, \\
\tilde{\mathbf{p}}_d &= \tilde{\mathbf{p}}_z.
\end{aligned}$$

We have introduced two vectors, $\mathbf{r}_z$ and $\mathbf{r}_c$, to account for the inexactness in the adjoint and forward solves. The following equations give the Lagrange multipliers, reduced gradient, and state and control steps for the exact (left column) and the inexact case (right column).

$$\begin{aligned}
\boldsymbol{\lambda} &= -\mathbf{A}_s^{-T}\mathbf{g}_s, & \tilde{\boldsymbol{\lambda}} &= \boldsymbol{\lambda} + \mathbf{A}_s^{-T}\mathbf{r}_z,\\
\mathbf{g}_z &= \mathbf{g}_d - \mathbf{A}_d^T\mathbf{A}_s^{-T}\mathbf{g}_s, & \tilde{\mathbf{g}}_z &= \mathbf{g}_z + \mathbf{A}_d^T\mathbf{A}_s^{-T}\mathbf{r}_z,\\
\mathbf{p}_d &= -\tilde{\mathbf{W}}_z^{-1}\mathbf{g}_z, & \tilde{\mathbf{p}}_d &= \mathbf{p}_d - \tilde{\mathbf{W}}_z^{-1}\mathbf{A}_d^T\mathbf{A}_s^{-T}\mathbf{r}_z,\\
\mathbf{p}_s &= \mathbf{A}_s^{-1}\mathbf{A}_d\tilde{\mathbf{W}}_z^{-1}\mathbf{g}_z - \mathbf{A}_s^{-1}\mathbf{c}, & \tilde{\mathbf{p}}_s &= \mathbf{p}_s + \mathbf{A}_s^{-1}\mathbf{A}_d\tilde{\mathbf{W}}_z^{-1}\mathbf{A}_d^T\mathbf{A}_s^{-T}\mathbf{r}_z + \mathbf{A}_s^{-1}\mathbf{r}_c.
\end{aligned}$$

Let us define the following constants : $\kappa_1 := \max(\|\tilde{\mathbf{W}}_z^{-1}\|)$, $\kappa_2 := \max(\|\mathbf{A}_s^{-1}\mathbf{A}_d\|)$, $\kappa_3 := \min(\sigma_{\min}(\tilde{\mathbf{W}}_z^{-1}))$, and $\kappa_4 := \max(\|\mathbf{A}_s^{-1}\|)$; σ denotes singular values and the min, max operations are across optimization iterations. We assume that these quantities are uniformly bounded.

4.2 l_1 merit function

The directional derivative of the merit function is given by

$$\begin{aligned}
\nabla\phi^T\mathbf{p}_x &= \nabla\phi^T\mathbf{p}_x + \mathbf{g}_s^T(\mathbf{A}_s^{-1}\mathbf{A}_d\tilde{\mathbf{W}}_z^{-1}\mathbf{A}_d^T\mathbf{A}_s^{-T}\mathbf{r}_z + \mathbf{A}_s^{-1}\mathbf{r}_c) - \mathbf{g}_d^T\tilde{\mathbf{W}}_z^{-1}\mathbf{A}_d^T\mathbf{A}_s^{-T}\mathbf{r}_z,\\
&= \nabla\phi^T\mathbf{p}_x + (\mathbf{A}_d^T\mathbf{A}_s^{-T}\mathbf{g}_s - \mathbf{g}_d)^T\tilde{\mathbf{W}}_z^{-1}\mathbf{A}_d^T\mathbf{A}_s^{-T}\mathbf{r}_z + (-\mathbf{A}_s^{-T}\mathbf{g}_s)^T\mathbf{r}_c,\\
&= -\mathbf{g}_z^T\tilde{\mathbf{W}}_z^{-1}\mathbf{g}_z - \boldsymbol{\lambda}^T\mathbf{c} - \rho_\phi\|\mathbf{c}\|_1 - \mathbf{g}_z^T\tilde{\mathbf{W}}_z^{-1}\mathbf{A}_d^T\mathbf{A}_s^{-T}\mathbf{r}_z - \boldsymbol{\lambda}^T\mathbf{r}_c.
\end{aligned}$$

To ensure that $\nabla\phi^T\mathbf{p}_x < 0$ we can set

$$\begin{aligned}
-\mathbf{g}_z^T\tilde{\mathbf{W}}_z^{-1}\mathbf{g}_z - \mathbf{g}_z^T\tilde{\mathbf{W}}_z^{-1}\mathbf{A}_d^T\mathbf{A}_s^{-T}\mathbf{r}_z &< 0,\\
-\boldsymbol{\lambda}^T\mathbf{c} - \boldsymbol{\lambda}^T\mathbf{r}_c - \rho_\phi\|\mathbf{c}\|_1 &< 0,
\end{aligned} \tag{18}$$

and therefore if we choose

$$\|\mathbf{r}_z\| < \frac{\kappa_3}{\kappa_1\kappa_2}\|\mathbf{g}_z\|, \tag{19}$$

we satisfy the first inequality in (18).

If we assume that the penalty parameter is given by (13) then

$$\begin{aligned}
&-\boldsymbol{\lambda}^T\mathbf{c} - \boldsymbol{\lambda}^T\mathbf{r}_c - \rho_\phi\|\mathbf{c}\|_1 < 0,\\
&-\boldsymbol{\lambda}^T\mathbf{c} - \boldsymbol{\lambda}^T\mathbf{r}_c - \|\boldsymbol{\lambda}\|_\infty\|\mathbf{c}\|_1 - \delta\|\mathbf{c}\|_1 <\\
&\qquad \|\boldsymbol{\lambda}\|_\infty\|\mathbf{c}\|_1 + \|\boldsymbol{\lambda}\|_\infty\|\mathbf{r}_c\|_1 - \|\boldsymbol{\lambda}\|_\infty\|\mathbf{c}\|_1 - \delta\|\mathbf{c}\|_1.
\end{aligned}$$

If we choose

$$\|\mathbf{r}_c\|_1 < \frac{1}{2}\frac{\delta}{\|\boldsymbol{\lambda}\|_\infty}\|\mathbf{c}\|_1, \tag{20}$$

then[1]

$$-\boldsymbol{\lambda}^T\mathbf{c} - \boldsymbol{\lambda}^T\mathbf{r}_c - \rho_\phi\|\mathbf{c}\|_1 < -\frac{1}{2}\delta\|\mathbf{c}\|_1.$$

For each iterate we therefore compute sufficient descent without having to increase the penalty parameter.

---

[1] In our implementation we use $\|\mathbf{r}_c\|_1 < \frac{1}{2}\delta/(\|\boldsymbol{\lambda}\|_\infty + 1)\|\mathbf{c}\|_1$ and the right hand side becomes $\delta/2(1/(1/\|\boldsymbol{\lambda}\|_\infty + 1) - 2)\|\mathbf{c}\|_1$. By noticing that $0 < \frac{1}{1/\|\boldsymbol{\lambda}\|_\infty + 1} < 1$ we get the sufficient descent condition.

4.3 Augmented Lagrangian merit function

For the inexact QN-RSQP method, the directional derivative of the augmented Lagrangian merit function becomes

$$\begin{aligned}\nabla\phi^T\tilde{\mathbf{p}} &= \tilde{\mathbf{p}}_x^T(\mathbf{g}+\mathbf{A}^T\tilde{\boldsymbol{\lambda}}+\rho_\phi\mathbf{A}^T\mathbf{c})+\mathbf{c}^T\tilde{\mathbf{p}}_\lambda,\\ &= -\mathbf{g}_z^T\tilde{\mathbf{W}}_z^{-1}\mathbf{g}_z-\rho_\phi\mathbf{c}^T\mathbf{c}+\mathbf{e}^T(\mathbf{g}+\mathbf{A}^T\tilde{\boldsymbol{\lambda}}+\rho_\phi\mathbf{A}^T\mathbf{c})+\mathbf{c}^T\tilde{\mathbf{p}}_\lambda+\\ &\quad+\mathbf{p}_x^T\mathbf{A}^T\mathbf{e}_\lambda+\mathbf{e}^T\mathbf{A}^T\mathbf{e}_\lambda,\end{aligned} \tag{21}$$

where

$$\begin{aligned}\mathbf{e}_s &:= \mathbf{A}_s^{-1}\mathbf{A}_d\tilde{\mathbf{W}}_z^{-1}\mathbf{A}_d^T\mathbf{A}_s^{-T}\mathbf{r}_z+\mathbf{A}_s^{-1}\mathbf{r}_c,\\ \mathbf{e}_d &:= -\tilde{\mathbf{W}}_z^{-1}\mathbf{A}_d^T\mathbf{A}_s^{-T}\mathbf{r}_z,\\ \mathbf{e}_\lambda &:= \mathbf{A}_s^{-T}\mathbf{r}_z.\end{aligned}$$

Now we examine the different terms of the gradient of the Augmented Lagrangian function. We use $\mathbf{g}+\mathbf{A}^T\boldsymbol{\lambda}=\{\mathbf{0}\quad\mathbf{g}_z\}^T$ and (21) becomes

$$\begin{aligned}\nabla\phi^T\tilde{\mathbf{p}} &= -\mathbf{g}_z^T\tilde{\mathbf{W}}_z^{-1}\mathbf{g}_z-\rho_\phi\mathbf{c}^T\mathbf{c}+\\ &\quad+\mathbf{g}_z^T\mathbf{e}_d+\mathbf{p}_x^T\mathbf{A}^T\mathbf{e}_\lambda+\mathbf{e}^T\mathbf{A}^T\mathbf{e}_\lambda+\rho_\phi\mathbf{e}^T\mathbf{A}^T\mathbf{c}+\mathbf{c}^T\tilde{\mathbf{p}}_\lambda.\end{aligned} \tag{22}$$

It is easy to check that the terms in the right hand of (22) simplify to:

$$\begin{aligned}\mathbf{g}_z^T\mathbf{e}_d &= -\mathbf{r}_z^T\mathbf{A}_s^{-1}\mathbf{c},\\ \mathbf{p}_x^T\mathbf{A}^T\mathbf{e}_\lambda &= -\mathbf{r}_z^T\mathbf{A}_s^{-1}\mathbf{c},\\ \rho_\phi\mathbf{e}^T\mathbf{A}^T\mathbf{c} &= \rho_\phi\mathbf{c}^T\mathbf{r}_c,\\ \mathbf{e}^T\mathbf{A}^T\mathbf{e}_\lambda &= \mathbf{r}_z^T\mathbf{A}_s^{-1}\mathbf{r}_c.\end{aligned}$$

Thus

$$\begin{aligned}\nabla\phi^T\tilde{\mathbf{p}} &= -\,\mathbf{g}_z^T\tilde{\mathbf{W}}_z^{-1}\mathbf{g}_z-\rho_\phi\mathbf{c}^T\mathbf{c}+\rho_\phi\mathbf{c}^T\mathbf{r}_c-\\ &\quad-\mathbf{r}_z^T(\mathbf{A}_s^{-1}\mathbf{c}+\mathbf{A}_s^{-1}\mathbf{A}_d\tilde{\mathbf{W}}_z^{-1}\mathbf{g}_z)+\mathbf{c}^T\tilde{\mathbf{p}}_\lambda+\mathbf{r}_z^T\mathbf{A}_s^{-1}\mathbf{r}_c.\end{aligned} \tag{23}$$

In the following we assume that $\mathbf{c}^T\tilde{\mathbf{p}}_\lambda$ is absorbed in the penalty parameter ρ_ϕ. We use the following inequality

$$\mathbf{g}_z^T\tilde{\mathbf{W}}_z^{-1}\mathbf{g}_z+\rho_\phi\mathbf{c}^T\mathbf{c}\le 2\max(\kappa_3\|\mathbf{g}_z\|^2,\rho_\phi\|\mathbf{c}\|^2)=:\gamma.$$

If we choose

$$\mathbf{r}_z^T(\mathbf{A}_s^{-1}\mathbf{c}+\mathbf{A}_s^{-1}\mathbf{A}_d\tilde{\mathbf{W}}_z^{-1}\mathbf{g}_z)<\eta\frac{1}{2}\gamma,\quad 0<\eta<1$$

then

$$\|\mathbf{r}_z\|<\frac{\eta}{2}\frac{\max(\kappa_3\|\mathbf{g}_z\|^2,\rho_\phi\|\mathbf{c}\|^2)}{\max(\kappa_4\|\mathbf{c}\|,\kappa_2\kappa_3\|\mathbf{g}_z\|)}. \tag{24}$$

Similarly if we choose

$$\mathbf{r}_c^T(\rho_\phi\|\mathbf{c}\| + \kappa_4\mathbf{r}_z) < \eta\frac{1}{2}\gamma, \quad 0 < \eta < 1$$

then

$$\|\mathbf{r}_c\| < \frac{\eta}{2}\frac{\max(\kappa_3\|\mathbf{g}_z\|^2, \rho_\phi\|\mathbf{c}\|^2)}{\max(\rho_\phi\|\mathbf{c}\|, \kappa_4\|\mathbf{r}_z\|)}. \tag{25}$$

By insisting that (24) and (25) are satisfied, we guarantee a descent direction without a penalty parameter significantly larger than that of the exact case. Of course these relations are not implemented in this form since we do not know $\|\mathbf{g}_z\|$. Instead of $\|\mathbf{g}_z\|$ we use the reduced gradient norm from the previous iteration—scaled by a standard inexact Newton forcing parameter η_N computed from

$$\eta_N^+ = \frac{\|\tilde{\mathbf{g}}_z^+ - \tilde{\mathbf{g}}_z - \alpha\tilde{\mathbf{p}}_d\|}{\|\tilde{\mathbf{g}}_z\|}$$

at the end of each SQP step.

Notice that the analysis is different from the case of the l_1 merit function. In the latter the errors $\mathbf{r}_z$ and $\mathbf{r}_c$ have been compared to $-\mathbf{g}_z^T\tilde{\mathbf{W}}_z^{-1}\mathbf{g}_z$ and $\rho_\phi\mathbf{c}^T\mathbf{c}$ respectively, whereas in the augmented Lagrangian both errors are compared to $\frac{1}{2}\gamma$. This allows a balance between feasibility and optimality. If, for example, we start very close to a feasible point but far from the optimum, an inexact Newton's criterion based only on the residual of the constraints (as in (20)) will oversolve, since far from the optimum we do not need the constraints to be satisfied. Of course it is very easy to extend the analysis for the augmented Lagrangian to the l_1 merit function.

4.4 The globalized inexact LNKS method

Algorithm 3 gives a high-level description of the globalized inexact version of LNKS. Some of the implementation details and heuristics are discussed in this section. We use the following notation: $\mathbf{q} = \{\mathbf{x} \ \ \boldsymbol{\lambda}\}^T$; $\phi(0) := \phi(\mathbf{q})$; $\mathbf{h}(0) := \mathbf{h}(\mathbf{q})$; $\phi(\alpha) := \phi(\mathbf{q} + \alpha\mathbf{p})$; and $\mathbf{h}(\alpha) := \mathbf{h}(\mathbf{q} + \alpha\mathbf{p})$.

The algorithm uses a three-level iteration. In the outer iteration the value of the continuation parameter (here symbolized by Re, the Reynolds number for viscous flows, following the examples of the next section) is gradually increased until the target value is reached. The middle iterations correspond to Lagrange-Newton linearizations of the optimality system for a fixed continuation number. Finally, the inner iteration consists of two core branches: the computation of a Newton direction and the computation of the search direction with QN-RSQP. The default branch is the Newton step. If this step fails to satisfy the line search conditions, we then switch to QN-RSQP. If QN-RSQP fails too, then we reduce the continuation parameter Re and return to the outer loop.

Algorithm 3 Globalized LNKS

1: Choose $\mathbf{x}_s$, $\mathbf{x}_d$, ρ_ϕ, t, δ_A, set $Re = Re_{start}$, $tol = tol_0$
2: $\mathbf{A}_s^T \boldsymbol{\lambda} + \mathbf{g}_s \approx \mathbf{0}$ — solve inexactly for $\boldsymbol{\lambda}$
3: while $Re \neq Re_{target}$ do
4:   loop
5:     Evaluate f, $\mathbf{c}$, $\mathbf{g}$, $\mathbf{A}$, $\mathbf{W}$
6:     $\mathbf{g}_z = \mathbf{g}_d + \mathbf{A}_d^T \boldsymbol{\lambda}$
7:     Check convergence: $\|\mathbf{g} + \mathbf{A}^T \boldsymbol{\lambda}\| \leq tol$ and $\|\mathbf{c}\| \leq tol$
8:     $\mathbf{P}^{-1}\mathbf{K}\mathbf{p} + \mathbf{P}^{-1}\mathbf{h} \approx \mathbf{0}$ — solve inexactly for $\mathbf{p}$
9:     Compute ρ_ϕ such that $\nabla\phi^T(0)\mathbf{p} \leq 0$
10:     Compute α s.t. $\phi(\alpha) \leq \phi(0) + \delta_A \alpha (\nabla\phi^T(0)\mathbf{p})$
11:     if Line search failed then
12:       Compute α s.t. $\|\mathbf{h}(\alpha)\| < t\|\mathbf{h}(0)\|$
13:     end if
14:     if Line search failed then
15:       $\tilde{\mathbf{W}}_z \mathbf{p}_d = -\mathbf{g}_z$ — solve inexactly for $\mathbf{p}_d$
16:       $\mathbf{A}_s \mathbf{p}_s + \mathbf{A}_d \mathbf{p}_d + \mathbf{c} \approx \mathbf{0}$ — solve inexactly for $\mathbf{p}_s$
17:       $\mathbf{A}_s^T \boldsymbol{\lambda}_+ + \mathbf{g}_s \approx \mathbf{0}$ — solve inexactly for $\boldsymbol{\lambda}_+$
18:       Compute α s.t. $\phi(\alpha) \leq \phi(0) + \delta_A \alpha (\nabla\phi^T(0)\mathbf{p})$
19:       if Line search failed then
20:         Reduce Re and go to step 5.
21:       end if
22:     end if
23:     $\boldsymbol{\lambda}+ = \boldsymbol{\lambda} + \mathbf{p}_\lambda$ — (only for LNKS step)
24:     $\mathbf{x}_+ = \mathbf{x} + \mathbf{p}_x$
25:   end loop
26:   $Re = Re + \Delta Re$
27:   Tighten tol
28: end while

The linear solves in Steps 8, 16 and 17 are performed inexactly (that is, by early termination of iterative solvers). In Step 8 we follow [10] in choosing the forcing term. In Steps 16 and 17 the forcing term is based on the formulas developed in Section 4.3.

In Step 6 we use the adjoint variables to update the reduced gradient. This is equivalent to $\mathbf{g}_z = \mathbf{g}_d - \mathbf{A}_d^T \mathbf{A}_s^{-T} \mathbf{g}_s$, if $\boldsymbol{\lambda}$ is computed by solving exactly $\mathbf{A}_s^T \boldsymbol{\lambda} + \mathbf{g}_s = \mathbf{0}$. When $\boldsymbol{\lambda}$ is taken from LNKS, it includes second order terms (which reduce to zero as we approach the solution), and when $\boldsymbol{\lambda}$ is taken from QN-RSQP it also introduces extra error since we never solve the linear systems exactly. In our numerical experiments this approximation has not caused problems.

We allow for non-monotone line searches. If the LNKS step is rejected by the merit function line search we do not switch immediately to QN-RSQP. Instead, we perform a line search (Step 12) on the KKT residual (as if we were treating the KKT conditions as nonlinear equations) and if the step is accepted we use it to update the variables for the next iteration. However, we do store the iterate and the merit function gradient, and we insist that some step satisfies the conditions of the merit line search (evaluated at the failure point) after a fixed number of iterations. Otherwise, we switch to QN-RSQP. This heuristic has been very successful. Typically, we permit two steps before we demand reduction of the merit function.

We use various heuristics to bound the penalty parameter and if possible reduce it. A new penalty parameter ρ_ϕ^+ is computed using the LNKS step and formula (14). If $\rho_\phi^+ > 4\rho_\phi$ we update the penalty parameter and we switch to QN-RSQP. If $\rho_\phi^+ < \rho_\phi/4$ we *reduce* the penalty parameter and set $\rho_\phi^+ = 0.5\rho_\phi$. We also reduce the penalty parameter after successful steps in the KKT residual.

We use the BFGS method for the quasi-Newton approximation of the reduced Hessian. To precondition $\mathbf{W}_z$ we use either BFGS or a matrix-free method we introduced in [6]. This preconditioner, which requires the action of $\mathbf{W}_z$ on a vector, can be also used as a driver for the reduced space globalization step. Although we have the luxury of second derivatives, computing the reduced Hessian exactly is very expensive. Instead we use an approximate reduced Hessian, given by

$$\tilde{\mathbf{W}}_z = \mathbf{W}_{dd} + \tilde{\mathbf{A}}_s^{-T}\mathbf{A}_d^T\mathbf{W}_{ss}\tilde{\mathbf{A}}_s^{-1}\mathbf{A}_d - \mathbf{W}_{ds}\tilde{\mathbf{A}}_s^{-1}\mathbf{A}_d - \mathbf{A}_d^T\tilde{\mathbf{A}}_s^{-T}\mathbf{W}_{sd}, \tag{26}$$

and $\tilde{\mathbf{A}}_s^{-1}$ is the preconditioner to the forward problem. We employ a Lanczos process to estimate the lower and upper eigenvalues of $\tilde{\mathbf{W}}_z$. In case of a negative eigenvalue (i.e. negative curvature) we can use a modified reduced Hessian, $\mu\mathbf{I} + \tilde{\mathbf{W}}_z$, where the parameter μ is chosen to shift the spectrum to the positive real axis.

5 Application to optimal boundary control of viscous flows

In this section we present results that typify the performance of the LNKS algorithm. The PDE-constrained optimization problem we consider is finding the optimal boundary control (suction/injection) that minimizes the rate of energy dissipation in a viscous flow. The flow (and hence state constraints) is described by the stationary incompressible Navier-Stokes equations. A survey and a number articles on flow control can be found in [15]. More on numerical approximation of the incompressible Navier-Stokes equations can be found in [14, 16]. We consider flow around a circular cylinder, which is anchored inside a rectangular duct, much like a numerical wind tunnel. A quadratic velocity profile is used as an inflow Dirichlet boundary condition and we prescribe a traction-free outflow. The control variables (i.e. decision variables) are the velocities $\boldsymbol{d}$ on the downstream portion of the cylinder surface. We use the velocity-pressure (*i.e.* $\boldsymbol{u}, p$) form of the incompressible steady state Navier-

Stokes equations. The objective function(al) is given by

$$\mathcal{J}(\boldsymbol{u}, \boldsymbol{d}) := \frac{\nu}{2} \int_{\Omega} \nabla \boldsymbol{u} \cdot \nabla \boldsymbol{u}^T + \frac{\rho}{2} \int_{\Gamma_d} \boldsymbol{d} \cdot \boldsymbol{d},$$

where the first term is the energy dissipation, and the second reflects the "cost" of the boundary velocity controls. The usual approach of defining a Lagrangian functional and requiring its stationarity with respect to Lagrange multipliers, state variables, and decision variables gives (the strong, infinite dimensional form of) the KKT optimality conditions, which consist of the *forward* problem

$$\begin{aligned} -\nu \nabla \cdot (\nabla \boldsymbol{u} + \nabla \boldsymbol{u}^T) + (\nabla \boldsymbol{u})\boldsymbol{u} + \nabla p = \boldsymbol{b} \quad &\text{in} \quad \Omega, \\ \nabla \cdot \boldsymbol{u} = 0 \quad &\text{in} \quad \Omega, \\ \boldsymbol{u} = \boldsymbol{u}_g \quad &\text{on} \quad \Gamma_u, \\ \boldsymbol{u} = \boldsymbol{d} \quad &\text{on} \quad \Gamma_d, \\ -p\boldsymbol{n} + \nu(\nabla \boldsymbol{u} + \nabla \boldsymbol{u}^T)\boldsymbol{n} = \boldsymbol{0} \quad &\text{on} \quad \Gamma_N, \end{aligned} \tag{27}$$

the *adjoint* problem

$$\begin{aligned} -\nu \nabla \cdot (\nabla \boldsymbol{\lambda} + \nabla \boldsymbol{\lambda}^T) + (\nabla \boldsymbol{u})^T \boldsymbol{\lambda} - (\nabla \boldsymbol{\lambda})\boldsymbol{u} + \nabla \mu = \nu \nabla \cdot (\nabla \boldsymbol{u} + \nabla \boldsymbol{u}^T) \quad &\text{in} \quad \Omega, \\ \nabla \cdot \boldsymbol{\lambda} = 0 \quad &\text{in} \quad \Omega, \\ \boldsymbol{\lambda} = \boldsymbol{0} \quad &\text{on} \quad \Gamma_u, \\ \boldsymbol{\lambda} = \boldsymbol{0} \quad &\text{on} \quad \Gamma_d, \\ -\mu \boldsymbol{n} + \nu(\nabla \boldsymbol{\lambda} + \nabla \boldsymbol{\lambda}^T)\boldsymbol{n} + (\boldsymbol{u} \cdot \boldsymbol{n})\boldsymbol{\lambda} = -\nu(\nabla \boldsymbol{u} + \nabla \boldsymbol{u}^T)\boldsymbol{n} \quad &\text{on} \quad \Gamma_N, \end{aligned} \tag{28}$$

and the *control* (i.e. decision) problem

$$\nu(\nabla \boldsymbol{\lambda} + \nabla \boldsymbol{\lambda}^T)\boldsymbol{n} + \nu(\nabla \boldsymbol{u} + \nabla \boldsymbol{u}^T)\boldsymbol{n} - \rho \boldsymbol{d} = \boldsymbol{0} \quad \text{on} \quad \Gamma_d. \tag{29}$$

Here $\nu = 1/Re$ and the decision variables are the velocities $\boldsymbol{d}$ on Γ_d; $\boldsymbol{\lambda}$ are the adjoint velocities and μ are the adjoint pressures. For a forward solve we need not distinguish between Γ_d and Γ_u. In the optimization problem, however, $\boldsymbol{u}_d$ is to be determined.

We discretize by the Galerkin finite element method, using tetrahedral Taylor-Hood elements (quadratic velocities, linear pressures). Our software is built on top of the PETSc library [1] and we use PETSc's block-Jacobi preconditioners with local ILU(0) for the domain decomposition approximation of the forward and adjoint operators. For the Krylov solves of the forward and adjoint problems, we use the quasi-minimum residual method (QMR) [12], and for the KKT Krylov solves we use a symmetric variant of QMR.

Figure 1 illustrates the optimization results for different Reynolds numbers. The optimal controls eliminate the recirculation region within the cylinder wake. This is achieved by injecting fluid on the downstream portion of the cylinder. We observe a tenfold relative reduction of the dissipation functional (which is proportional to drag on the cylinder).

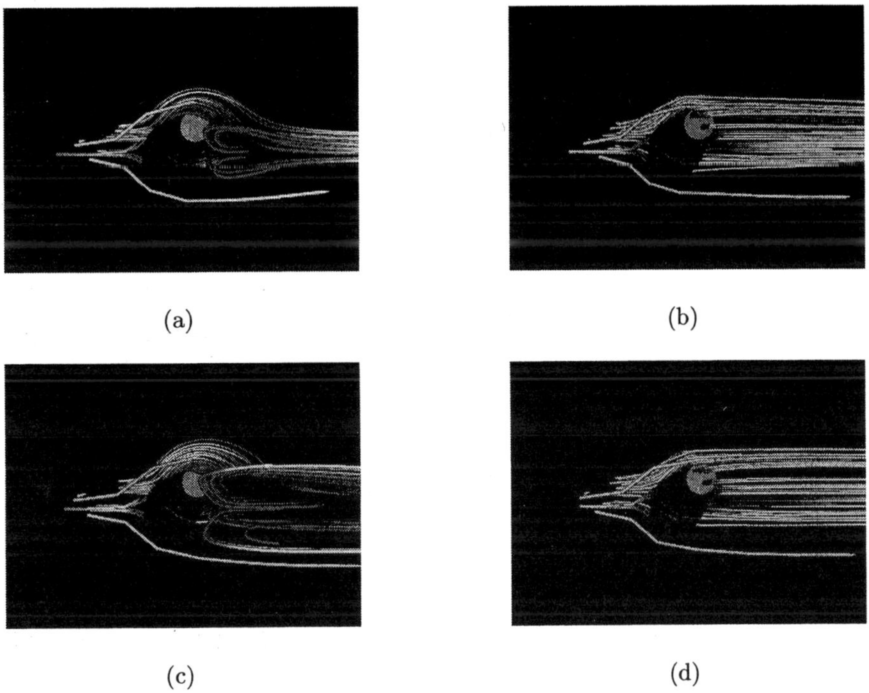

Fig. 1. PDE-constrained optimal control problem. The constraints are the steady three-dimensional incompressible Navier-Stokes equations modeling viscous flow around a cylinder. The objective is to minimize a linear combination of the energy dissipation in the fluid and the cost of the controls. The controls are injection/suction velocities on the downstream portion of the cylinder surface. The left images depict streamtubes for the uncontrolled flow at Re 20 (top) and 40 (bottom). The right images depict streamtubes of the optimally-controlled flow (same Reynolds numbers). Injecting fluid entirely eliminates recirculation within the wake of the cylinder, thus minimizing dissipation. The optimization problem was solved on 256 processors of the Cray T3E-900 at the Pittsburgh Supercomputing Center.

Table 1 shows results for 32, 64, and 128 processors of a Cray T3E-900 for a roughly doubling of problem size. We compare QN-RSQP (exact solves), with LNKS (exact solves) and IN-LNKS (inexact solves). Continuation was used for the initial guess at Re 60 by using the solution from Re 30 as the initial guess. The reduced Hessian preconditioner is a BFGS approximation, initialized using several iterations of a 2-step stationary iterative method [6]. For this problem, QN-RSQP successfully converged but only after a significant amount of time.[2] LNKS does much better—4 to 5 times faster than

[2] Here we used the l_1 merit function with second order correction line search. In past experiments we used standard l_1 and after 48 hours QN-RSQP was terminated with just two orders of magnitude reduction in the reduced gradient.

Table 1. Scalability of QN-RSQP and LNKS algorithms for solution of the optimal flow control problem on 32, 64, and 128 processors of the Cray T3E-900. Results correspond to a roughly doubling of problem size for each doubling of number of processors. **QN-RSQP** is quasi-Newton reduced-space SQP; in **LNKS** we terminate the KKT Krylov iterations when the Euclidean norm of residual is less than 0.9×10^{-7}; in **IN-LNKS** we use a inexact Newton method on the KKT conditions; **N or QN iter** is the number of Newton or quasi-Newton steps; **KKT iter** is the number of inner iterations averaged across the outer iterations; **time** is wall-clock time in hours. Continuation was used for Re=60.

$Re = 30$

states **controls**	**method**	**N or QN iter**	**KKT iter**	**time**
117,048	QN-RSQP	161	—	32.1
2,925	LNKS	6	1,367	5,7
(32 procs)	IN-LNKS	11	163	1.4
389,440	QN-RSQP	189	—	46.3
6,549	LNKS	6	2,153	15.7
(64 procs)	IN-LNKS	13	238	3.8
615,981	QN-RSQP	204	—	53.1
8,901	LNKS	6	3,583	16.8
(128 procs)	IN-LNKS	12	379	4.1

$Re = 60$

states **controls**	**preconditioning**	**Newton iter**	**average KKT iter**	**time (hours)**
117,048	QN-RSQP	168	—	33.4
2,925	LNKS	7	1,391	6,8
(32 procs)	IN-LNKS	11	169	1.5
389,440	QN-RSQP	194	—	49.1
6,549	LNKS	7	2,228	18.9
(64 procs)	IN-LNKS	15	256	4.8
615,981	QN-RSQP	211	—	57.3
8,901	LNKS	8	3,610	13.5
(128 procs)	IN-LNKS	16	383	5.1

QN-RSQP. The most notable finding in Table 1 is the dramatic acceleration of LNKS that is achieved by allowing inexactness (IN-LNKS). The inexactness did not interfere at any point with the merit function and in all cases we observed quadratic convergence. For both Re 30 and 60, IN-LNKS converges over an order of magnitude more quickly than QN-RSQP.

In Table 2 we compare LNKS with the inexact version of the QN-RSQP method. In these numerical tests we have chosen Reynolds numbers in which the steady state model of the flow is not correct physically (i.e. a steady state does not exist for this value of Re). This was done to increase the nonlinearity of the problem, to stress the globalizations used in LNKS. BFGS is chosen as the quasi-Newton update in both the QN-RSQP method and the QN-RSQP preconditioner for LNKS.

The effect of inexact computations is examined in Table 2. As can be seen from the iterations column (**N or QN itr**), the number of outer iterations depends very mildly on the nonlinearity of the problem. In the fourth column (**ls failed**) we measure the success of the line search algorithms. Its meaning is overloaded; for the IN-QN-RSQP method it indicates how many times the penalty parameter increased to greater than twice that of the maximum penalty parameter encountered in the exact QN-RSQP solves; for the LNKS methods it indicates failures of the merit function line search. If the latter happens we switch to a line search on the KKT residual. The fifth column (**KKT failed**) indicates the number of times that this approach failed (and as a result we had to backtrack the outer iteration and switch to a quasi-Newton step). For this numerical experiment we observe that the number of excessive penalty parameter increments within the inexact QN-RSQP is relatively small. Overall the inexact methods are decidedly faster than exact versions, as shown in the last column, which gives CPU timings.

Table 2. Results for three different Reynolds numbers for the 117,048 states problem on 32 processors. Exact and inexact variants of both QN-RSQP and LNKS algorithms are presented. **QN-RSQP** is quasi-Newton reduced-space SQP; **IN-QN-RSQP** is the inexact QN-RSQP method; in **LNKS** we terminate the KKT Krylov iterations when the Euclidean norm of residual is less than 0.9×10^{-7}; in **IN-LNKS** we use an inexact Newton method for the KKT conditions; **itr** is the number of Newton (or quasi-Newton) steps; for IN-QN-RSQP, **ls failed** indicates the number of excessive increases of the merit function penalty parameter ρ_ϕ; for the two LNKS methods, it indicates the number of unsuccessful augmented Lagrangian line search attempts; **KKT failed** indicates the number of iterations in which the KKT steps had to be rejected; **time** is wall-clock time in hours on the T3E-900. In this example we did not employ continuation.

Reynolds	method	N or QN itr	ls failed	KKT failed	time
90	QN-RSQP	181	-	-	35.4
	IN-QN-RSQP	184	5	-	22.1
	LNKS	9	1	1	7.2
	IN-LNKS	14	4	0	1.5
120	QN-RSQP	185	-	-	36.1
	IN-QN-RSQP	192	5	-	23.2
	LNKS	10	2	2	8.1
	IN-LNKS	15	6	1	2.3
150	QN-RSQP	184	-	-	36.3
	IN-QN-RSQP	194	6	-	25.1
	LNKS	11	2	2	8.6
	IN-LNKS	15	6	2	2.9

For the LNKS methods we see that for exact solves switching to a line search on the KKT residual does not help (in fact it slows down the algorithm). On the contrary, for inexact solves this approach helps, and the relatively expensive quasi-Newton steps are usually avoided.

6 Conclusions

We presented the basic algorithmic components of the LNKS method, and considered inexact variants that speed up the method without compromising convergence. We considered an application to the optimal control of a viscous flow around a cylinder by boundary suction/injection. Our tests illustrate that LNKS is a robust and scalable algorithm for PDE-constrained optimization. It exhibits the well-known mesh-independence convergence properties of Newton methods, which combined with the inner Krylov-Schur iteration results in a very fast method. Often, the optimal solution is found in a small multiple of the cost of a single forward problem solve.

Inexactness is introduced into the LNKS method through inexact Lagrange-Newton solves within continuation loops, early termination of Krylov-Schur iterations to compute the Newton direction, and a BFGS approximation of the reduced Hessian based on inexact state Jacobian solves (used both as a reduced space preconditioner within the KKT solves as well as within the QN-RSQP globalization). The numerical experiments on the effects of inexactness are of limited scope, yet give an indication of effectiveness of inexact computations in reducing wall-clock time while retaining robustness. Typically, introducing inexactness into LNKS led to a factor of 3 to 4 reduction in cost relative to exact LNKS. Moreover, the augmented Lagrangian globalization we tested performed robustly and we did not have convergence problems for the (highly) nonlinear problem of separated flow around a cylinder. The results reveal at least an order of magnitude improvement in time over conventional quasi-Newton methods, rendering tractable some problems that otherwise would have required unacceptable amounts of parallel supercomputing time.

References

1. Satish Balay, William D. Gropp, Lois Curfman McInnes, and Barry F. Smith. PETSc home page. `http://www.mcs.anl.gov/petsc`, 1999.
2. Lorenz T. Biegler, Jorge Nocedal, and Claudia Schmid. A reduced Hessian method for large-scale constrained optimization. *SIAM Journal on Optimization*, 5:314–347, 1995.
3. George Biros. *Parallel Algorithms for PDE-Constrained Optimization and Application to Optimal Control of Viscous Flows.* PhD thesis, Carnegie Mellon University, Pittsburgh, PA, August 2000.
4. George Biros and Omar Ghattas. Parallel Newton-Krylov algorithms for PDE-constrained optimization. In *Proceedings of SC99*, The SCxy Conference series, Portland, Oregon, November 1999. ACM/IEEE.

5. George Biros and Omar Ghattas. Parallel preconditioners for KKT systems arising in optimal control of viscous incompressible flows. In D. E. Keyes, A. Ecer, J. Periaux, and N. Satofuka, editors, *Parallel Computational Fluid Dynamics 1999*. North-Holland, 1999.
6. George Biros and Omar Ghattas. Parallel Lagrange-Newton-Krylov-Schur methods for PDE-constrained optimization. Part I: The Krylov-Schur solver. Technical report, Laboratory for Mechanics, Algorithms, and Computing, Carnegie Mellon University, 2000.
7. George Biros and Omar Ghattas. Parallel Lagrange-Newton-Krylov-Schur methods for PDE-constrained optimization. Part II: The Lagrange Newton solver, and its application to optimal control of steady viscous flows. Technical report, Laboratory for Mechanics, Algorithms, and Computing, Carnegie Mellon University, 2000.
8. John Dennis E., Jr., Mahmoud El-Alem, and Maria C. Magicl. A global convergence theory for general trust-region-based algorithms for equality constrained optimization. *SIAM Journal on Optimization*, 7(1):177–207, 1997.
9. Stanley C. Eisenstat and Homer F. Walker. Globally convergent inexact Newton methods. *SIAM Journal on Optimization*, 4(2):393–422, 1994.
10. Stanley C. Eisenstat and Homer F. Walker. Choosing the forcing terms in an inexact Newton method. *SIAM Journal on Scientific Computing*, 17(1):16–32, 1996.
11. Roger Fletcher. *Practical Methods of Optimization*. John Wiley and Sons, second edition, 1987.
12. Roland W. Freund and Noël M. Nachtigal. An implementation of the QMR method based on coupled two-term recurrences. *SIAM Journal of Scientific Computing*, 15(2):313–337, March 1994.
13. Omar Ghattas and Jai-Hyeong Bark. Optimal control of two- and three-dimensional incompressible Navier-Stokes flows. *Journal of Computational Physics*, 136:231–244, 1997.
14. Max D. Gunzburger. *Finite Element for Viscous Incompressible Flows*. Academic Press, 1989.
15. Max D. Gunzburger, editor. *Flow Control*, volume 68 of *IMA Math. Appl.* Springer-Verlag, New York, 1995.
16. Max D. Gunzburger and Roy A. Nicolaides, editors. *Incompressible Computational Fluid Dynamics*. Cambridge University Press, 1993.
17. Matthias Heinkenschloss and Luis N. Vicente. Analysis of inexact trust-region SQP algorithms. Technical Report TR99-18, Rice University, Department of Computational and Applied Mathematics, 1999.
18. C. T. Kelley and David E. Keyes. Convergence analysis of pseudo-transient continuation. *SIAM Journal on Numerical Analysis*, 35:508–523, 1998.
19. C.T. Kelley and Ekkehard W. Sachs. Truncated Newton methods for optimization with inaccurate functions and gradients. *SIAM Journal on Optimization*, 10(1):43–55, 1999.
20. F. Leibritz and E. W. Sachs. Inexact SQP interior point methods and large scale optimal control problems. *SIAM Journal on Control and Optimization*, 38(1):272–293, 1999.
21. Ivan Malčević. Large-scale unstructured mesh shape optimization on parallel computers. Master's thesis, Carnegie Mellon University, 1997.
22. Stephen G. Nash and Ariela Sofer. *Linear and Nonlinear Programming*. McGraw-Hill, 1996.

23. Jorge Nocedal and Stephen J. Wright. *Numerical Optimization.* Springer, 1999.
24. T. Steihaug. The conjugate gradient method and trust regions in large scale optimization. *SIAM Journal on Numerical Analysis*, 20:626–637, 1983.

Part IV

Sensitivities for PDE-based Optimization

Solution Adapted Mesh Refinement and Sensitivity Analysis for Parabolic Partial Differential Equation Systems*

Shengtai Li[1,2], Linda R. Petzold[1], and James M. Hyman[2]

[1] University of California, Santa Barbara, CA 93106, USA
[2] T-7, Los Alamos National Laboratory, Los Alamos, NM 87544

Abstract. We have developed a structured adaptive mesh refinement (SAMR) method for parabolic partial differential equation (PDE) systems. Solutions are calculated using the finite-difference or finite-volume method in space and backward differentiation formula (BDF) integration in time. The combination of SAMR in space and BDF in time is designed for problems where the fine-scale profile of sharp fronts in space should be resolved and implicit integration in time is necessary to improve the efficiency of the computation. Methods for forward sensitivity analysis on the adaptive mesh are presented.

1 Introduction

Structured adaptive mesh refinement (SAMR) has been used extensively to solve partial differential equations (PDEs) [3, 11, 15]. SAMR uses a hierarchical block data structure where each block (called *patch*) can be solved as a single grid. Most implementations of SAMR have used an explicit time integration, and refined time as well as space by taking local smaller time steps for finer grids. The time stepsize for explicit integration is limited by the Courant-Friedrichs-Lewy (CFL) condition [9]. Explicit methods are appropriate for hyperbolic systems, where the CFL number is proportional to $\Delta t/\Delta x$. However, for a parabolic system the CFL number is proportional to $\Delta t/(\Delta x)^2$. Hence the time step for an explicit integration needs to be very small to ensure stability. It is desirable to solve this type of problem with implicit time integration. Implicit time integration is also preferred for solving steady-state and slow-transient problems, because the stepsize restrictions are less stringent than for explicit schemes (there may not be any). Although SAMR has been available for more than a decade, its combination with implicit time integration is still in its infancy. Before describing our algorithm and implementation, we first discuss two available adaptive grid implementations for parabolic PDE systems.

Verwer et al. [17] designed a local uniform grid refinement (LUGR) with second order backward differentiation formula (BDF) method in time. Instead

* This work was partially supported by DOE contract number DE-FG03-00ER25430, NSF grant CCR-9896198, NSF/ARPA PC-239415, and NSF ACI-0086061.

of using a hierarchical block data structure, LUGR uses a data structure specially designed for its algorithm. A standard second-order finite difference is used in the spatial discretization, central on the internal domain and one-sided at the boundaries. A fixed second-order two-step implicit BDF method with variable stepsizes is used for the time integration. The resulting system of nonlinear equations in each time step is solved by a modified Newton method and (preconditioned) iterative linear solver. LUGR is not flexible with respect to changes in the spatial discretization or time integration method.

Flaherty et al. [10] designed an adaptive overlapping grid (AOG) method using Galerkin's method with a piecewise polynomial basis in space and a singly implicit Runge-Kutta (SIRK) integration method in time. A tree-based hierarchical data structure is used. The grid refinement strategy is based on error estimates for Galerkin's method. Due to the frequent stops and starts that are needed in conjunction with the refinement/coarsening process, AOG opted for single-step SIRK methods for the time integration. Unfortunately, these methods often proved to be more costly than a multistep method. The integration of each local patch is done separately, and a Schwarz alternation iteration is used to obtain satisfactory accuracy in overlapping regions. AOG also uses overlapping and rotated grids in the refinement processing, which proved to be disadvantageous for general problems.

In this paper, we study how to combine the SAMR method with variable order and variable stepsize BDF time integration. The BDF methods are implicit multistep methods for solving systems of ordinary differential equations (ODEs) or differential-algebraic equations (DAEs). An efficient BDF code DASSL, developed by Petzold [4], has been widely used. In DASSL, the implicit system is solved by a modified Newton iteration. The linear system at each Newton iteration is solved by a dense or banded direct solver. An extension of DASSL, DASPK, was developed by Brown, Hindmarsh and Petzold [5]. DASPK has an additional option of using a preconditioned incomplete GMRES method to solve the linear system at each Newton iteration, which is particularly effective in the method of lines (MOL) solution of time-dependent PDEs in two or three dimensions.

Sensitivity analysis is important in many engineering and scientific applications. The information contained in the sensitivity trajectories is useful for parameter estimation, optimization, model reduction and experimental design. A DASPK package (DASPK3.0) with forward sensitivity analysis capability was developed by the authors [13]. DASPK3.0 also incorporated many other new features for the efficient integration of the ODE/DAE.

Compared with explicit integration, applying implicit time integration with an AMR system has a lot of difficulties. First, each patch cannot be integrated independently in one time step, because it may share boundary points with other patches. Second, the implicit system needs to be solved in each time step. Therefore at least one linear system must be solved in each time step. How to solve the linear system efficiently is essential to the success of the PDE solver. Related issues of how to reuse the previously-evaluated

Jacobian or preconditioner for the current time step, and how to get the initial guess for the Newton iteration are also important. Third, it is more difficult to take a local time step for the finer grid for an implicit method than for an explicit method. The time stepsize at each level is usually determined by accuracy instead of stability for the implicit method. Thus, the ratio of the stepsizes for different levels usually is not an integer(which is required for the hierarchical AMR data structure). There are some other difficulties, such as storing and updating the Jacobian or other linear system information for each level/patch if it is solved separately, restarting the integration after each refinement, etc.

In the following sections, we propose some strategies to overcome or circumvent these difficulties. We also study how to efficiently compute the sensitivities of adaptive solutions for the PDEs. Difficulties related to adaptive data structure and discretization are addressed.

2 AMR with Hierarchical Block Structure

Our SAMR method makes use of the AMR data structure and refinement strategy in [11]. In order to be read independently, we outline the method here.

2.1 Hierarchical block structure

To efficiently manage the data on each level in the AMR algorithm, the points are grouped (clustered) into logically-rectangular blocks called patches. These patches are the building blocks for the hierarchical grid structure and are the basic data unit for refining the grid in space. A patch is treated as a single grid with all the attributes of a single grid.

We use an indexed linear array hierarchical data structure in [11]. The hierarchical grid data structure $G{=}|n|G_1|G_2|...|G_n|$ contains the number of levels of the grid and pointers to the grid on each of the lower levels. The data structure on the i-th level $G_i{=}|m_i|p_1|G_{i,1}|p_2|G_{i,2}|...|p_{m_i}|G_{i,m_i}|$ contains information on the patches, where m_i denotes the number of patches. The data structure $G_{i,j}$ contains information for the j-th patch on the i-th level. For a 1-D grid, the pointer p_j is the index of the patch in the coarse grid that is the parent of the patch $G_{i,j}$. For a 2-D and 3-D grid, the variable p_j contains the number of parent coarse grids for the patch $G_{i,j}$. An auxiliary array is used to store the indices of the parent grids.

2.2 Refinement strategy

The core of the AMR algorithm is in choosing how to cover the regions that need refinement with a finer grid. The *Remesh* stage to cover subdomains with higher resolution patches is the most algorithmically complex AMR

```
Remesh(level)
begin
    maxlevel = the maximum level allowable, flevel = the finest level existing;
    flevel = max(flevel + 1, maxlevel);
    while (flevel-1 needs no refining) decrease flevel by 1;
    // Readapt the current grid
    for slevel = flevel-1 downto level do
        Refine(slevel)
            Select (slevel): flag the inaccurate points which need refining;
            Expand (slevel): add buffer zones around the flagged region;
            Cluster (slevel): group the flagged points into clusters;
    for slevel = level upto flevel-1 do
        Regrid (slevel +1): define the solution values for the readapted grid;
    // Refine to generate new finer grid
    while (flevel < maxlevel and flevel needs refining) do
        Refine (flevel);
        Regrid (flevel + 1);
        increase flevel by 1;
end
```

Fig. 2.1. Pseudo-code for AMR remeshing algorithm.

operation in the refinement process. The remesh stage is divided into two processes (see Fig. 2.1): *readapt* (including refine and coarsen) the current grid and *refine* to generate a new finer grid. Both processes have two small steps (see Fig. 2.1): *refine* and *regrid.*

The readaptation must be designed to capture the features that appear in the finer levels but would not be identified if the process started with the solution on the coarsest grid and then adapted the grid to the finer levels. Therefore, we initiate the mesh readaptation on the finest level possible. Note that this is different from the local uniform grid refinement (LUGR) method [17]. This grid is then coarsened or refined based on the selection algorithm. This process continues until all of the indicated levels have been readapted. The regridding step (see Fig. 2.1) (defining the solution values for the readapted grid) is done in reverse. It starts from the coarsest level possible. After the first process, if the finest level available does not reach the maximum level allowable and needs further refinement, we start the second process to refine and generate finer level patches.

We adopt the monitor function proposed by Verwer et al. [17] to identify the regions to coarsen or refine. The monitor function is defined for each grid point (i, j). We initiate a level refinement if there is a point where the monitor function exceeds the tolerance. In order to ensure proper nesting, if the current level grid has grandchildren, those points are also flagged. To be flexible, our software has an option to allow users to provide monitor functions [11].

For some applications, the monitor function may fail to identify all the regions that need to be refined. Also, there are situations where we may only be interested in the final steady state solution. In these cases, the efficiency can be improved if the user has control over the AMR process. In an extreme situation, a user may want complete control to guide the refinement process at any time and any place.

We incorporate several options in our software for user control of the refinement. The user can force a refinement through a grid file and modify the refinement parameters at any time [11].

3 Time Integration

It would be appealing to take a local time step for a local finer grid, as is done in conjunction with explicit time integration [11]. However, due to the difficulties mentioned in Section 1, we decided to synchronize the time step for all the grids. In fact, for a parabolic problem solved by an implicit method, the time step is determined by accuracy rather than stability considerations, and the difference in stepsize between grids in different refinement levels is generally small.

As pointed out in [17], the solution injected from a finer grid is in general not a solution of the PDE system discretized on a coarser grid and hence can cause convergence problems in the Newton iteration if it is used as the initial guess. Therefore, we solve the whole AMR system simultaneously. That is, the entire AMR hierarchical structure is transformed into one linear structure used by the implicit time integration solver.

To design software that is easy to use by an application scientist, we use the method of lines (MOL) approach. That is, the PDEs are first discretized into ODEs/DAEs, and then existing ODE/DAE software is used in the time integration. Our implementation has no restriction on the spatial discretization and time integration solver. We use DASPK3.0 for the time integration because of its capabilities for DAEs, implicit time integration and sensitivity analysis. In the next subsection, we describe the transformation between the DASPK3.0 and AMR data structures.

3.1 Transformation between DASPK and AMR data structures

The hierarchical data structure of AMR provides us a possibility to integrate each level or patch separately. However, the difficulty of such a separate computation is synchronization of the time step. Because the time stepsize is computed inside the time integration solver (DASPK in our case), we cannot expect that different levels or patches would use the same time stepsize. Another problem with separate computation is that the Schwarz alternation iteration must be used.

To avoid these complications and difficulties, we integrate the whole system as one big ODE/DAE system. Thus, we must transform the hierarchical data structure into a flat structure that can be used by DASPK. In order that the equations or residuals can be evaluated patch by patch in the AMR hierarchical system and the solutions visualized easily, the transformation must also be done in the reverse.

To eliminate redundancy and inconsistency, we require that any point or cell in the AMR system be evaluated only once. The transformation is illustrated in Fig. 3.1 and is designed as follows. Beginning with the finest level, each point in each patch in a level is copied to a linear array and marked after it is copied; if a point is marked by a previous patch or level, it is skipped. This process is done level by level until the base grid is finished. The inverse transformation is a little more complex. After the inverse copying from the linear array to the AMR hierarchical structure is done, the uninitialized points in the AMR hierarchical structure are collected by copying from the sibling grid and finer children grids.

For the inverse transformation, the ghost boundaries are also required in evaluation of the equations or during refinement. The ghost boundaries for each patch must be collected from three sources. First they are calculated from external boundary conditions if any of its boundaries reaches the external boundary. Then they are copied from the sibling internal grid points. Finally, if there are still uninitialized ghost boundary points, interpolation from the parent coarse grid is used.

3.2 Warm restart after refinement

After a new grid is generated and the solution has been interpolated from the old mesh to the new one, the simplest approach would be to restart the time integrator as though solving a new problem. This is called a *full restart* by Berzins et al.[2] and is appropriate for single-step time integration methods such as singly implicit Runge-Kutta (SIRK) [6] integration. For multistep methods, a full restart would cause the ODE/DAE solver to choose the lowest order single-step method and to reduce the time step size to satisfy the error tolerance of the lowest order method.

In a warm restart (or *flying restart* [2]) the history array used by the ODE/DAE solver is also interpolated to the new mesh, and the integration is continued with almost the same step size and order as would have been used had the remeshing not taken place. Because the number of equations may have changed during the remeshing and the Jacobian matrix (preconditioner in our case) is difficult to interpolate accurately, we always reevaluate the Jacobian matrix in a warm restart.

Compared with a global rezone method [12], the interpolation from an old grid to a new one in AMR involves less error, because the most interesting portion of the fine grid overlaps with that of the old grid, and the overlapping part can just be copied from the old mesh. If the refinement in the AMR

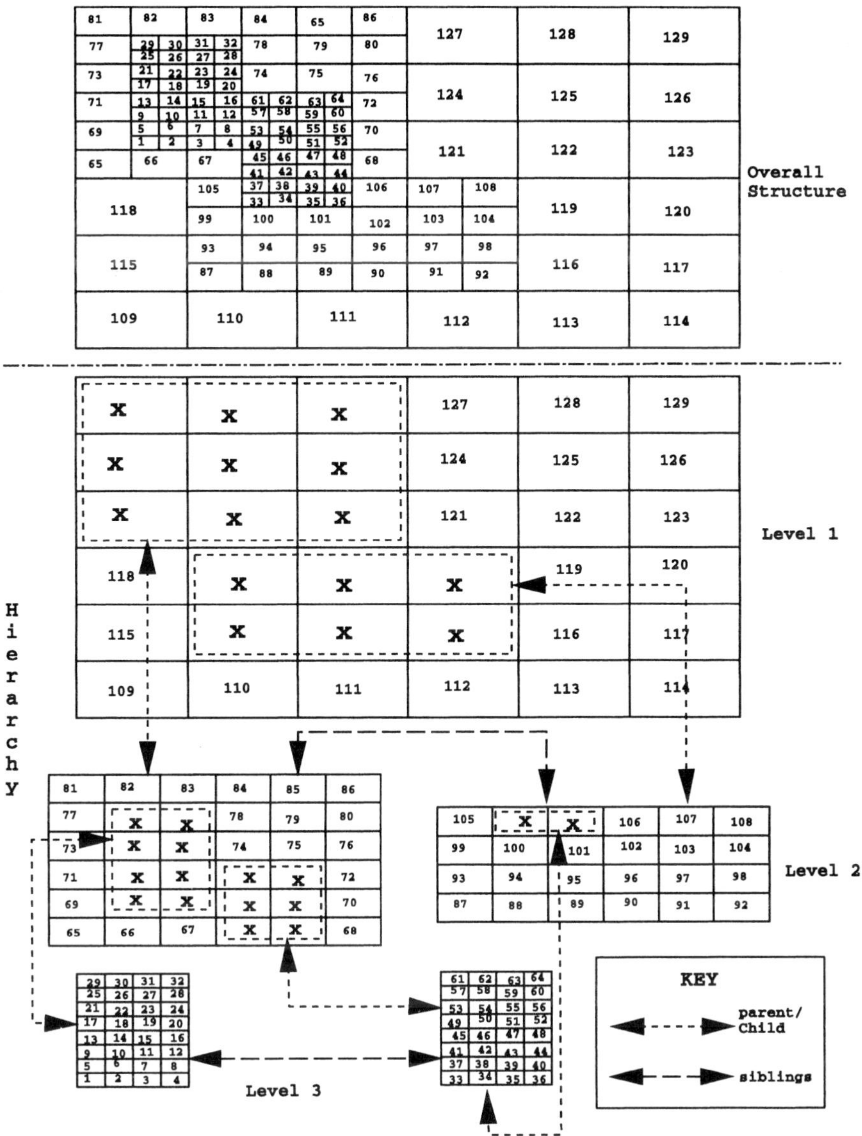

Fig. 3.1. The hierarchical data structure for AMR and its transformation to the DASPK flat structure. The number inside the mesh is the order number in the DASPK flat structure. The cells with "x" inside have been defined by other patches. A total of 129 cells is copied from the AMR hierarchical data structure to the DASPK flat structure.

system is timely, the interpolation occurs only near the coarse-fine interface, where the new grid points are generated by the refinement process. Since the discretization error near the coarse-fine interface is generally much smaller than at the internal points, linear interpolation is sufficient in most cases. When a new grid has more refinement levels than the old one, bivariate cubic interpolation can be applied.

The interpolation errors in a warm restart may not be sufficiently small and may cause the ODE/DAE solver to reduce the stepsize and/or order. In our experience, even if the ODE/DAE solver eventually restarts with the first-order method, the time-step size is much larger than that for a full restart.

3.3 Reducing the overhead related to mesh adaptation

Even with the warm restart, the overhead of the mesh adaptation is relatively high. The most significant cost is evaluation of the Jacobian. The adaptation process (including refining and regridding steps) and interpolations for the history array actually take less time than the evaluation of the Jacobian. Since the Jacobian must be evaluated after each adaptation, it is important to reduce the number of adaptations and/or reduce the computational work of evaluation of the Jacobian.

The number of adaptations is determined by the number of time steps k_{amr} between two adjacent refinements. k_{amr} is affected by many factors. One of them is the number of buffer zones (k_{buf}) added during the refinement. We note in our experiments that if k_{buf} increases, k_{amr} can be larger. However, when k_{buf} increases, the finer grid becomes larger and requires more time to solve. In our numerical tests, we found that $k_{buf} = 2$ and $k_{amr} \approx 10$ have good performance for most problems. How to choose k_{amr} dynamically, as we did in [12] in the case of explicit integration, is under investigation.

Replacing with the new grid adaptively An alternative approach to reduce the overhead of the adaptations is to replace the old mesh with the new adaptive one only when the variance is big enough. A fixed k_{amr} can be taken for this approach.

As we have mentioned, the new adaptive mesh shares the most interesting part with the old one. In our implementation, we calculate the number of grid cells shared by both the new grid and the old one after each adaptation. A *ratio* that measures the shared percentage is calculated as follows,

$$ratio = \frac{2 * N_{share}}{N_{old} + N_{new}}, \tag{1}$$

where N_{share} is the number of shared grid cells, N_{old} is the number of cells in the old grid and N_{new} is the number of cells in the new adaptive grid. The old mesh is replaced by the new one only if $ratio < 0.92$. This number was determined experimentally to give the best performance over a wide range of problems. Otherwise, the old mesh is used as if no adaptation has occurred. This strategy can sharply reduce the number of warm restarts and the number of Jacobian evaluations.

ILU preconditioner and ADIFOR To evaluate the Jacobian efficiently, we can choose a simple preconditioner that requires less computational work, such as a block-diagonal or block-Jacobi preconditioner. However, the performance for these simple preconditioners was not very promising during our numerical experiments. We opted for an incomplete LU (ILU) factorization preconditioner [16] in our software.

For an ILU preconditioner, the Jacobian should be evaluated and stored in each evaluation of the preconditioner. Unlike a single nonadaptive grid, the bandwidth of the Jacobian in our system can be very large because the solution in one patch/level can be related to the solution in another patch/level and they can be far away in different locations in the flat DASPK structure after transformation. Note that in Fig. 3.1, the cell at position 33 relates to the cells at positions 100 and 105.

The cost of evaluating the Jacobian J (C_{eval}) via ADIFOR, if the sparse forward mode is used, is related to the cost of residual evaluation (C_{fun}) by $C_{eval} \simeq a \cdot m \cdot C_{fun}$, where $a = 3$ for the basic forward mode of automatic differentiation, and m is the maximum number of nonzero entries in any row of the Jacobian. However, when the finite-difference method is used, the cost is $C_{eval} \simeq b \cdot C_{fun}$, where b is the bandwidth of the Jacobian matrix. For a PDE in a 2-D domain, m is usually small ($m = 5$ if central difference in space is used for a scalar PDE) but b is large. Therefore, we recommend using ADIFOR to evaluate the Jacobian whenever possible.

Reducing the computation of data structure transformation The transformation between the AMR hierarchical data structure and the DASPK linear structure is frequently used during function evaluations, mesh refinements and warm restarts. To reduce the overhead of the adaptations, we must reduce the computation time of each interface transformation.

We can go through the matching process proposed in Section 3.1 each time to do the transformation or inverse transformation. Noting that the matching process is the same if the mesh does not change, we propose to do the matching process only once for a new mesh.

The algorithm is as follows. When a new mesh is generated, we go through the AMR hierarchical data structure and the DASPK linear structure and match them one by one using the algorithm of Section 3.1. During the processing, we can take advantage of the indexed linear array implementation for the hierarchical data structure. We actually do the matching between two linear structures instead of one hierarchical (tree) structure and one linear structure. The solutions in the AMR hierarchical data structure contain data at the ghost boundaries and shared internal points among different patches. Hence there are more of them than those in the DASPK linear structure. Two index arrays are used to store the locations of the solutions in each linear structure during the matching process. One is used to store the indices of the elements in the AMR structure for the elements in the DASPK structure. The other is used to store the indices of the already marked elements in the AMR structure and their corresponding elements in the DASPK structure.

For the transformation from the AMR structure to the DASPK structure, we use only the first pointer array. For the inverse transformation, both pointer arrays are used. The ghost boundary data for the AMR structure, if needed, are collected separately. The transformations can be done much faster with the help of the two arrays. Since these two arrays are computed only once for a new mesh, the total computational efficiency is improved.

4 Sensitivity Analysis for PDEs

DASPK3.0 has a capability for forward sensitivity analysis[13]. The sensitivity equations for the DAEs have many good properties which can be taken advantage of. First they are linear with respect to the sensitivity variables. Second, the Jacobian matrix for the sensitivities is the same as for the original DAEs. We would like to make use of this sensitivity analysis in the SAMR solution of PDEs.

4.1 Sensitivity ODEs vs. sensitivity PDEs

There are two possibilities for evaluating the sensitivity residuals of a PDE system. First, we can use the MOL approach and transform the PDE system into an ODE/DAE system. Then the sensitivity methods in DASPK3.0 can be used. The sensitivity equations can be evaluated by several options in DASPK3.0, such as the finite-difference or ADIFOR options. This approach does not require any modification of the PDE discretization codes.

The other approach is to solve the sensitivity PDEs coupled with the original physical PDEs directly. They are simultaneously discretized in space and then the coupled ODE/DAE system is solved by DASPK3.0. Similar to the sensitivity DAEs, the sensitivity PDEs are linear with respect to the sensitivity variables. Since the PDE system is usually much simpler before discretization, the sensitivity PDEs can be easily obtained. Some special discretizations or transformations used for the state PDEs can be reused in the sensitivity PDEs. Therefore, it is usually more efficient and accurate to evaluate the sensitivity equations by this approach than by the first approach. In fact, we have found that if a nonlinear spatial discretization scheme (e.g. upwinding scheme) was used, the first approach might produce incorrect sensitivities.

For an implicit solver like DASPK, the cost of the Newton iteration for solving the nonlinear system of equations often dominates the computation.

It is easy to solve the coupled system without distinguishing the state and sensitivity variables in the second approach. However, it is much more efficient if we evaluate the Jacobian/preconditioner only for the state variables, and reuse them in solving for the sensitivities. DASPK3.0 has an option for the user to input the residual for the state and sensitivity equations respectively. Distinguishing the state and sensitivity variables in DASPK also allows the user to exclude the sensitivity variables from the stepsize control, which in our experience has led to better performance as well as accurate sensitivities.

For an adaptive grid solver, we must decide whether the selection of mesh refinement should be based only on the state PDEs or on both the state and sensitivity PDEs. We observed in our applications that the sensitivity PDEs for the sensitivity parameters that appears in the PDEs shared the same refinement regions as the state PDEs. Therefore, we can exclude the sensitivity equations from the monitor function evaluations for efficiency considerations for those applications. This might not be true if the sensitivity parameters are in the initial or boundary conditions.

4.2 Sensitivity analysis with AMR hierarchical structure

In DASPK3.0, the sensitivity variables are stored separately right after the state variables, whereas a sensitivity variable is taken as a PDE variable and all of the variables in a patch are stored together in the AMR data structure. This causes some difficulty in transformation from the AMR hierarchical data structure to the DASPK flat structure and in the residual evaluations of DASPK.

The transformation between the AMR and DASPK data structures proceeds in two steps. In the first step, we do the transformation only for the state variables. In the second step, we transform one by one for the sensitivity variables. If the sensitivity variables are not needed, the second step is skipped. In sensitivity analysis using DASPK3.0, the Krylov iteration uses only the residual evaluations of the state variables. The number of residual evaluations for the state variables is much larger than that for the sensitivity variables. Therefore, the overhead in the transformation can be much reduced by the two-step technique.

5 Numerical Experiments

In this section, we give an example to illustrate the effectiveness of our algorithm and software. The number of steps k_{amr} between two adaptations is chosen to be 6. The refinement ratio is chosen to be 2, and the number of buffer zones (k_{buf}) is 2 unless it is specified otherwise. Central differencing discretization in space is used. The error tolerance in DASPK is chosen to be RTOL=ATOL=10^{-5}. All of our computations are done in double precision on a 450HZ PC with the Linux operating system. For comparison, we give some key statistics of our computation.

NWR	Number of warm restarts
NTS	Number of time steps
NRE	Number of residual evaluations
NJE	Number of Jacobian evaluations
NETF	Number of error test failures
MXEQ	Maximum number of equations in DASPK format
CPU	Total CPU time taken to solve the problem

5.1 SAMR solution

This example of reaction-diffusion type is described in Zegeling [18]. The PDE is given by

$$u_t = \Delta u + D(2-u)\exp(-d/u), \quad \text{on the domain } \Omega = (0,1)\times(0,1) \quad (2)$$

$$u|_{t=0} = 1, \qquad \frac{\partial u}{\partial n} = 0, \text{ at } x = 0, y = 0, \quad \text{and } u = 1, \text{ at } x = 1, y = 1,$$

where Δ is the Laplacian operator and $D = Re^d/d, R = 5, d = 20$. We output the solution at $t = 0.30$. Since the solution before $t = 0.25$ is very smooth, we turned off the refinement and used only the base grid before $t = 0.25$. The tolerance for grid refinement was chosen to be TOLS=0.001.

The full restart is extremely slow. The warm restart, however, is much faster. We note that after refinement the solver DASPK3.0 uses almost the same order as before the refinement. The first refinement takes place at about $t = 0.25$. We chose $k_{amr} = 8$ for the later refinements. The comparison for methods with different refinement levels and different buffer zones are shown in Table 1. The contour plots and refinement patches are displayed in Fig. 5.1.

Table 1. Comparison of different methods for reaction-diffusion problem (2).

Base grid	level	k_{buf}	NWR	NTS	NRE	NJE	NETF	MXEQ	CPU
201×201	1	N/A	0	147	687	15	18	40401	155
101×101	2	2	17	188	742	37	22	17450	56
51×51	3	2	18	209	802	38	29	12637	32
51×51	3	1	19	220	854	41	32	11557	31

The warm restart does have some adverse effect on the time step selection, which is shown in Fig. 5.1-c. We suspect that this is due to the interpolation errors from the old grid to the new grid.

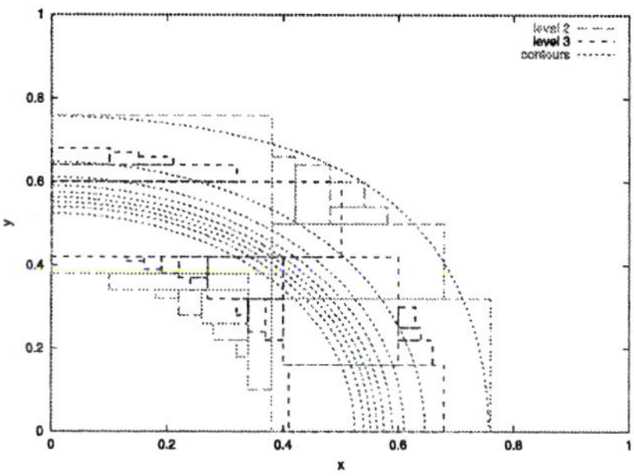

Fig. 5.1-a. AMR with 50×50, 3 refinement levels and 2 buffer zones. 9 contours between 1.1 and 1.9 are plotted.

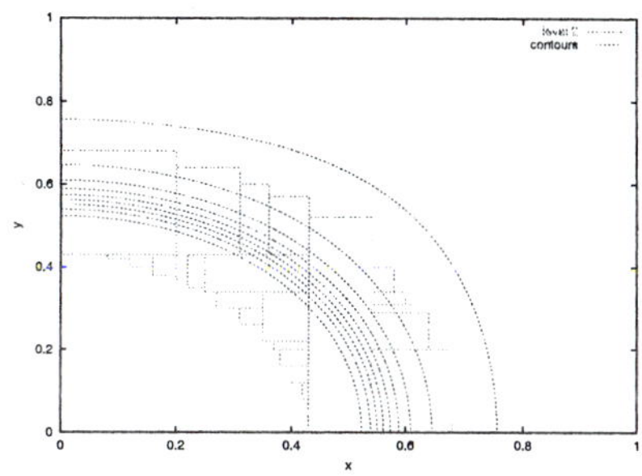

Fig. 5.1-b. AMR with 100×100, 2 refinement levels and 2 buffer zones. 9 contours between 1.1 and 1.9 are plotted.

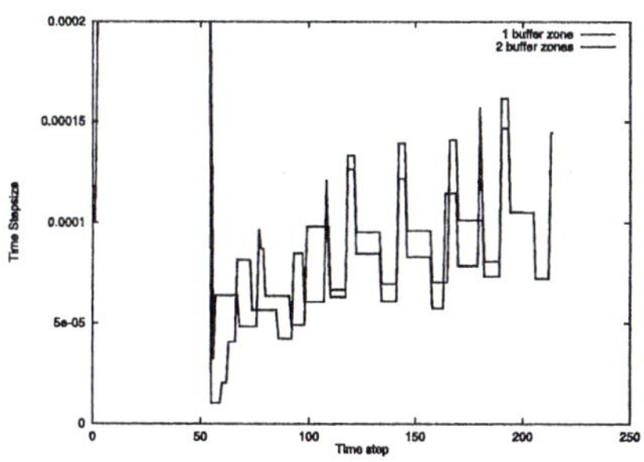

Fig. 5.1-c. The warm restart can cause the DAE solver to reduce the time step after each refinement. Three-level refinement and a 50×50 base grid is used.

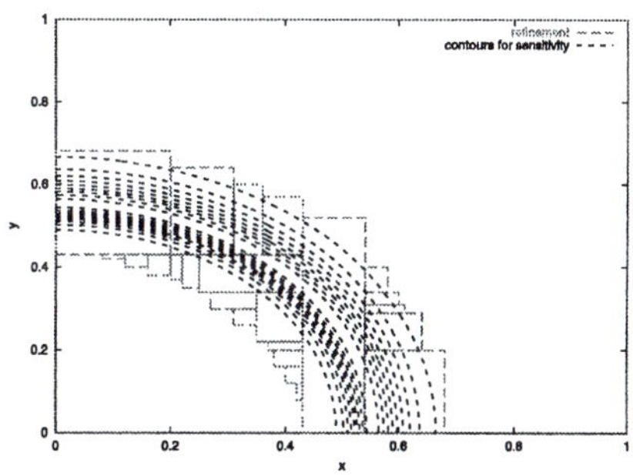

Fig. 5.1-d. Contour plots for the sensitivity with respect to R. 9 contours between 2 and 18 are used. The contours increase from both sides to the middle of the refinement.

Figure 5.1: Results for reaction-diffusion problem (2) (see Plate 1 on page 345).

5.2 Sensitivity analysis

We also computed the sensitivity with respect to the parameter R in Eq. (2). The sensitivity PDE is given by

$$s_t = \Delta s + D/R(2-u)\exp\left(-\frac{d}{u}\right) - sD\exp\left(-\frac{d}{u}\right)\left(1 - \frac{d(2-u)}{u^2}\right). \quad (3)$$

We used two refinement levels and a 51×51 base grid. The error tolerance for the sensitivity variables was the same as that for the state variables. The partial error test (excluding the sensitivity variables from the error test) and staggered corrector method option was used in DASPK3.0. The other parameters were the same as those without considering sensitivity.

Two options for sensitivity evaluation in DASPK3.0 were tested: input analytically and by ADIFOR with seed matrix. They produced the same results, which is not surprising since no special technique is used during the spatial discretization. The efficiency for the two options was almost the same. We also solved Eqs. (2) and (3) without using the sensitivity techniques of DASPK3.0 (see "taken as PDEs" method in Table 2). In this method, we cannot exclude the sensitivity variables from the temporal error test in DASPK3.0, and the warm-restart after each refinement does not work well after $t = 0.29$. We suspect the reason is that the sensitivity changes too rapidly. We should mention that the accuracy of the sensitivity did not improve much by including the sensitivity variables in the error test. The contour plots are almost the same.

Table 2. Comparison of different methods for sensitivity analysis of reaction-diffusion problem (2). *This number includes evaluations of the sensitivity equations.

Sensitivity Evaluation Method	NST	NRE	NJE	NETF	CPU
Input analytically	191	3817	39	24	79
ADIFOR with seed matrix	191	3817	39	24	82
Taken as PDEs	1315	4336*	491	211	329

6 Conclusion

We have presented our implementation of AMR with the implicit DAE solver DASPK, for parabolic problems where implicit time integration is best suited. Several difficulties have been described when AMR is combined with implicit integration. We have provided some strategies to overcome and/or circumvent the difficulties. Numerical results demonstrate that these strategies are effective.

We have also discussed how to combine the sensitivity analysis with the AMR hierarchical data structure. An interface was designed between the AMR hierarchical structure and the DAE solver flat structure to facilitate the use of the DAE solver and data visualizations.

A large-scale sparse linear system must be solved in the implicit time integration. How to improve the efficiency of the linear solver is key to the success of our combination of AMR and DASPK. We provide an ILU preconditioner which is evaluated by ADIFOR. Other kinds of preconditioners, such as additive Schwarz alternation preconditioners, are under investigation.

Although the warm restart technique can greatly improve the efficiency, the restart/interpolation process still has an adverse effect on the time step selection. We think this is due to the interpolation error after the refinement. How to improve the time step selection after each restart is an open problem.

The forward sensitivity method described in this paper is attractive when there are relatively few sensitivity parameters. When a large number of sensitivity parameters and only a few derived functions are involved, the adjoint sensitivity method may be more advantageous. With the help of the recent results on the adjoint method for DAEs [7, 8], we have studied several theory and implementation issues for the adjoint sensitivity method on an adaptive grid for partial differential-algebraic equations (PDAE) [14].

References

1. C. Bischof, A. Carle, G. Corliss, A. Griewank and P. Hovland, *ADIFOR–Generating derivative codes from Fortran programs,* Scientific Programming (1992).
2. M. Berzins, P. J. Capon and P. K. Jimack, On spatial adaptivity and interpolation when using the method of lines, *Appl. Numer. Math.*, **26** (1998) 117-133.
3. M. J. Berger and P. Colella, Local adaptive mesh refinement for shock hydrodynamics, J. Comput. Phys. **82** (1989) 64-84.
4. K. E. Brenan, S. L. Campbell and L. R. Petzold, Numerical Solution of Initial-Value Problems in Differential-Algebraic Equations, Second Edition, SIAM, 1995.
5. P. N. Brown, A. C. Hindmarsh and L. R. Petzold, *Using Krylov methods in the solution of large-scale differential- algebraic systems*, SIAM J. Sci. Comput., **15** (1994) 1467-1488.
6. J.C. Butcher, *On the implementation of Runge-Kutta methods*, BIT, **16** (1976), 237-240.
7. Y. Cao, S. Li, L. Petzold, Adjoint sensitivity analysis for differential-algebraic equations: Algorithms and Software, to appear, J. Comp. Appl. Math.
8. Y. Cao, S. Li, L. Petzold and R. Serban, *Adjoint sensitivity analysis for differential-algebraic equations: The adjoint DAE system*, submitted.
9. R. Courant, K.O.Friedrichs, and H. Lewy, *Über die partiellen differenzengleichungen der mathematicschen physik.*, Mathematische Annalen, **100** (1928), 32-74.
10. J. E. Flaherty, P. K. Moore and C. Ozturan, Adaptive overlapping grid methods for parabolic systems, in Adaptive Methods for Partial Differential Equations, J. E. Flaherty, P.J. Paslow, M. S. Shephard, and J. D. Vasilakis, eds., Society for Industrial and Applied Mathematics, Philadelphia, PA, 1989.
11. J. M. Hyman and S. Li, Solution Adapted Nested Grid Refinement for 2-D PDEs, Los Alamos National Lab. Report, LA-UR-98-5463 (1998).
12. J. M. Hyman, S. Li and L. R. Petzold, An Adaptive Moving Mesh Method with Static Rezoning for Partial Differential Equations, Los Alamos National Laboratory Report (1998).
13. S. Li and L. R. Petzold, Software and algorithms for sensitivity analysis of large-scale differential-algebraic systems, J. Comp. and Appl. Math., **125** (2001) 131-145.
14. S. Li and L. R. Petzold, Adjoint sensitivity analysis for partial differential-algebraic equations, in preparation.
15. J. Quirk, An Adaptive Grid Algorithm for Computational Shock Hydrodynamics, Ph.D thesis (1991), College of Aeronautics, Cranfield Institute of Tech.
16. Y. Saad, Iterative Methods for Sparse Linear Systems, PWS Publishing Company, Boston, 1996.

17. J. G. Verwer, J. G. Blom, VLUGR2: A Vectorized Local Uniform Grid Refinement Code for PDEs in 2D, Report NM-R9307 (1993), CWI, Amsterdam.
18. P. A. Zegeling, Moving Finite-Element Solution of Time-Dependent Partial Differential Equations in Two Space Dimensions. Department of Numerical Mathematics, CWI, Amsterdam, Report NM-R9206 (1992).

Challenges and Opportunities in Using Automatic Differentiation with Object-Oriented Toolkits for Scientific Computing

Paul Hovland[1], Steven Lee[2], Lois McInnes[1], Boyana Norris[1], and Barry Smith[1]

[1] Mathematics and Computer Science Division, Argonne National Laboratory, 9700 S. Cass Ave., Argonne, IL 60439-4844.
[2] Center for Applied Scientific Computing, Lawrence Livermore National Laboratory, Box 808, L-560, Livermore, CA 94551.

Abstract. The increased use of object-oriented toolkits in large-scale scientific simulation presents challenges and new opportunities for the use of automatic (or algorithmic) differentiation (AD) techniques, especially in the context of optimization. Because object-oriented toolkits use well-defined interfaces and data structures, there is potential for simplifying the AD process. Furthermore, derivative computation can be improved by exploiting high-level information about numerical and computational abstractions. However, challenges to the successful use of AD with these toolkits also exist. Among the greatest challenges is balancing the desire to limit the scope of the AD process with the desire to minimize the work required of a user. We discuss our experiences in integrating AD with the PETSc, PVODE, and TAO toolkits and our plans for future research and development in this area.

1 Introduction

The ever-increasing complexity of advanced computational science applications has led to an increase in the use of object-oriented software practices in the development of scientific applications and toolkits. A good object-oriented design increases productivity by allowing developers to focus on a small component of a complex system. Furthermore, increased code reuse provides justification for expending significant effort in the development of highly optimized object-oriented toolkits.

Many high-performance numerical toolkits include components designed to be combined with an application-specific nonlinear function. Examples include optimization components, nonlinear equation solvers, and differential algebraic equation solvers. Often the numerical methods implemented by these components also require first and possibly second derivatives of the function. Frequently, the toolkit is able to approximate these derivatives by using finite differences; however, the convergence rate and robustness are often improved if the derivatives are computed analytically.

Developing correct parallel code for computing the analytic derivatives of a complicated nonlinear function can be an onerous task, especially when second derivatives are required. An automated alternative such as automatic differentiation (AD) [15, 16] is therefore very attractive as a mechanism for providing these analytic derivatives. Furthermore, because object-oriented toolkits provide well-defined interfaces for nonlinear functions, the AD process can potentially be simplified.

We examine the use of AD in conjunction with object-oriented numerical toolkits. In particular, we describe the use of AD to provide first and, where appropriate, second derivatives in conjunction with the Portable, Extensible Toolkit for Scientific Computing (PETSc), the Toolkit for Advanced Optimization (TAO), and a parallel ODE solver for computing sensitivities (SensPVODE).

This paper is organized as follows. Section 2 provides a brief introduction to automatic differentiation. Section 3 introduces the toolkits considered in this paper. Section 4 discusses the challenges and opportunities that arise in using AD with these toolkits. Section 5 provides a synopsis of experimental results. Section 6 summarizes our experiences and describes our expectations for using AD in the context of PDE-constrained optimization.

2 Automatic Differentiation

Automatic, or algorithmic, differentiation is a technique for augmenting arbitrarily complex computer subprograms with instructions for the computation of derivatives. The technique combines rules for analytically differentiating the finite number of elemental functions in a programming language with the chain rule of differential calculus.

The two principal approaches to implementing AD are operator overloading and compiler-based source transformation. Each method has its advantages and disadvantages [7]. There are also two basic modes of AD, the forward mode and the reverse mode. The forward mode is particularly appropriate when the number of independent variables is small or when a small number of directional derivatives are required. The reverse mode is attractive when the number of dependent variables is small or when a Jacobian-transpose-vector product is required. We focus on the source transformation approach using the forward mode, but most of the issues discussed in this paper are also applicable to the operator overloading approach and the reverse mode. More details of how AD works and various methods for exploiting chain rule associativity can be found in [16].

Figure 1 shows a simple example of the code generated by the ADIFOR [6] source transformation tool for Fortran 77. While this example is not indicative of the power of automatic differentiation, which is equally applicable to applications spanning thousands or millions of lines of code and using complex control structures, it does illustrate some important concepts.

```
y(1) = 10.0 * (x2-x1*x1)
y(2) = 1.0 - x1
```

```
d2_b = dble(10.0)
d5_b = (-d2_b) * x1 + (-d2_b) * x1
do g_i_ = 1, g_p_
  g_y(g_i_, 1) = d5_b * g_x1(g_i_) +
+                d2_b * g_x2(g_i_)
enddo
y(1) = dble(10.0) * (x2 - x1 * x1)

do g_i_ = 1, g_p_
  g_y(g_i_, 2) = -g_x1(g_i_)
enddo
y(2) = 1.0d0 - x1
```

Fig. 1. A simple example of automatic differentiation. The code on the right was generated by the ADIFOR tool from the code fragment on the left.

Each scalar variable in the original program has associated with it a derivative vector (sometimes generically referred to as a derivative object). In the case of ADIFOR-generated code, this association is by name. For example, the derivative vector associated with `x1` is `g_x1`. In the case of an AD tool for C, such as ADIC [8], association by name is not possible because of aliasing. Instead, association must be by address. ADIC accomplishes this association by converting all floating-point variables to a new datatype, `DERIV_TYPE`.

```
typedef struct {
        double value;
        double  grad[ad_GRAD_MAX];
} DERIV_TYPE;
```

The `value` field carries the original floating-point value, while the `grad` field contains the associated derivative vector.

The derivative vectors associated with independent variables are collectively referred to as the seed matrix. The seed matrix can be initialized such that upon completion of the computation the derivative vectors associated with the dependent variables contain the Jacobian, a Jacobian-vector product, or an arbitrary Jacobian-matrix product. For instance, if in our simple example we initialize `g_x1` to $[1.0, 0.0]$ and `g_x2` to $[0.0, 1.0]$, then `g_y(i,j)` will contain $\partial y_j / \partial x_i$.

3 Toolkits

We have investigated the use of AD in conjunction with three object-oriented toolkits for scientific computing: PETSc, TAO, and SensPVODE. All of these toolkits employ an object-oriented design and are implemented in C/C++. In the next sections, we briefly describe these toolkits and the role derivatives play.

3.1 Portable, Extensible Toolkit for Scientific Computing

PETSc [2, 3] is a suite of data structures and routines for the scalable solution of scientific applications modeled by partial differential equations. The software integrates a hierarchy of components that range from low-level distributed data structures for vectors and matrices through high-level linear, nonlinear, and timestepping solvers. The algorithmic source code is written in high-level abstractions so that it can be easily understood and modified. This approach promotes code reuse and flexibility and, in many cases, helps to decouple issues of parallelism from algorithm choices.

PETSc provides a suite of nonlinear solvers that couple a Newton-based method, offering the advantage of rapid convergence when an iterate is near to a problem's solution, with a line search, trust region, and pseudo-transient continuation strategy to extend the radius of convergence of the Newton techniques. The linearized systems are typically solved inexactly with preconditioned Krylov methods. The basic Newton method requires the Jacobian matrix, $J = F'(u)$, of a nonlinear function $F(u)$. Matrix-free Newton-Krylov methods require Jacobian-vector products, $F'(u)v$, and may require an approximate Jacobian for preconditioning.

PETSc also provides components for managing computations on structured grids, including hierarchies of grids for use in multigrid methods. Among the functions provided by these components are generalized gather-scatter operations for communicating ghost values, colorings for use in finite difference (and AD) Jacobian computations, and simplified facilities for mapping between local and global indices.

3.2 Toolkit for Advanced Optimization

TAO [4, 5] focuses on scalable optimization software, including nonlinear least squares, unconstrained minimization, bound-constrained optimization, and general nonlinear optimization. The TAO optimization algorithms use high-level abstractions for matrices and vectors and emphasize the reuse of external tools where appropriate, including support for using the linear algebra components provided by PETSc and related tools.

Many of the algorithms employed by TAO require first and sometimes second derivatives. For example, unconstrained minimization solvers that require the gradient, $f'(u)$, of an objective function, $f(u)$, include a limited-memory variable metric method and a conjugate gradient method, while solvers that require both the gradient, $f'(u)$, and Hessian, $f''(u)$ (or Hessian-vector products), include line search and trust region variants of Newton methods. In addition, algorithms for nonlinear least squares and constrained optimization often require the Jacobian of the constraint functions.

3.3 SensPVODE

PVODE [10] is a high-performance ordinary differential equation solver for the types of initial value problems (IVPs) that arise in large-scale computational simulations.

$$y'(t) = f(t, y, p), \quad y(t_0) = y_0(p), \quad y \in \mathbf{R}^N, \quad p \in \mathbf{R}^m. \tag{1}$$

Often, one wants to compute sensitivities with respect to certain parameters p_i in the IVP, $s_i = \partial y / \partial p_i$. SensPVODE [22] is a variant of PVODE that simultaneously solves the original ODE and the sensitivity ODEs.

$$s_i'(t) = \frac{\partial f}{\partial y} s_i(t) + \frac{\partial f}{\partial p_i}, \quad s_i(t_0) = \frac{\partial y_0(p)}{\partial p_i}, \quad i = 1, \cdots, m. \tag{2}$$

Thus, SensPVODE requires the derivatives $\frac{\partial f}{\partial y} s_i(t) + \frac{\partial f}{\partial p_i}$.

4 Using Automatic Differentiation with Object-Oriented Toolkits

Automatic differentiation can be used in conjunction with object-oriented toolkits at many different levels. At the highest level, AD can be applied to a toolkit to facilitate sensitivity analysis and optimization of models constructed with the toolkit. Another option is to use AD to provide the derivatives required by the toolkit. Specifically, AD can be applied directly to the parallel nonlinear function or to the building blocks of the nonlinear function, such as the nonlinear function on a local subdomain or an element or vertex function. In the following sections we consider some of the challenges and the opportunities in applying AD at these various levels.

4.1 Toolkit Level

Applying AD to an object-oriented toolkit for scientific computing offers the opportunity to compute derivatives (also called sensitivities) of scientific applications that use the toolkit. These derivatives can be used for sensitivity analysis, to understand the sensitivity of the simulation results to uncertainties in model parameters, or for optimization. Applying AD to a toolkit also provides an opportunity to employ semi-automatic differentiation techniques exploiting high-level mathematical and algorithmic features. Using such techniques can also circumvent some of the challenges that arise in applying AD to a toolkit, including the fact that functions and derivatives may not converge at the same rate in an iterative method and that certain numerical methods, including ODE solvers with adaptive stepsize control, introduce feedback into a computation that may seriously influence the derivatives.

One simple example of a technique based on mathematical insight arises in the differentiation of a linear solver. Rather than differentiate through a preconditioner and iterative Krylov solver, one can solve directly a linear system with multiple right-hand sides, perhaps using block Krylov methods [9, 25] or a projection method [12]. We have conducted preliminary research in this area in developing a differentiated version of PETSc [19]. A more sophisticated example of semi-automatic differentiation, applicable to other types of iterative solvers, is described elsewhere in this volume [17]. We also note that the development of SensPVODE may be interpreted as applying a combination of analytic and automatic differentiation techniques to the PVODE toolkit.

4.2 Parallel Nonlinear Function Level

Automatic differentiation is often not quite "automatic." Often, the user of an AD tool needs to specify the independent and dependent variables and possibly initialize the seed matrix to indicate exactly which derivatives are to be computed. The use of AD in conjunction with a numerical toolkit promises to make full automation possible because of the use of well-defined interfaces. For example, the nonlinear solver in PETSc requires a nonlinear function adhering to the interface

```
int Function(SNES, Vec, Vec, void *);
```

while SensPVODE requires a function adhering to the interface

```
void function(integer, Real, N_Vector, N_Vector, void *);
```

In both cases, all of the required information regarding the derivatives to be computed is known in advance, making automation of the AD process possible. Other work [11, 13, 24] has also demonstrated the benefits of well-defined interfaces for automating the AD process.

However, applying AD directly to the parallel nonlinear function required by the toolkit is not without challenges. The function may include many calls to toolkit support functions. Thus, differentiated versions of these functions must be developed using automatic or semi-automatic differentiation techniques. In many cases, these support functions are used for communicating ghost values or performing similar problem setup or data movement functions. Consequently, the differentiated versions of the functions may end up performing unnecessary work. Parallelism is also an issue. The nonlinear function may use OpenMP or make calls to MPI functions, so the AD tool must support the parallel programming paradigm being used. Furthermore, the parameters to the function are likely parallel objects, as is the case with the `Vec` and `N_Vector` objects in the examples above. Therefore, care must be taken in the way seed matrices are initialized; additional communication may also be required.

An important issue in the differentiation of application functions arises from the benign-looking `void *` that appears as the final argument in both of the examples above. This argument is used to pass a pointer to an application-specific data structure that contains various data objects of use to the application. Figure 2 contains several examples of these data structures.

As discussed earlier, in order to provide association by address between variables and their derivative vectors, ADIC changes the datatypes of all floating-point variables. The consequence for these application-specific data structures is that if AD is naively applied, the type of several fields within the data structure will be changed. Therefore, all application functions that access the data structure, not just the nonlinear function, must be modified to use the new datatypes.

An alternative is to use two data structures, one with the original data types and one with the derivative datatypes, and to copy data between them. As the final example in Figure 2 illustrates, however, some care must be taken with this approach. If the data structure includes workspace, such as the `uext` field in this example, it may be inefficient to copy such unneeded data. Furthermore, it is usually not possible to generate code automatically for copying from one data structure to another. Therefore, at this point in our work coupling ADIC with SensPVODE, we require the user to provide the routine for copying data between the application-specific data structures and their differentiated counterparts.

4.3 Local Subdomain Level

As alluded to in the preceding section, many parallel nonlinear functions follow the pattern gather/scatter ghost values, compute function on local subdomain, and assemble/map function from local to global indices. Therefore, an approach that avoids many of the complications associated with applying AD to the full parallel nonlinear function is to apply AD only to the local subdomain function, with manual modifications to the setup and assembly phases. We have examined this approach in conjunction with PETSc and TAO by manually extracting the local subdomain function, as illustrated in Figure 3 and described in [1, 20].

Recent work has examined how we can use the structured grid components to simplify the process. The setup and assembly phases are essentially the same for most nonlinear functions on a structured grid. Therefore, instead of extracting the subdomain computation from the parallel nonlinear function, we ask the user to provide only the subdomain function. Instead of providing a nonlinear function with the interface

```
int Function(SNES, Vec, Vec, void *);
```

the user provides, for example, a local subdomain function with the interface

```
int LocalFormFunction2d(Field  **, Field  **, Coords, void *);
```

```
typedef struct {
   double      param;          /* test problem parameter */
   int         mx,my;          /* discretization in x, y directions */
   Vec         localX,localF;  /* ghosted local vectors */
   DA          da;             /* distributed array data structure */
   int         rank;           /* processor rank */
} AppCtx;

typedef struct {
   double     lidvelocity,prandtl,grashof;
   PetscTruth draw_contours;
} AppCtx;

typedef struct {
  real om, dx, dy, q4;
  real uext[NVARS*(MXSUB+2)*(MYSUB+2)];
  integer my_pe, isubx, isuby, nvmxsub, nvmxsub2;
  real *p;
  real *pbar;
  MPI_Comm comm;
} *UserData;
```

Fig. 2. Examples of application-specific data structures used in PETSc, TAO, and SensPVODE applications.

where `Coords` contains information about the corners of the subdomain and `Field` is a structure with a number of scalar fields corresponding to the number of degrees of freedom at each vertex. The setup and assembly are handled by the structured grid component. Given a differentiated version of `LocalFormFunction2d`, the structured grid component can also perform the necessary setup, including seed matrix initialization, and assembly of the Jacobian. The setup and assembly phases use coloring [14] to reduce the length of the derivative vectors to the stencil size times the number of degrees of freedom. Future research will extend this work to Hessian computations and unstructured meshes.

4.4 Element or Vertex Function Level

A final option is to differentiate the element or vertex function and then assemble the full Jacobian from the element (or vertex) Jacobians. This is common practice in finite element computations, especially when the element Jacobian is simple to derive by hand. For complex element or vertex functions, or for higher-order derivatives, AD can be employed. This approach eliminates the need for a matrix coloring, reduces the memory requirements, and is easily extended to second and higher-order derivatives.

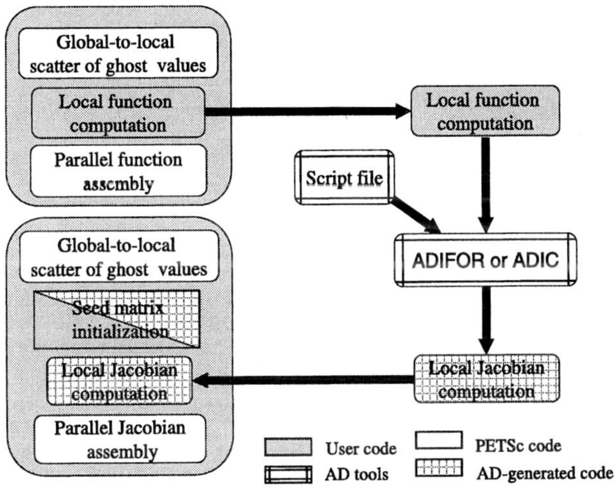

Fig. 3. Schematic diagram of the use of automatic differentiation tools to generate the Jacobian routine for a nonlinear PDE computation.

The principal impediment to this approach is that not all functions are easily decomposed to this level. It may be computationally more efficient to precompute fluxes for all of the edges/faces in the subdomain, then use these fluxes in computing the element functions. Furthermore, boundary conditions may introduce many special cases. It is computationally more efficient to handle these special cases separately, then loop over the remaining elements, than to test every element to determine whether a special case applies.

5 Experimental Results

We briefly summarize experimental results from the use of AD in conjunction with the PETSc and SensPVODE toolkits. More detailed descriptions of these and related experiments can be found in [1, 19–21, 23].

5.1 Toolkit Level

In [19] we describe research into the application of automatic differentiation to PETSc. Here we briefly discuss some of the relevant experimental results.

We have tested the differentiated version of PETSc, and in particular its linear solver component, with an example involving the solution of a linear system of equations $Ax = b$ where A is the 256×256 matrix whose sparsity pattern corresponds to a five-point stencil discretization of a 16×16 computational domain. In all of the experiments, we have used a GMRES solver in combination with an incomplete LU factorization preconditioner.

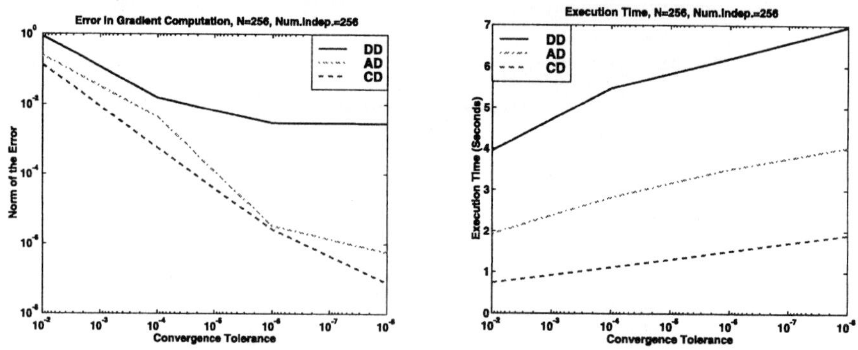

Fig. 4. Gradient error and execution time with varying convergence tolerances.

Figure 4 shows the accuracy and performance results for various convergence tolerances. DD (for "divided differences") designates finite difference approximation, AD designates black-box automatic differentiation, and CD indicates a combination of analytic and automatic differentiation using successive linear solves on the multiple right-hand sides. The termination condition of the Krylov subspace methods is based on the relative decrease of the l_2-norm of the residual and the convergence tolerance value, which is plotted along the x-axis. The y-axis of the accuracy plot is the l_2-norm of the matrix representing the difference between the derivatives produced by the various approaches and the actual solution, $\nabla x = A^{-1}b$, which we compute separately up to machine precision for verification purposes. For the finite difference and AD approaches the convergence tolerance refers to the convergence of x, whereas in the combined approach (semi-automatic differentiation) it refers to the convergence of ∇x. In this example, the semi-automatic differentiation approach exhibits significant performance improvement over finite differences and AD.

5.2 Parallel Nonlinear Function Level

We applied SensPVODE to a simple test case, a two-species diurnal kinetics advection-diffusion system in two space dimensions. The PDEs can be written as

$$\frac{\partial c_i}{\partial t} = K_h \frac{\partial^2 c_i}{\partial x^2} + V \frac{\partial c_i}{\partial x} + \frac{\partial}{\partial y}\left(K_v(y)\frac{\partial c_i}{\partial y}\right) + R_i(c_1, c_2, t) \quad (i = 1, 2),$$

where the subscripts i are used to distinguish the chemical species. The reaction terms are given by

$$R_1(c_1, c_2, t) = -q_1 c_1 c_3 - q_2 c_1 c_2 + 2q_3(t)c_3 + q_4(t)c_2 \text{ and}$$
$$R_2(c_1, c_2, t) = q_1 c_1 c_3 - q_2 c_1 c_2 - q_4(t)c_2,$$

and $K_v(y) = K_0 e^{(y/5)}$. The scalar constants for this problem are $K_h = 4.0 \times 10^{-6}$, $V = 10^{-3}$, $K_0 = 10^{-8}$, $q_1 = 1.63 \times 10^{-16}$, $q_2 = 4.66 \times 10^{-16}$, and $c_3 = 3.7 \times 10^{16}$. The diurnal rate constants are

$$\begin{aligned} q_i(t) &= e^{[-a_i/\sin \omega t]} && \text{for } \sin \omega t > 0, \\ q_i(t) &= 0 && \text{for } \sin \omega t < 0, \end{aligned}$$

where $i = 3$ and 4, $\omega = \pi/43200$, $a_3 = 22.62$, and $a_4 = 7.601$. The time interval of integration is $[0, 86400]$, representing 24 hours measured in seconds.

The problem is posed on the square $0 \leq x \leq 20$, $30 \leq y \leq 50$ (all in km), with homogeneous Neumann boundary conditions. The PDE system is treated by central differences on a uniform 100×100 grid, with simple polynomial initial profiles. See [22] for more details. For the purpose of sensitivity analysis, we identify the following eight parameters associated with this problem: $p_1 = q_1$, $p_2 = q_2$, $p_3 = c_3$, $p_4 = a_3$, $p_5 = a_4$, $p_6 = K_h$, $p_7 = V$, and $p_8 = K_0$.

We varied the number of sensitivities from 1 to 8 and compared the use of AD with three finite difference methods to compute the derivatives $\frac{\partial f}{\partial y} s_i(t) + \frac{\partial f}{\partial p_i}$ required by SensPVODE. In the combined central difference method, p_i and y are perturbed simultaneously to obtain both terms in the derivative expression. In the other finite difference methods, the terms are approximated separately. See [22] for a complete description of the finite difference strategies.

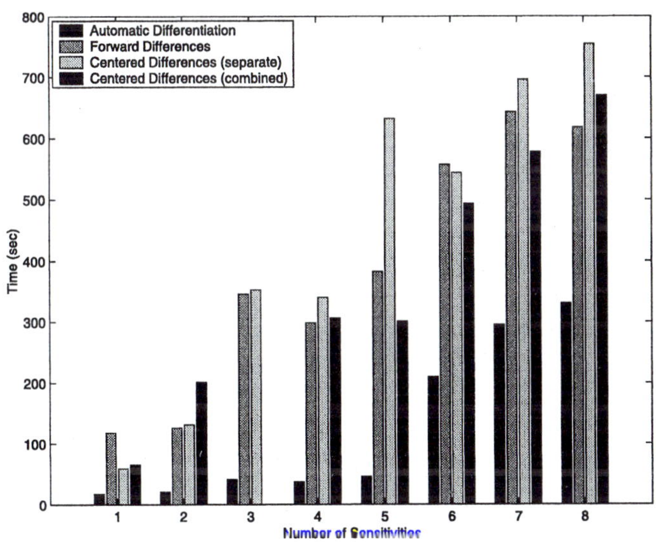

Fig. 5. Comparison of SensPVODE performance (total time to solution) for various derivative computation strategies. Results are the average of three runs on 16 processors of a Linux cluster (see Plate 2 on page 346).

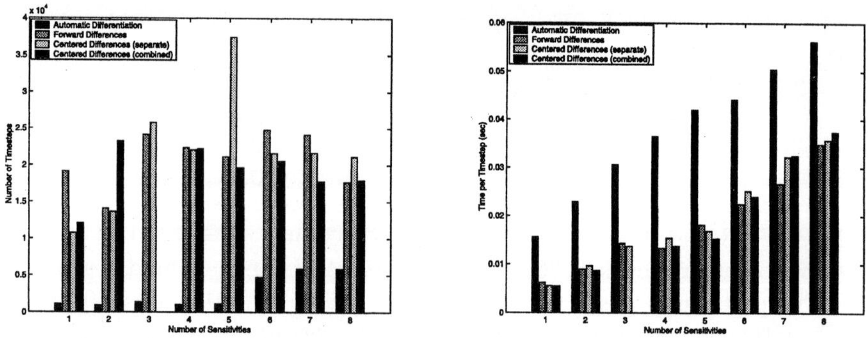

Fig. 6. Number of timesteps and time per timestep for various derivative computation strategies in SensPVODE. Results are the average of three runs on 16 processors of a Linux cluster (see Plate 3 on page 346).

Figures 5 and 6 summarize our results on 16 nodes of a Linux cluster. Each node in the cluster has two 550 MHz Pentium III processors (only one processor per node was used) and Myrinet interconnect. AD shows a significant performance advantage over the finite difference methods. However, as Figure 6 illustrates, the performance improvements are due to a reduction in the number of timesteps required; the runtime per timestep actually increases.

5.3 Local Subdomain Level

We used AD to provide the directional derivatives required by PETSc to solve the steady-state, three-dimensional compressible Euler equations on mapped, structured meshes using a second-order, Roe-type finite volume discretization. We solved in parallel a nonlinear system, using matrix-free Newton-Krylov-Schwarz algorithms with pseudo-transient continuation to model transonic flow over an ONERA M6 airplane wing. See [18] for details about the problem formulation and algorithmic approach. The linearized Newton correction equations were solved by using restarted GMRES preconditioned with the restricted additive Schwarz method with one degree of overlap.

As discussed in depth in [21] and summarized in Figure 7, our results indicate that, for matrix-free Newton-Krylov methods, AD offers significantly greater robustness and provides better algorithmic performance than do finite difference approximations (FD). Because the directional derivatives required by a matrix-free method can be computed less expensively by FD than by AD, however, AD does not always provide a performance advantage in terms of runtime [21]. We are therefore investigating hybrid AD-FD strategies that combine the robustness of AD with the reduced cost of FD.

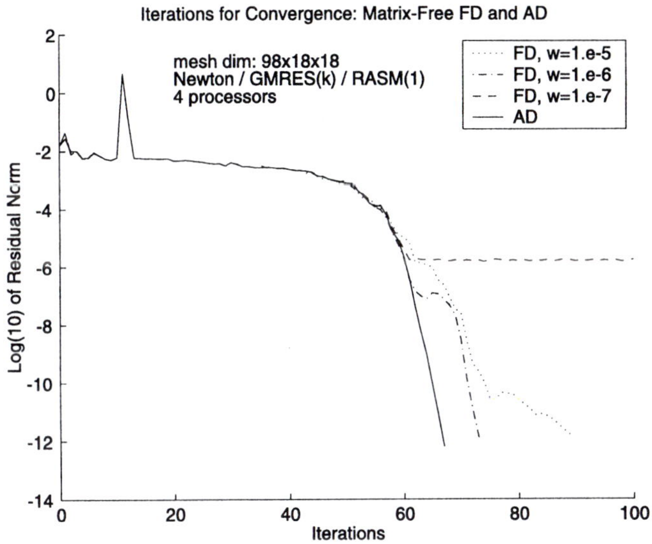

Fig. 7. Algorithmic performance of automatic differentiation derivatives versus finite difference approximations.

6 Conclusions and Expectations

The combination of AD and object-oriented toolkits has proven to be an effective instrument for scientific computing. The analytic derivatives of AD enhance robustness and accelerate the convergence of Newton methods. The well-defined interfaces and data encapsulation of object-oriented toolkits simplify the AD process. There are many options as to the level at which AD can be applied. Each level has its advantages and disadvantages.

AD offers great promise as a useful tool in PDE-constrained optimization. Analytic derivatives are often necessary to ensure robust and efficient convergence to the true minimum. For complex PDE-based simulations, however, developing analytic derivatives, especially second derivatives, by hand is often intractable. Judicious use of AD can overcome these obstacles. Furthermore, the reverse mode of AD enables gradients, Jacobian-transpose-vector, and Hessian-vector products to be computed at significantly lower cost than is possible with finite difference approximations. The ability to compute accurately and efficiently a full suite of first- and second-derivative matrices and directional derivatives should facilitate the algorithmic experimentation necessary for the advancement of PDE-constrained optimization. Thus, AD can play an important role in advancing both the science and the practice of PDE-constrained optimization.

Acknowledgments

This work was supported by the Mathematical, Information, and Computational Sciences (MICS) subprogram of the Office of Advanced Scientific Computing Research, U.S. Department of Energy, under Argonne and Lawrence Livermore Contracts W-31-109-Eng-38 and W-7405-Eng-48, respectively.

We thank Peter Brown, Alan Hindmarsh, David Keyes, Matt Knepley, Jorge Moré, and Linda Petzold for informative discussions and Gail Pieper for proofreading an early draft of this manuscript.

References

1. J. Abate, S. Benson, L. Grignon, P. Hovland, L. McInnes, and B. Norris. Integrating automatic differentiation with object-oriented toolkits for high-performance scientific computing. In G. Corliss, C. Faure, A. Griewank, L. Hascoët, and U. Naumann, editors, *Automatic Differentiation of Algorithms: From Simulation to Optimization*, pages 173–178. Springer, New York, 2002.
2. S. Balay, W. D. Gropp, L. C. McInnes, and B. F. Smith. Efficient management of parallelism in object oriented numerical software libraries. In E. Arge, A. M. Bruaset, and H. P. Langtangen, editors, *Modern Software Tools in Scientific Computing*, pages 163–202. Birkhauser Press, 1997.
3. S. Balay, W. D. Gropp, L. C. McInnes, and B. F. Smith. PETSc 2.0 users manual. Technical Report ANL-95/11 - Revision 2.0.28, Argonne National Laboratory, March 2000. See `http://www.mcs.anl.gov/petsc`.
4. S. Benson, L. C. McInnes, and J. Moré. TAO users manual. Technical Report ANL/MCS-TM-242, Mathematics and Computer Science Division, Argonne National Laboratory, 2000. See `http://www.mcs.anl.gov/tao`.
5. S. Benson, L. C. McInnes, and J. Moré. A case study in the performance and scalability of optimization algorithms. *ACM Transactions on Mathematical Software*, 27:361–376, 2001.
6. C. Bischof, A. Carle, P. Khademi, and A. Mauer. ADIFOR 2.0: Automatic differentiation of Fortran 77 programs. *IEEE Computational Science & Engineering*, 3(3):18–32, 1996.
7. C. Bischof and A. Griewank. Tools for the automatic differentiation of computer programs. In *ICIAM/GAMM 95: Issue 1: Numerical Analysis, Scientific Computing, Computer Science*, pages 267–272, 1996. Special Issue of Zeitschrift für Angewandte Mathematik und Mechanik (ZAMM).
8. C. Bischof, L. Roh, and A. Mauer. ADIC — An extensible automatic differentiation tool for ANSI-C. *Software—Practice and Experience*, 27(12):1427–1456, 1997.
9. C. H. Bischof, H. M. Bücker, and P. D. Hovland. On combining computational differentiation and toolkits for parallel scientific computing. In A. Bode, T. Ludwig, W. Karl, and R. Wismüller, editors, *Euro-Par 2000 – Parallel Processing, Proceedings of the 6th International Euro-Par Conference, Munich, Germany, August/September 2000*, volume 1900 of *Lecture Notes in Computer Science*, pages 86–94, Berlin, 2000. Springer.
10. G. D. Byrne and A. C. Hindmarsh. PVODE, an ODE solver for parallel computers. *Int. J. High Perf. Comput. Appl.*, 13:354–365, 1999.

11. M. C. Ferris, M. P. Mesnier, and J. J. Moré. NEOS and Condor: Solving optimization problems over the Internet. *ACM Trans. Math. Software*, 26(1):1–18, Mar. 2000.
12. P. Fischer. Projection techniques for iterative solution of $Ax = b$ with successive right-hand sides. *Comput. Methods Appl. Mech. Engng.*, 163:193–204, 1998.
13. M. Gertz, 2000. Personal communication.
14. D. Goldfarb and P. L. Toint. Optimal estimation of Jacobian and Hessian matrices that arise in finite difference calculations. *Mathematics of Computation*, 43:69–88, 1984.
15. A. Griewank. On automatic differentiation. In *Mathematical Programming: Recent Developments and Applications*, pages 83–108, Amsterdam, 1989. Kluwer Academic Publishers.
16. A. Griewank. *Evaluating Derivatives: Principles and Techniques of Algorithmic Differentiation.* SIAM, Philadelphia, 2000.
17. A. Griewank and C. Faure. Piggyback differentiation and optimization. In L. T. Biegler, O. Ghattas, M. Heinkenschloss, and B. van Bloemen Waanders, editors, *Large-Scale PDE-Constrained Optimization*, volume 30 of *Lecture Notes in Computational Science and Engineering.* Springer-Verlag, Heidelberg, 2003.
18. W. D. Gropp, D. E. Keyes, L. C. McInnes, and M. D. Tidriri. Globalized Newton-Krylov-Schwarz algorithms and software for parallel implicit CFD. *Int. J. High Performance Computing Applications*, 14:102–136, 2000.
19. P. Hovland, B. Norris, L. Roh, and B. Smith. Developing a derivative-enhanced object-oriented toolkit for scientific computations. In *Proceedings of the SIAM Workshop on Object Oriented Methods for Inter-operable Scientific and Engineering Computing*, pages 129–137. SIAM, October 1998.
20. P. Hovland, B. Norris, and B. Smith. Making automatic differentiation truly automatic: Coupling PETSc with ADIC. In P. M. A. Sloot, C. J. K. Tan, J. J. Dongarra, and A. G. Hoekstra, editors, *Computational Science – ICCS 2002, Proceedings of the International Conference on Computational Science, Amsterdam, The Netherlands, April 21–24, 2002. Part II*, volume 2330 of *Lecture Notes in Computer Science*, pages 1087–1096, Berlin, 2002. Springer.
21. P. D. Hovland and L. C. McInnes. Parallel simulation of compressible flow using automatic differentiation and PETSc. *Parallel Computing*, 27:503–519, 2001.
22. S. L. Lee, A. C. Hindmarsh, and P. N. Brown. User documentation for SensPVODE, a variant of PVODE for sensitivity analysis. Technical Report UCRL-MA-140211, Lawrence Livermore National Laboratory, 2000.
23. S. L. Lee and P. D. Hovland. Sensitivity analysis using parallel ODE solvers and automatic differentiation in C: SensPVODE and ADIC. In G. Corliss, C. Faure, A. Griewank, L. Hascoët, and U. Naumann, editors, *Automatic Differentiation of Algorithms: From Simulation to Optimization*, pages 223–229. Springer, New York, 2002.
24. S. Li and L. Petzold. Design of new DASPK for sensitivity analysis. Technical Report TRCS99-28, University of California at Santa Barbara, 1999.
25. D. P. O'Leary. The block conjugated gradient algorithm and related methods. *Linear Algebra Appl.*, 29:293–322, 1980.

Piggyback Differentiation and Optimization

Andreas Griewank[1] and Christèle Faure[2]

[1] Technical University Dresden, Institute of Scientific Computing D-01062 Dresden, Germany
[2] Polyspace Technologies, Paris, FRANCE

Abstract. In design optimization and parameter identification objective, or response, function(s) are typically linked to the actually independent variables through equality constraints, which we will refer to as state equations. Our key assumption is that it is impossible to form and factor the corresponding constraint Jacobian, but one has instead some fixed point algorithm for computing a feasible state, given any reasonable value of the independent variables. The ultimate goal is to derive from a given state equation solver and a function for evaluating the objective(s) an iterative procedure that achieves primal and (some kind of) dual feasibility as well as optimality more or less simultaneously. Ideally, the cost should only amount to a handful of simulation runs to (re-)gain primal feasibility. So far we have concentrated on ways of obtaining dual feasibility in a piggyback fashion. By this we mean the simultaneous solution of the adjoint equations by an iterative procedure that is obtained by automatic differentiation from the state residual evaluation code. It is shown in particular that the Lagrange function exhibits superconvergence to the reduced function value.

Keywords. fixed point iteration, derivative convergence, algorithmic or automatic differentiation, implicit functions, reduced gradient, reduced Hessian, Q- and R-linear convergence.

1 Introduction

Let us consider a parameter-dependent system of nonlinear equations

$$w \equiv F(z,x) = 0 \quad \text{with} \quad F : \mathbb{R}^l \times \mathbb{R}^n \mapsto \mathbb{R}^l \tag{1}$$

where x represents the vector of independent variables or parameters with respect to which we wish to optimize. The goal is to find some desirable value of the "response" function

$$y = f(z,x) \quad \text{with} \quad f : \mathbb{R}^l \times \mathbb{R}^n \mapsto \mathbb{R}^m \tag{2}$$

that evaluates a few key quantities of the "state vector" z. Although this is not necessary for our theoretical statements one may usually assume that the dimension l of the state space is orders of magnitudes larger than the number n of parameters x and the dimension m of y, which we may think of as vector of objectives and soft constraints. Often some of the design parameters do not enter directly into the state equations but determine some intermediate parameters via a grid generation process. From a conceptual point of view this preprocessing does not effect our basic approach very much, though it may raise some serious implementation issues.

For example, consider the design optimization of an aircraft, one of the prime challenges in scientific computing. Here x parameterizes the shape of the wing and fuselage and its structural realization; z may represent the three dimensional flow field and the elastic deformations of the aircraft in flight; and y could be a handful of performance indices like lift, drag, weight, and cost of production etc. With h the grid size in one spatial dimension one may then assume for a full three dimensional model that n is of order h^{-1} or h^{-2}, l of order h^{-3} and m is a small number independent of h.

To arrive at desirable response values by varying x we need the reduced function value $y = y(x)$ and its Jacobian dy/dx possibly post- or pre-multiplied by vectors $\dot{x}$ or $\bar{y}$, respectively. In order to effectively eliminate the variables z and w at least theoretically we need the following assumption.

Assumption 1 (JR Jacobian Regularity). *On some neighbourhood of a solution (z_*, x) with $F(z_*, x) = 0$ the Jacobians of $F(z, x)$ and $f(z, x)$ with respect to $z \in \mathbb{R}^l$ and $x \in \mathbb{R}^n$ are once Lipschitz-continuously differentiable and the square matrix*

$$F_z(z, x) \equiv \frac{\partial}{\partial z} F(z, x) \in \mathbb{R}^{l \times l}$$

is nonsingular at all points in that neighbourhood with $\|F_z(z, x)^{-1}\| \leq \Gamma$.

Consequences of the Implicit Function Theorem

By the implicit function theorem equation (1) implies at regular roots $z_* \in \mathbb{R}^l$

$$\frac{dz}{dx} \equiv -\left(\frac{\partial w}{\partial z}\right)^{-1} \frac{\partial w}{\partial x} = -F_z(z_*, x)^{-1} F_x(z_*, x) ,$$

where we abbreviate $F_x(z, x) \equiv \partial F(z, x)/\partial x \in \mathbb{R}^{l \times n}$.

Then it follows by the chain rule for the total derivative of the responses or objectives y with respect to the design parameters x that

$$\frac{dy}{dx} = \frac{\partial y}{\partial x} + \frac{\partial y}{\partial z}\frac{dz}{dx} = f_x(z_*, x) - f_z(z_*, x) F_z(z_*, x)^{-1} F_x(z_*, x) . \quad (3)$$

Here the subscripts of f_x and f_z denote differentiation with respect to x and z, respectively. As we will see below the triple matrix product on the right can be computed in two different ways related to the forward and reverse mode of automatic differentiation. The latter method is more efficient if $m < n$ as we may assume and a combination of both yields better accuracy.

Optimization Aspects

In optimization calculations with a single objective function $y = f(z,x) \in \mathbb{R}$ the vector $dy/dx \in \mathbb{R}^n$ is called the reduced gradient. At least theoretically, accurate and affordable values of the reduced gradient allow us to treat the optimization of y as an unconstrained problem. This approach leads to a class of schemes for equality constrained optimization problems that are unsurprisingly named "reduced gradient methods". In a certain sense all rapidly converging constrained optimization methods are ultimately reduced gradient methods, but the wisdom of maintaining more or less exact feasibility earlier on is debatable. On one hand it means that a feasible solution with a somewhat reduced objective function is available, whenever the optimization calculation stops, possibly because the computing resources or the patience of management has been exhausted.

On the other hand, if the optimization can be carried out until the end, then allowing earlier on significant infeasibilities (in the sense of large residuals in the state equation) may reduce the overall runtime significantly. This effect is especially likely to occur when the nature of the state equation $F(z,x) = 0$ is such that (re-)gaining feasibility is a rather slow iterative process. Then one might also be interested in determining quite rapidly whether a suggested change in the design variables x actually leads to a desirably low value of the reduced objective function. The right compromise between feasibility and optimality is the hallmark of a good merit function for judging whether an optimization step has been successful or not. In the context of design optimization gaining feasibility and optimality at the same time has been proposed as one-shot approach by S. Ta'asan [15] and also employed successfully by A. Jameson [11]. In this paper we will not debate these wider issues but concentrate on the task of arriving as fast as possible at accurate values of the reduced function and its first derivatives.

Note that we have assumed here and throughout that there is a given partitioning of all variables into a set of state variables z and a set of design variables x with the former being considered as dependent on the latter via the state equation. In the terminology of MINOS [12] and similar nonlinear programming tools the state variables are always basic and the design variables always superbasic. This fixed partition is only possible due to the regularity Assumption JR and because we have excluded the possibility of inequality constraints. If the latter are also present our analysis applies only locally, once all active constraints have been identified.

Preview of Contents

In the second section of the paper we set up direct and adjoint sensitivity equations whose solutions can be interpreted as feasible directions and Lagrange multipliers, respectively. Either of them yield immediately the desired reduced derivatives. However, in view of their size in typical applications the sensitivity equations can usually not be solved exactly and we must instead accept approximate solutions obtained by iterative solvers. In Section 3 we apply standard results on fixed point iterations to the direct or forward differentiation of the original iteration loop and then we analyse the new approach of *iterated adjoints*. It differs completely from the iteration that would be obtained by mechanically applying the reverse mode to the original iteration. In particular the memory requirement does not grow with the number of steps taken. As a result we obtain estimates of the reduced function that converge twice as fast as the underlying state space iterates. Our approach is numerically validated on an Euler Code in two dimensions. The paper concludes with the customary summary and outlook in Section 4.

2 Approximating Reduced Functions and Derivatives

For the remainder of this paper it is of crucial importance that the triple matrix product $-f_z\big[F_z\big]^{-1}F_x$ on the right-hand side of (3) can be bracketed in two different ways. More specifically, we face the alternative of computing first

$$\frac{dz}{dx} = -F_z\big(z_*,x\big)^{-1}F_x(z_*,x) \in R^{l\times n} \tag{4}$$

or alternatively

$$\frac{\partial y}{\partial w} = f_z(z_*,x)F_z\big(z_*,x\big)^{-1} \in R^{m\times l} \tag{5}$$

and then multiplying these matrices by $f_z(z_*,x)$ from the left or $-F_x(z_*,x)$ from the right, respectively. The notation $\partial y/\partial w = (\partial y/\partial z)(\partial z/\partial w)$ is appropriate since $\partial z/\partial w = F_z(z,x)^{-1}$ for fixed x by the inverse function theorem.

The Direct and Adjoint Sensitivity Equation

We will find below that (4) is best calculated in the forward mode and (5) can be obtained by something akin to the reverse mode. In the forward mode we may restrict the differentiation to a single direction vector $\dot{x} \in \mathbb{R}^n$ and in the reverse mode to a single adjoint vector $\bar{y} \in \mathbb{R}^m$. Hence we compute or approximate the *feasible direction*

$$\dot{z}_* \equiv -F_z\big(z_*,x\big)^{-1}F_x(z_*,x)\,\dot{x} \in R^l \tag{6}$$

and the *multiplier vector*

$$\bar{w}_* \equiv -\bar{y}\,f_z(z_*,x)F_z\big(z_*,x\big)^{-1} \in R^l\,. \tag{7}$$

In other words we wish to solve the *direct sensitivity equation*

$$\dot{w} \equiv \dot{F}(z_*,x,\dot{z}_*,\dot{x}) \equiv F_z(z_*,x)\,\dot{z}_* + F_x(z_*,x)\,\dot{x} = 0 \tag{8}$$

or the *adjoint sensitivity equation*

$$\bar{z} = \bar{F}(z_*, x, \bar{w}_*, \bar{y}) \equiv \bar{w}_* F_z(z_*, x) + \bar{y} f_z(z_*, x) = 0 . \tag{9}$$

The adjoint sensitivity equation has been analysed and used occasionally in the engineering literature [4, 9, 13].

For given vectors $\dot{x}$ or $\bar{y}$ the resulting vectors $\dot{z}_*$ or $\bar{w}_*$ yield the following reduced quantities

$$\begin{aligned} \dot{y}_* &= f_x\dot{x} + f_z\dot{z}_* \in \mathbb{R}^m && \text{Reduced Tangent} \\ \bar{x}_* &= \bar{y}f_x + \bar{w}_*F_x \in \mathbb{R}^n && \text{Reduced Gradient} \\ \bar{y}\,\dot{y}_* &= \bar{x}_*\dot{x}\ , \in \mathbb{R} && \text{Directional Derivative} \end{aligned}$$

where all derivatives are evaluated at (z_*, x). In particular, when $\dot{x}$ or $\bar{y}$ are Cartesian basis vectors one obtains a column $\dot{y}_*$ or a row $\bar{x}_*$ of the reduced Jacobian dy/dx, respectively. Naturally, one can compute several such column or row vectors simultaneously, possibly to compute the whole reduced Jacobian at once. While this tends to make sense from the practical point of view, it does not reduce the computational complexity by more than a small factor. Hence we prefer to couch our analysis in terms of the derivative vectors $\dot{z}_*$ or $\bar{w}_*$, the resulting $\dot{y}_*$, $\bar{x}_*$ and the reduced partial $\bar{y}\,\dot{y}_* = \bar{x}_*\dot{x}$. The last scalar quantity is listed separately since it can be approximated with a higher accuracy than the reduced tangent $\dot{y}_*$ and the reduced gradient $\bar{x}_*$ whose intersection it represents in a certain sense.

At least theoretically $\dot{z}_*$ and $\bar{w}_*$ can be computed by solving a linear system involving the Jacobian $F_z(z_*, x)$ and its transpose, respectively. The corresponding right-hand sides $\dot{F}(z_*, x, 0, -\dot{x}) \in R^l$ and $\bar{F}(z_*, x, 0, -\bar{y}) \in R^l$ as defined in (8) and (9) can be evaluated by a single forward or reverse sweep on the evaluation procedure for (F, f), respectively. Hence, in either case the operations count for the right hand side is a small multiple of that for F itself. The same applies to the memory requirement for the adjoint $\bar{F}(z_*, x, 0, -\bar{y}) \in R^l$, which is also proportional to the basic operations count, unless more sophisticated versions of the reverse mode with checkpointing are employed. We expect this to be necessary only in rather exceptional cases where the evaluation of F itself involves time-like evolutions.

In many applications the main obstacle to solving either sensitivity equation is that the Jacobian cannot be formed and factored at a reasonable cost. We will certainly make this assumption here, since otherwise one may also perform Newton steps and the whole idea of extracting extra information from the users fixed point iteration becomes moot. The same would still be true if the Jacobian could be preconditioned well enough such that a suitable iterative solver could find Newton steps or the solutions to our sensitivity equations quite rapidly, even when high accuracy was required. Hence we will assume that the iterative solution of the state equation is a rather drawn out process, possibly effected by a legacy code including various tricks of the trade that even the current user may not fully be aware of. While this is the scenario to which our approach is applicable in principle, we will in fact inch back to the Newton-like scenario sketched above when it comes to establishing convergence at certain asymptotic rates. However, there is numerical evidence that the approach works still in cases where the assumptions of our theory are not satisfied, or at least not easily verified. For example, this is the case for the Euler code from which we obtained the numerical results plotted at the end of Section 5.

Whatever methods one uses to generate approximate solutions $\dot{z}$ and $\bar{w}$ to either or both sensitivity equations their quality can be gauged by evaluating the *derivative residuals* $\dot{F}(z_*, x, \dot{z}, \dot{x})$ and $\bar{F}(z_*, x, \bar{w}, \bar{y})$ defined in (8) and (9). These vectors can be obtained just as cheaply as the right hand sides mentioned above and bound the derivative errors according to Lemma 11.2 in [10] as follows

Proposition 1 (Forward and Reverse Consistency Check).
Under Assumption JR and with $\dot{x} \in \mathbb{R}^n$ or $\bar{y} \in \mathbb{R}^m$ fixed there exist constants $\delta > 0$ and $\gamma < \infty$ such that with $\dot{F}$ defined in (8)

$$\|\dot{z} - \dot{z}_*\| \;\leq\; \gamma\,(\|F(z,x)\| + \|\dot{F}(z,x,\dot{z},\dot{x})\|)$$

or with $\bar{F}$ defined in (9)

$$\|\bar{w} - \bar{w}_*\| \;\leq\; \gamma\,(\|F(z,x)\| + \|\bar{F}(z,x,\bar{w},\bar{y})\|)$$

for all z with $\|z - z_*\| < \delta$ and $\dot{z} \in \mathbb{R}^l$ or $\bar{w} \in \mathbb{R}^l$ arbitrary.

The constant γ in Proposition 1 is a function of local Lipschitz constants and a bound Γ on the size of the inverse $F_z(z,x)^{-1}$. Like always in nonlinear equation solving good estimates for these quantities are hard to come by. As one can see in Proposition 1 both derivative vectors $\dot{z}$ and $\bar{w}$ are usually affected by error in the underlying z, which explains why there is often a time-lag in their convergence. The delay has been observed on most iterative schemes other than Newton's method. One significant contribution of this paper is the observation that any such delay can be avoided by combining the results of forward and reverse differentiation appropriately.

As an immediate consequence of Proposition 1 we obtain the following approximation result

Corollary 2 (Reduced Function Estimate).
Under the assumptions of Proposition 1 and with $\bar{w}$ as defined above we have

$$\begin{aligned}\Big|\big[\bar{y}f(z,x) + \bar{w}F(z,x)\big] - \bar{y}f(z_*,x)\Big| \\ \leq \Gamma\,\|F(z,x)\|\,\|\bar{F}(z,x,\bar{w},\bar{y})\| + \mathcal{O}(\|F(z,x)\|^2)\end{aligned}$$

Proof. The assertion follows from the Taylor expansions

$$
\begin{aligned}
f(z_*, x) &= f(z, x) + f_z(z, x)(z_* - z) + \mathcal{O}(\|z - z_*\|^2) \quad , \\
0 = F(z_*, x) &= F(z, x) + F_z(z, x)(z_* - z) + \mathcal{O}(\|z - z_*\|^2)
\end{aligned}
$$

with the definitions of $\bar{F}$ and Γ as an upper bound on the inverse Jacobian norm.

As we will see in Section 5 one can expect that within an iterative procedure the corrected estimate $\bar{y}f(z,x) + \bar{w}F(z,x)$ converges roughly twice as fast to the actual reduced function value $\bar{y}f(z_*,x)$ as $\bar{y}f(z,x)$ by itself. Moreover, we have an approximate bound on the remaining error that can be obtained at virtually no extra cost. This technique for the doubling of the order of the estimate is related to the methods of Becker and Rannacher [1] for obtaining superconvergent approximations to integral quantities from the solution of partial differential equations.

Reduced Gradients and Jacobians

By combining several directions $\dot{x}$ into a matrix $\dot{X} \in \mathbb{R}^{n\times p}$ we obtain instead of the vector sensitivity equation (8) the matrix identity

$$\dot{F}(z_*, x, \dot{Z}_*, \dot{X}) \;\equiv\; F_z(z_*, x)\dot{Z}_* + F_x(z_*, x)\dot{X} \;=\; 0 \;\in \mathbb{R}^{l\times p}$$

with $\dot{Z}_* \in \mathbb{R}^{l\times p}$. Analogously, combining several weight vectors $\bar{y}$ into a matrix $\bar{Y} \in \mathbb{R}^{q\times l}$ we obtain instead of (9) the adjoint sensitivity equation

$$\bar{F}(z_*, x, \bar{W}_*, \bar{Y}) \;\equiv\; \bar{W}_* F_z(z_*, x) + \bar{Y} f_z(z_*, x) \;=\; 0 \;\in \mathbb{R}^{q\times l}$$

with $\bar{W}_* \in \mathbb{R}^{q\times l}$. For approximations $\dot{Z} \approx \dot{Z}_*$ and $\bar{W} \approx \bar{W}_*$ Proposition 1 and its Corollary 2 apply column- or row-wise, respectively. In particular we obtain for $\dot{X} = I$ with $p = n$ and $\bar{Y} = I$ with $q = m$ the reduced Jacobian approximations

$$\dot{Y} \equiv f_x(z, x) + f_z(z, x)\dot{Z} \quad \text{and} \quad \bar{X} \equiv f_x(z, x) + \bar{W}F_x(z, x) \tag{10}$$

as well as the synergetic estimate

$$f_x(z, x) + f_z(z, x)\dot{Z} + \bar{W}F_x(z, x) + \bar{W}F_z(z, x)\,. \tag{11}$$

All these are $m\times n$ matrices, whose values coincide if $\dot{Z} = \dot{Z}_*$ and $\bar{W} = \bar{W}_*$ are exact. If the weighting $\bar{y}$ has been fixed or there is only one response function in the first place only the forward differentiation needs to be performed in vector mode and we obtain the reduced gradient

$$\bar{y}f_x(z, x) + \bar{y}f_z(z, x)\dot{Z} + \bar{w}F_x(z, x) + \bar{w}F_z(z, x)\,. \tag{12}$$

Leaving off the last term and either one of the middle two terms leads back to the simple approximations based on $\bar{w}$ or $\dot{Z}$ alone. It can be shown [7], that the estimates (11) and (12) exhibit superconvergence in a certain sense. Another benefit of computing or approximating both $\dot{Z}_*$ and $\bar{w}_*$ is that they allow the calculation of the reduced Hessian, in fact they are indispensable for that purpose.

Two Stage Scenario

Often some of the design parameters do not enter directly but via a grid generator, so that we have effectively the partitioned system

$$\text{Optimize} \quad f(z,x) \quad s.t. \quad G(x,v) = 0 \in \mathbb{R}^n \quad \text{and} \quad F(z,x) = 0 \in \mathbb{R}^l \ .$$

Now the dimension $n = dim(x)$ of the 'grid' is typically orders of magnitude larger than $\nu = dim(v)$, the number of the underlying optimization parameters. From an abstract point of view we may consider $\tilde{F} \equiv (G, F) = 0 \in \mathbb{R}^{n+l}$ as state equation in the combined variable vector $\tilde{z} \equiv (z, x) \in \mathbb{R}^{l+n}$ and apply the results of this section accordingly. The only problem with this approach is that typically the two conditions $G(x,v) = 0$ and $F(z,x) = 0$ are solved in a two stage procedure. Hence, while solving the first equation iteratively we have no approximation to the gradient of the final objective function. Such approximations are needed if we wish to apply the adjoint fixed point iteration to be described in the following section, which avoids any storage of intermediate results along the iteration trajectory. Hence we advocate the following approach for the two stage scenario.

Solve $u \equiv G(x,v) = 0$ approximately as before carrying forward only the feasible direction approximations $\dot{X} \in \mathbb{R}^{n\times\nu}$ that are derived from $\dot{V} = I \in \mathbb{R}^{\nu\times\nu}$. Then solve $F(z,x) = 0$ approximately and compute simultaneously approximations to $\dot{Z} \in \mathbb{R}^{l\times\nu}$ and/or $\bar{w} \in \mathbb{R}^l$ that corresponds to some $\bar{y} \in \mathbb{R}^m$. Then forward differentiation through the objective as given in (10) yields the full reduced Jacobian $\dot{Y} = \partial y/\partial v$, whose total cost would be of course proportional to the number of design variables $\nu = dim(v)$.

Alternatively, we may now apply the analysis of this section to the first stage with the response function given by $\tilde{f}(x) = f(z_*(x), x)$ or rather its linearization about the current point as given by its gradient $\bar{x} = \bar{y}f_x + \bar{w}F_x = d(\bar{y}\tilde{f})/dx$. As a result we obtain an adjoint vector $\bar{u}$ solving at least approximately the sensitivity equation

$$\bar{u}\, G_x + \bar{x} \quad = \quad 0 \quad .$$

When the grid condition $u = G(x,v) = 0$ is also solved iteratively this approach requires a second iteration with adjoint calculation. To this end one should strive to identify a number of steps at the end of the original iteration over which it can be expected to be nicely contractive. This cycle should then be rerun once or several times until the adjoint sensitivity equation for

$\bar{u}$ is satisfied with acceptable accuracy. Finally, assuming that v impacts y only via x and z we obtain the reduced gradient $\bar{v} = \bar{u}\, G_v$, which allows the optimization of the weighted objective $\bar{f}$ with respect to the actual design variables v. The two-stage scenario will not be considered further in this paper.

3 Derivative Recurrences and their Convergence

The iterates $z_k \in \mathbb{R}^l$ generated by many practical methods for approximating a solution $z_* = z_*(x)$ with $F(z_*, x) = 0$ satisfy a recurrence of the form

$$z_{k+1} \; = \; H_k(z_k, x) \; \equiv \; z_k \; - P_k F(z_k, x) \, . \tag{13}$$

Here the preconditioner P_k is some $l \times l$ matrix that approximates the inverse of the Jacobian $F_z(z_k, x)$. The closer that approximation, the more the iteration resembles Newton's methods with its excellent local convergence properties. As long as $F(z_k, x) \neq 0$ any new iterate z_{k+1} can be written in the form (13) since we have not yet imposed any conditions on P_k. To ensure stable convergence from within a vicinity of z_* we make the following assumption

Assumption 2 (CP CONTRACTIVE PRECONDITIONING). *The preconditioners P_k satisfy*

$$|\!|\!| I - P_k \, F_z(z_*, x) |\!|\!| \; \leq \; \rho \; < 1 \quad \textit{for all} \quad k \tag{14}$$

with respect to some induced matrix norm $|\!|\!| \cdot |\!|\!|$.

Because the norm $|\!|\!| \cdot |\!|\!|$ must be independent of the iteration counter k this hypothesis is a little stronger than the condition that the spectral radius ($\equiv$ largest eigenvalue modulo) of all $[I - P_k F_z(z_*, x)]$ is uniformly bounded below 1. According to Ostrowsky's Theorem (see Propositions 10.1.3 and 10.1.4 in [14]) it follows from Assumption CF that all initial guesses z_0 whose distance to z_* is less than some bound lead to convergence with

$$Q\{z_k - z_*\}_{k \in \mathrm{N}} \; \equiv \; \limsup_k \frac{|\!|\!| z_{k+1} - z_* |\!|\!|}{|\!|\!| z_k - z_* |\!|\!|} \; \leq \; \rho \, . \tag{15}$$

Here the vector norm $|\!|\!| \cdot |\!|\!|$ must be consistent with the matrix norm used in Assumption CP.

Quotient and Root Convergence Factors

The chosen norm strongly influences the so-called Q-factor $Q\{z_k - z_*\}$ defined by (15) for any iteration sequence $\{z_k - z_*\}$ with $z_k \neq z_*$ for all k. In contrast it follows from the equivalence of all norms on finite dimensional spaces that the R-factor

$$R\{z_k - z_*\}_{k\in\mathbb{N}} \equiv \limsup_k \sqrt[k]{\|z_k - z_*\|} \leq Q\{z_k - z_*\}_{k\in\mathbb{N}} \leq \rho \quad , \tag{16}$$

The last inequality holds by (15) and the other one is established as Proposition 9.3.1 in [14]. In both (15) and (16) we may replace the uniform bound $\rho < 1$ from Assumption CP by the corresponding limit superior

$$\rho_o \equiv \limsup_k \|I - P_k\, F_z(z_k, x)\| \leq \rho \quad . \tag{17}$$

Except on academic test functions one can normally not compute the solution error $\|z_k - z_*\|$ and must therefore be content to gauge the quality of the current approximation z_k in terms of the residual $F_k \equiv F(z_k, x)$. Under our Assumptions JR one may view $\|F(z_k, x)\|$ as an equivalent norm to $\|z_k - z_*\|$ since there must be constants $\delta > 0$ and $1 \leq \gamma < \infty$ such that

$$\frac{1}{\gamma} \leq \frac{\|F(z, x)\|}{\|z - z_*\|} \leq \gamma \quad \text{for} \quad \|z - z_*\| < \delta \, .$$

This implies for any sequence z_k converging to, but never attaining exactly z_* that

$$R\{F_k\}_{k\in\mathbb{N}} = R\{z_k - z_*\}_{k\in\mathbb{N}} \leq Q\{F_k\}_{k\in\mathbb{N}} \leq \gamma^2\, Q\{z_k - z_*\}_{k\in\mathbb{N}} \, .$$

To succinctly indicate convergence with the R-factor ρ_* we will write

$$z_k = z_* + \tilde{\mathcal{O}}(\rho_*^k) \quad \text{and} \quad F(z_k, x) = \tilde{\mathcal{O}}(\rho_*^k) \, .$$

Direct Derivative Recurrence

Now we attempt to derive from the original fixed point iteration extra recurrences that simultaneously compute the desired derivative quantities in a piggy-back fashion. First we consider more or less straight forward differentiation. Suppose the preconditioner matrix P_k is for each k at least locally a smooth function of (z, x); often it will even be constant. Moreover, let x vary along the straight line $x = x(t) = x(0) + t\dot{x}$ as a function of the scalar parameter $t \approx 0$. Then it follows by induction on k that when $z_k = z_k(t)$ is differentiable in t so is z_{k+1} and the derivatives $\dot{z}_k = \dot{z}_k(t)$ must satisfy the recurrence

$$\dot{z}_{k+1} = \dot{z}_k - P_k\, \dot{F}(z_k, x, \dot{z}_k, \dot{x}) - \dot{P}_k\, F(z_k, x) \, . \tag{18}$$

Here the matrix $\dot{P}_k \equiv d\, P_k(z_k(t), x(t))/dt$ is the derivative of the preconditioner with respect to t, which exists under the assumption made above. The derivative residual $\dot{F}(z_k, x, \dot{z}_k, \dot{x})$ is defined by (8).

The last term $\dot{P}_k\, F(z_k, x)$ is in some way the most interesting. If the preconditioner is fixed so that (13) reduces to a simple substitution method, the last term vanishes since clearly $\dot{P}_k \equiv 0$. Even if the $\dot{P}_k$ are nonzero but their size is uniformly bounded, the term $\dot{P}_k\, F(z_k, x)$ disappears gradually as $F_k = F(z_k, x)$ converges to zero. This happens for example in Newton's method where $P_k = F_z(z_k, x)^{-1}$ is continuously differentiable in (z_k, x), provided F itself is at least twice continuously differentiable. However, second derivatives should not really come into it at all as the implicit derivative $\dot{z}_*$ is according to the explicit representation (6) uniquely defined by the extended Jacobian of F. Hence we may prefer to simply drop the last term and use instead the *simplified recurrence*

$$\dot{z}_{k+1} \;=\; \dot{z}_k - P_k\, \dot{F}(z_k, x, \dot{z}_k, \dot{x})\,. \tag{19}$$

The implementation of this recurrence requires the *deactivation* of P_k when this preconditioner depends on x as it usually does. By this we mean that the dependence on x is suppressed so that it looks as through P_k consists of real entries that have fallen from the sky. Whether and how this can be done depends on the particular AD tool. It should not be made too easy, because the unintentional suppression of active dependencies can lead to wrong derivative values. A comparison of the simplified recurrence with black box differentiation was performed in the computational study [2]. For our analysis we make the following assumption

Assumption 3 (**UL** LIPSCHITZ CONTINUOUS DEPENDENCE). *For the convergent sequence $z_k \to z_*$ the preconditioners P_k are all differentiable with respect to (z, x) on some neighbourhood of (z_k, x) such that the derivative tensors*

$$\frac{\partial}{(\partial z, \partial x)} P_k(z, x) \in \mathbb{R}^{l \times l \times l \times n} \tag{20}$$

are uniformly bounded over all k.

Under this assumption both derivative recurrences lead to the same convergence. As shown in [6] we have as a generalization of a result by Jean Charles Gilbert [8].

Proposition 3 (ROOT CONVERGENCE FACTOR OF DERIVATIVES). *Under Assumptions CP and UL we have for all z_0 sufficiently close to z_* and arbitrary $\dot{z}_0$ applying, either (18) or (19)*

$$R\{\dot{z}_k - \dot{z}_*\}_{k \in \mathbb{N}} \;\leq\; \rho_0 \tag{21}$$

where ρ_o is defined in (17).

According to Proposition 3 the complete (18) and the simplified (19) derivative recurrence yield R-linear convergence to the exact derivative $\dot{z}_*$. In both cases the root convergence factor is bounded by the limiting spectral radius ρ_0, which bounds the quotient convergence factor of the iterates z_k themselves. If $\rho_0 = 0$ we have Q-superlinear convergence of the z_k and the slightly weaker property of R-superlinear convergence for the $\dot{z}_k$. This result applies in particular for Newton's method, where the z_k converge in fact quadratically.

In [6] a considerable effort was made to extend the result to quasi-Newton methods where the sequence P_k is obtained by secant updating, which leads to Assumption UL being violated in general. Nevertheless, it could be shown that due to classical convergence characteristics of secant updating methods the decline of the residuals $F(z_k, x)$ is just fast enough that the second term in (18) still gradually disappears. We will avoid this complication here since secant updating methods do not seem to be used a lot on the class of design optimization problems we are mainly interested in. Of course, for many other methods like conjugate gradient type schemes and of course multi-level approaches our assumptions are not not easily verified either.

Our simplified piggyback iteration for solving simultaneously the equations $F(z) = 0$ and $\dot{F}(z, \dot{z}) = 0$ is displayed in Table 1.

Table 1. Direct Fixed Point Iteration

fix $x, \dot{x} \in \mathbb{R}^n$
initialize $z_0, \dot{z}_0 \in \mathbb{R}^l$
for $k = 0, 1, 2, \ldots$
  $w_k = F(z_k, x)$
  $\dot{w}_k = \dot{F}(z_k, x, \dot{z}_k, \dot{x})$
  stop if $\|w_k\|$ and $\|\dot{w}_k\|$ are small
  $z_{k+1} = z_k - P_k\, w_k$
  $\dot{z}_{k+1} = \dot{z}_k - P_k\, \dot{w}_k$
$y_k = f(z_k, x)$
$\dot{y}_k = f_z(z_k, x)\dot{z}_k + f_x(z_k, x)\dot{x}$

Table 2. Adjoint Fixed Point Iteration

fix $x \in \mathbb{R}^n, \bar{y} \in \mathbb{R}^m$
initialize $z_0, \bar{w}_0 \in \mathbb{R}^l$
for $k = 0, 1, 2, \ldots$
  $[w_k, y_k] = [F(z_k, x), f(z_k, x)]$
  $\bar{z}_k = \bar{F}(z_k, x, \bar{w}_k, \bar{y})$
  stop if $\|w_k\|$ and $\|\bar{z}_k\|$ are small
  $z_{k+1} = z_k - P_k\, w_k$
  $\bar{w}_{k+1} = \bar{w}_k - \bar{z}_k\, P_k$
$y_k = f(z_k, x), \quad \sigma_k = \bar{y}\, y_k + \bar{w}_k\, w_k$
$\bar{x}_k = \bar{w}_k\, F_x(z_k, x) + \bar{y}\, f_x(z_k, x)$

Adjoint Fixed Point Iteration

Except for the omission of the term involving $\dot{P}_k$ and the suggested modification of the stopping criterion the scheme listed in Table 1 could have been obtained by simply differentiating the original fixed point iteration in the forward mode. This black box approach often yields virtually identical results and it alleviates the need for any code modification by hand. Unfortunately, things are not nearly as convenient in the reverse mode.

Suppose one has a code for executing the update (13) a certain number of T times and subsequently evaluating the response function $y = f(z, x)$ at the final z. If one then applies an adjoint generator nominating x as independent and y as dependent variables the resulting code will have some pretty undesirable features. The main crux is that it will save all intermediate states on the way forward, which means a T-fold increase in memory relative to the original fixed point iteration. The reason for this apparent waste of storage is that AD tools cannot detect automatically whether an iteration represents a genuine evolution whose complete trajectory is important for the adjoints to be calculated, or whether the early stages are only of passing interest as the trajectory homes in on a fixed point later. In the latter case, which is of interest here, the return sweep of the reverse mode regresses every sense of the word from good information in the vicinity of the solution point to much earlier iterates where function and derivative values have little to do with the desired implicit derivatives at the limit. Moreover, in contrast to the constructive tests given in Proposition 1, we have no way of gauging the quality of the approximation $\bar{x} \approx \bar{x}_*$ that finally pops out of the black box adjoint procedure. Even if we could test its quality, there would be no apparent way of refining the approximation other than by rerunning both sweeps with an increased T. Some of these arguments against mechanical adjoining can be made even in the linear case as was done in [9]. As the reader might suspect by now this gloomy description only sets the stage for the following enlightenment.

To compute $\bar{w}_*$ one may of course solve the adjoint sensitivity equation

$$F_z(z_*, x)^T \, \bar{w}^T \;=\; -f_z(z_*, x)^T \bar{y}^T \;=\; \bar{F}(z_*, x, 0, -\bar{y})^T \tag{22}$$

obtained by transposing (9). The transposed Jacobian $F_z(z_*, x)^T$ has the same size, spectrum, and sparsity characteristics as $F_z(z_*, x)$ itself. Hence the task of solving the adjoint sensitivity equation (22) is almost exactly equivalent to the task of solving the direct sensitivity, equation (8). Because of the similarity relation

$$P_k^T \, F_z(z_k, x)^T \;=\; P_k^T \left[P_k F_z(z_k, x)\right]^T \, P_k^{-T} \tag{23}$$

the square matrices $I - P_k F_z(z_k, x)$ and $I - P_k^T F_z(z_k, x)^T$ have the same spectrum. Hence the latter has by Assumption CP a spectral norm less or equal to $\rho < 1$ and we may solve the adjoint sensitivity equation (22) by the iteration

$$\bar{w}_{k+1}^T = \bar{w}_k^T - P_k^T \left[F_z(z_*, x)^T \bar{w}_k^T + f_z(z_*, x)^T \bar{y}^T\right] = \bar{w}_k^T - P_k^T \bar{F}(z_*, x, \bar{w}_k, \bar{y})^T . \tag{24}$$

where $\bar{F}$ is defined as in (9). Starting from any $\bar{w}_0$ one obtains again Q-linear convergence with

$$R\{\bar{w}_k - \bar{w}_*\}_{k \in \mathrm{N}} \;\leq\; Q\{\bar{w}_k - \bar{w}_*\}_{k \in \mathrm{N}} \;\leq\; \rho_o \tag{25}$$

where the Q-factor is defined in terms of the consistent norm for which Assumption CP holds. The recurrence (24) was apparently first analyzed by Christianson [3], albeit with a fixed final preconditioner P_k.

Now the question arises whether the adjoint sensitivity calculations can also be performed in a piggyback fashion, i.e. without setting up a second phase iteration. One can easily check that (25) remains true when (24) is applied with z_* replaced by z_k. This means that we can propagate the adjoint vectors $\bar{w}_k$ forward without the need to record the intermediates z_k and the corresponding preconditioner s. Only each coupled evaluation $(F(z_k, x), f(z_k, x))$ must be reverted to yield the adjoint residual $\bar{F}(z_k, x, \bar{w}_k, \bar{y})^T \in \mathbb{R}^l$ at a comparable computational effort. However, it should be noted that the response function f must now be evaluated at each iterate rather than just at the end. This extra effort that is typically quite small and sometimes required anyway for the adjustment of boundary conditions [5]. The size of the resulting adjoint residual $\bar{F}$ should be included in the overall stopping criterion, which yields the iteration displayed in Table 2.

What we call here "adjoint fixed point iteration" has been referred to as "iterative incremental form" of the adjoint sensitivity equation in the aerodynamic literature [13]. It must be stressed that contrary to what one might except the vectors $\bar{w}_k$ and $\bar{z}_k$ are not the adjoints of the intermediate values w_k and z_k in the usual sense. The concept of an adjoint is normally only defined for evaluation procedures that involve an a priori fixed sequence of elemental operations. In the linear case one may interpret the quantities $\bar{z}_k$ and $\bar{w}_k$ still as adjoints but not of z_k and w_k respectively. Substituting Ostrowsky's Theorem, Propositions 3 and equation (25) into Proposition 1 and its Corollary 2 we obtain the following list of R-linear convergence results

$$\begin{aligned}
z_k - z_* &= \mathcal{O}(\|F(z_k)\|) &&= \tilde{\mathcal{O}}(\rho_*^k) \\
y_k - y_* &= \mathcal{O}(\|F(z_k)\| &&= \tilde{\mathcal{O}}(\rho_*^k) \\
\dot{z}_k - \dot{z}_* &= \mathcal{O}(\|F(z_k)\| + \|\dot{F}(z_k, \dot{z}_k)\|) &&= \tilde{\mathcal{O}}(\rho_0^k) \\
\dot{y}_k - \dot{y}_* &= \mathcal{O}(\|F(z_k)\| + \|\dot{F}(z_k, \dot{z}_k)\|) &&= \tilde{\mathcal{O}}(\rho_0^k) \\
\bar{w}_k - \bar{w}_* &= \mathcal{O}(\|F(z_k)\| + \|\bar{F}(z_k, \bar{w}_k)\|) &&= \tilde{\mathcal{O}}(\rho_0^k) \\
\bar{x}_k - \bar{x}_* &= \mathcal{O}(\|F(z_k)\| + \|\bar{F}(z_k, \bar{w}_k)\|) &&= \tilde{\mathcal{O}}(\rho_0^k) \\
\sigma_k - \bar{y}\, y_* &= \mathcal{O}(\|F(z_k)\|\,\|\bar{F}(z_k)\| + \|\bar{F}(z_k)\|^2) &&= \tilde{\mathcal{O}}(\rho_0^k\,\rho_*^k)
\end{aligned}$$

where

$$\sigma_k \equiv \bar{y} f(z_k) + \bar{w}_k F(z_k) \quad .$$

Here we have again omitted the constant vectors $x, \dot{x}$ and $\bar{y}$ as arguments of F, $\dot{F}$ and $\bar{F}$. The following figures were obtained on a 2D Euler code for an aircraft wing

Inspection of Fig. 1 one sees that on our aerodynamical test problem the three residuals $F(z_k)$, $\dot{F}(z_k, \dot{z}_k)$, $\bar{F}(z_k, \bar{w}_k)$ and hence the iterates z_k, $\dot{z}_k$ and $\bar{w}_k$ all converge at about the same asymptotic rate. At iteration 1000 we reset the derivatives $\dot{z}_k$ and $\bar{w}_k$ to zero, which almost led to a repetition of

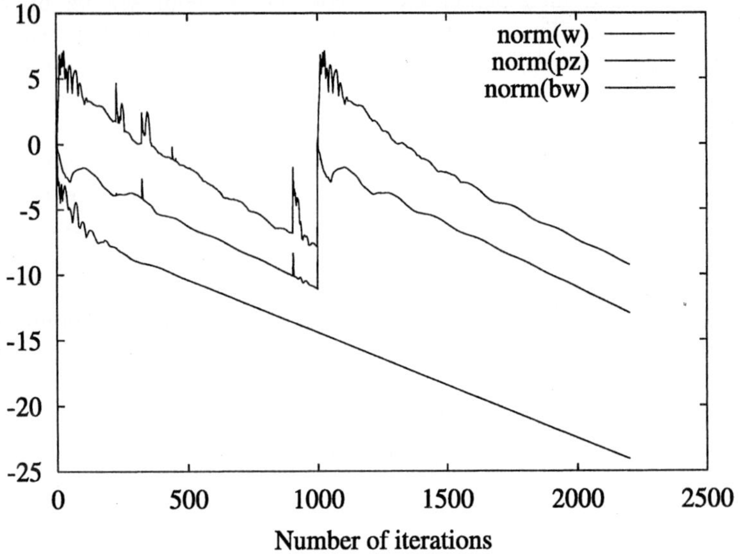

Fig. 1. Convergence of Residuals $w = F$, $p_w = \dot{F}$ and $b_z = \bar{F}$

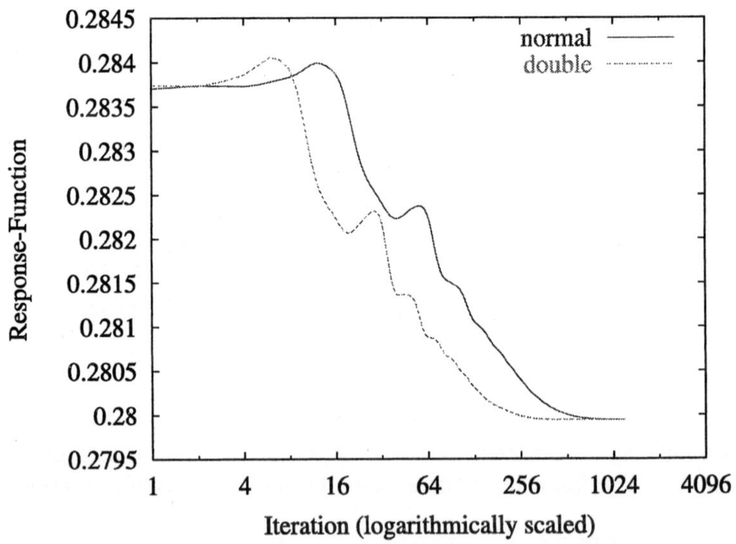

Fig. 2. Convergence of Lift Coefficient and its Corrected Value

the earlier residual reduction as far as the derivatives is concerned. Hence we may conclude that the derivative recurrences were nearly linear in that the shifts in the underlying iterates z_k were small compared to the updates of the derivatives $\dot{z}_k$ and $\bar{w}_k$ themselves. In Fig. 2 we see that the corrected lift coefficient "double" converges in agreement with Corollary 2, indeed significantly faster, and arrives after some 128 iterations at an approximate value that should already be close enough to decide whether the current design parameter settings are competitive or whether this constellation should be rejected.

4 Summary and Outlook

On the basis of a given fixed point iteration for solving a state equation $F(z,x) = 0$ with respect to z we wish to calculate rapidly converging estimates for the reduced function $f(z(x),x)$ and its total derivatives with respect to x. This reduced gradient and a corrected version of the reduced function $f(z(x),x)$ is obtained from the results of an adjoint fixed point iteration that can be grafted onto the original, user supplied solver. The corrected function estimate is is shown to converge twice as fast as the underlying fixed point iterations. The algorithmic development and convergence analysis was confirmed by test calculations on a $2D$ Euler code.

The same can be shown [7] for a hybrid estimate of reduced first derivatives that are obtained by a combination of adjoint and direct differentiation. The latter process generates feasible state space direction, which yield in combination with the Lagrange multipliers generated by the adjoint differentiation also reduced Hessians and other second derivative information. The rapid availability of accurate estimates for the reduced response function and its derivatives should facilitate the development of piggy-back design optimization methods that achieve optimality in addition to feasibility at little extra cost.

5 Acknowledgment

The authors are indebted to Vittorio Selmin, Alenia for supplying the Euler code.

References

1. R. Becker and R. Rannacher, *An optimal control approach to error control and mesh adaptation in finite element methods*, Acta Numerica 2001 (A. Iserles, ed.), Cambridge University Press, 2001, A. Iserles, ed.
2. A. Carle and M. Fagan, *Improving derivative performance for CFD by using simplified recurrences*, Computational Differentiation: Techniques, Applications, and Tools (Martin Berz, Christian Bischof, George Corliss, and Andreas Griewank, eds.), SIAM, Philadelphia, Penn., 1996, pp. 343–351.

3. B. Christianson, *Reverse accumulation and attractive fixed points*, Optim. Methods Softw. **3** (1994), 311–326.
4. D.G. Cacuci, C.F. Weber, E.M. Oblow, and J.H. Marable, *Sensitivity theory for general systems of nonlinear equations*, Nuclear Science & Engineering **88** (1980), 88–110.
5. S.A. Forth and T.P. Evans, *Aerofoil Optimisation via Automatic Differentiation of a Multigrid Cell-Vertex Euler Flow Solver*, Proceedings of Automatic Differentiation 2000: From Simulation to Optimization (Berlin), Springer Verlag, 2000.
6. A. Griewank, C. Bischof, G. Corliss, A. Carle, and K. Williamson, *Derivative convergence of iterative equation solvers*, Optimiz. Methods Softw. **2** (1993), 321–355.
7. A. Griewank and Ch. Faure, *Reduced functions, gradients and hessians from fixed point iterations for state equations*, Numerical Algorithms **30** (2002), no. 2, 113–139.
8. J.Ch. Gilbert, *Automatic differentiation and iterative processes*, Optim. Methods Softw. **1** (1992), 13–21.
9. M.B. Giles, *On the iterative solution of adjoint equations*, Proceedings of Automatic Differentiation 2000: From Simulation to Optimization (Berlin), Springer Verlag, 2000.
10. A. Griewank, *Evaluating Derivatives, Principles and Techniques of Algorithmic Differentiation*, Frontiers in Appl. Math., no. 19, SIAM, Philadelphia, 2000.
11. A. Jameson, *Optimum aerdynamic design using cfd and control theory*, 12th AIAA Computational Fluid Dynamics Conference, AIAA Paper 95-1729 (San Diego, CA), American Institute of Aeronautics and Astronautics, 1995.
12. B.A. Murtagh and M.A. Saunders, *MINOS 5.4 User's Guide*, Report, Systems Optimization Laboratory, Stanford Uinversity, December 1983, Report SOL 83-20R, (revised February 1995).
13. P.A. Newman, G.J.-W. Hou, H.E. Jones, A.C. Taylor, and V.M. Korivi, *Observations on computational methodologies for use in large-scale, gradient-based, multidisciplinary design incorporating advanced CFD codes*, Technical Memorandum 104206, NASA Langley Research Center, February 1992, AVSCOM Technical Report 92-B-007.
14. J.M. Ortega and W.C. Rheinboldt, *Iterative solution of nonlinear equations in several variables*, Academic Press, New York, 1970.
15. S. Ta'asan, G. Kuruvila, and M.D. Salas, *Aerdynamic design and optimization in one shot*, 30th AIAA Aerospace Sciences Meeting and Exhibit, AIAA Paper 91-0025 (Reno, Nevada), American Institute of Aeronautics and Astronautics, 1992.

Part V

NLP Algorithms and Inequality Constraints

Assessing the Potential of Interior Methods for Nonlinear Optimization

José Luis Morales[1], Jorge Nocedal[2], Richard A. Waltz[2], Guanghui Liu[3], and Jean-Pierre Goux[2]

[1] Departamento de Matemáticas, Instituto Tecnológico Autónomo de México, Río Hondo 1, Col Tizapán San Angel, México D.F. CP 01000, México. †
[2] Electrical and Computer Engineering Department, Northwestern University, Evanston IL 60208, USA. ‡
[3] Industrial Engineering and Management Sciences Department, Northwestern University, Evanston IL 60208, USA.‡

Abstract. A series of numerical experiments with interior point (LOQO, KNITRO) and active-set sequential quadratic programming (SNOPT, filterSQP) codes are reported and analyzed. The tests were performed with small, medium-size and moderately large problems, and are examined by problem classes. Detailed observations on the performance of the codes, and several suggestions on how to improve them are presented. Overall, interior methods appear to be strong competitors of active-set SQP methods, but all codes show much room for improvement.

1 Introduction

The goal of this paper is to evaluate the efficiency and robustness of interior methods for nonlinear programming, and to assess their potential in the solution of very large problems. The hope is that this study will highlight some of the strengths and weaknesses of interior methods, suggest avenues for improving their performance, and identify classes of problems for which they appear to be well suited. In order to conduct this evaluation we will perform numerical tests using two established interior point packages, LOQO and KNITRO, and will compare their performance with the state-of-the-art sequential quadratic programming codes, SNOPT and filterSQP. This will also allow us to make some new observations about the numerical behavior of these two active-set SQP algorithms.

The nonlinear programming problem will be written as

$$\min_x \quad f(x) \tag{1a}$$

$$\text{s.t.} \quad h(x) = 0 \tag{1b}$$

$$g(x) \leq 0 \tag{1c}$$

† This author was supported by CONACyT, Asociación Mexicana de Cultura AC, the Fulbright Comission, and National Science Foundation grant CCR 9907818.
‡ These authors were supported by National Science Foundation grant CCR 9907818, and by Department of Energy grant DE-FG02-87ER25047-A004.

We will consider only problems where the objective $f : \mathbb{R}^n \to \mathbb{R}$ and constraint functions $h : \mathbb{R}^n \to \mathbb{R}^m, \;\; g : \mathbb{R}^n \to \mathbb{R}^q$ are smooth. Our testing will be done primarily on small, medium-sized and moderately large problems from the CUTE collection. By this we mean problems that have up to 10,000 variables or constraints.

A variety of interior (or barrier) methods for nonlinear programming have been proposed in the last few years [1, 5, 7, 12, 15, 17, 25, 26], but only a few implementations are currently available for public use. We have chosen to experiment with the software packages LOQO [25] and KNITRO [5], which are available through the NEOS system [9], as well as in machine executable form. These packages implement two distinct interior point approaches: LOQO is a line search algorithm that has much in common with interior algorithms for linear and convex quadratic programming, whereas KNITRO is a trust region method which uses sequential quadratic programming methodology to treat the barrier sub-problems. LOQO and KNITRO also differ significantly in the linear algebra techniques used for the step computation.

One of the main objectives is to try to determine if the interior methods implemented in these two packages are competitive with established codes for large-scale nonlinear programming. Several production packages, such as LANCELOT [8], MINOS [20], and CONOPT [11] would have served our purposes, but we have chosen SNOPT [18] and filterSQP [14] because we have experience with both codes, and because they constitute two complementary approaches to SQP algorithms: SNOPT is a line search method using convex models and filterSQP is a trust region method that allows the model to be non-convex.

Before proceeding, we must clarify what this study can and cannot achieve. First, the software packages used in our tests are regularly being updated and their performance could change significantly after a new release. We can expect this situation to continue in the near future; even SNOPT, which follows a well established active-set SQP approach, is still being modified in significant ways. Therefore we warn the reader against using our results to rank the codes. Not only is such a ranking dubious given that it is based on a particular set of problems, but even within this testing environment relative performance of the codes can change at any time.

Second, our goal is to assess the effectiveness of *optimization methods*, more specifically interior methods and active-set SQP methods, and not simply to evaluate the performance of specific *software implementations*. Of course, it is not possible to entirely isolate the performance of a method from its particular implementation because implementation aspects (e.g. sparse elimination techniques) impact performance in significant ways. Moreover, algorithms within a given class of methods can vary due to various lower-level algorithmic decisions, such as the line search strategy and the choice of merit function.

Nevertheless, by judiciously controlling the testing environment, we will be able to make some observations and state a few conjectures about the methods of interest. To focus on algorithmic features we separate the problems into three classes: (i) unconstrained, (ii) equality constrained, and (iii) general constrained problems, excluding linear and quadratic programs.

2 Overview of the Codes

We now outline the algorithms implemented in the four packages under consideration. More details about them will be presented in subsequent sections as we analyze their performance on various classes of problems.

2.1 LOQO

LOQO 4.05 [25] originated as an interior point code for linear and quadratic programming, and has been extended to solve nonlinear problems. Our interest here is only in its behavior as a nonlinear solver. LOQO implements a line search primal-dual interior point method. It uses second derivatives of the objective function and constraints.

The nonlinear programming problem (1) is replaced by a sequence of barrier problems of the form

$$\min_{x,s} \quad f(x) - \mu \sum_{i=1}^{q} \ln s_i \tag{2a}$$

$$\text{s.t.} \quad g(x) + s = 0, \tag{2b}$$

where $\mu > 0$ is the barrier parameter, s is a vector of slacks, and q denotes the number of inequality constraints. (For simplicity of exposition, we assume in this section that all the constraints are inequalities.) The algorithm computes a search direction d by factoring a system of the form

$$\begin{bmatrix} -\nabla^2 L(x,y) & A(x)^T \\ A(x) & SY^{-1} \end{bmatrix}, \tag{3}$$

where $A(x)$ stands for the Jacobian of the constraints $g(x)$, y is the vector of Lagrange multipliers, and $L(x,y)$ denotes the Lagrangian, $L(x,y) = f(x) - y^T g(x)$. Here we use the notation that S is a diagonal matrix with $S_{ii} = s_i$ (and similarly for Y^{-1}).

Once the search direction is computed, a backtracking line search is performed to guarantee sufficient decrease in the ℓ_2 penalty function,

$$\psi(x,s;\mu,\nu) = f(x) - \mu \sum_{i=1}^{q} \ln s_i + \nu \|g(x) + s\|_2^2, \tag{4}$$

where $\nu > 0$ is the penalty parameter. This is not an exact merit function, since for any given ν, a minimum of ψ is normally not a KKT point for (1). Moreover, a heuristic is used to determine the value of the penalty parameter

ν; it attempts to set $\nu = 0$, and increases it as necessary to ensure that the direction d is a descent direction for ψ.

A dynamic rule (see (11)) is used to update the value of the barrier parameter μ at each iteration. The KKT system (3) is solved using a sparse LDL^T factorization [22]. In order to handle non-convex problems, LOQO monitors the Cholesky factors of

$$\nabla^2 L(x, y) + A(x)^T S^{-1} Y A(x),$$

and if the factorization implies that this matrix is not sufficiently positive definite, a multiple λ of the identity matrix is added to $\nabla^2 L$. Since a suitable value of λ is not known in advance, several trial values and refactorizations may be needed.

There is no global convergence theory for the algorithm implemented in LOQO, and it would be difficult to establish one without restrictive assumptions. The main obstacles to proving convergence are the use of a dynamic rule for changing the barrier parameter μ, and the use of the penalty function (4) as a merit function.

2.2 KNITRO

KNITRO 1.00 [5] implements a primal-dual interior point method with trust regions. It makes use of second derivative information and, as in LOQO, employs slack variables to formulate barrier subproblems of the form (2). KNITRO differs from most interior methods proposed in the literature in that it does not factor the primal-dual matrix (3). Instead it applies an SQP-type method for equality constrained optimization to compute steps d that approximately minimize the barrier problem (2). When this minimization is completed, the barrier parameter μ is decreased by a fixed factor of five.

Each step is computed as the sum $d = v + t$ of a normal component v whose objective is to improve feasibility, and a tangential component t that aims toward optimality [3, 21]. The main cost of the iteration lies in the computation of the tangential component, which is performed using a projected conjugate gradient (CG) iteration. Each CG iteration involves two main computations: (i) a projection, which makes use of the factors of the matrix

$$\begin{bmatrix} I & 0 & A(x)^T \\ 0 & I & S \\ A(x) & S & 0 \end{bmatrix}, \tag{5}$$

and, (ii) a product of the Hessian $\nabla^2 L$ times a vector. The normal step v and the Lagrange multipliers y are inexpensive to compute, since they each require only one backsolve using the factors of (5).

After the step d is computed, the algorithm tests whether it provides sufficient decrease in the (non-differentiable) merit function

$$\phi(x, s; \mu; \nu) = f(x) - \mu \sum_{i=1}^{q} \ln s_i + \nu \|g(x) + s\|_2, \qquad \nu > 0. \tag{6}$$

If there is no decrease, the step is rejected and the trust region radius is reduced. Second-order correction steps are computed under certain circumstances to prevent the rejection of steps. The penalty parameter is chosen at each iteration so that the step d is a descent direction for the merit function (6).

A global convergence theory of an algorithm similar to that implemented in KNITRO is presented in [4], whereas rate of convergence results are given in [6].

2.3 SNOPT

SNOPT 5.3-4 [18] is a line search SQP method in which the search direction is determined by an activer-set method for convex quadratic programming. The line search enforces decrease of an augmented Lagrangian function, which in the case when all the constraints are inequalities, takes the form

$$\mathcal{M}(x, s, y) = f(x) - y^T(g(x) + s) + \tfrac{1}{2}(g(x) + s)^T D(g(x) + s), \tag{7}$$

where D is a diagonal matrix of penalty parameters. The code requires only first derivatives of the objective function and constraints, and maintains a dense BFGS approximation to the reduced Hessian of the *modified* Lagrangian,

$$\mathcal{L}(x, x_k, y_k) = f(x) - y_k^T \left[g(x) - g_k - A_k^T(x - x_k) \right].$$

When the number of variables is large the algorithm automatically switches to a limited memory BFGS mode.

All iterates satisfy the linear constraints in the problem. To promote global convergence, the iteration may enter *elastic mode* in which the constraints are relaxed using slack variables. "Linear elastic mode" is invoked to determine if the linear constraints are compatible, and "nonlinear elastic mode" is used when the quadratic subproblem is infeasible or unbounded, or when the Lagrange multiplier estimates y become large.

To ensure that the quasi-Newton approximations are positive definite, SNOPT modifies the correction vectors used in the update, if necessary; under certain conditions it may even skip the update. Global convergence results are difficult to establish for SNOPT, mainly because the merit function (7) is treated as a function of x, s, y, and is therefore not minimized by a solution point.

2.4 FilterSQP

The FilterSQP 1.0 [14] package implements a trust region SQP method that makes use of second derivatives of the objective function and constraints. The step is computed by means of an active-set method for indefinite quadratic programming (QP). If a direction of negative curvature is detected, the QP solver follows it until a constraint or the trust region boundary is encountered. This process continues until a local minimum of the quadratic model is obtained. A novel feature is the use of a "filter", a step-acceptance mechanism that eliminates the need for a merit function [14]. If a step is not accepted by the filter, a second-order correction step is typically attempted. The trust region is defined by an ℓ_∞ norm, and its update strategy is simpler than in trust region methods that use merit functions.

The algorithm contains a *feasibility restoration phase* that is invoked when the constraints of the subproblem are infeasible, or when the filter mechanism requires that progress on the constraints be made. The feasibility restoration phase is similar to the procedure used by SNOPT in elastic mode. The main differences are: the use of a trust region, and the fact that the objective function f is ignored during the feasibility restoration phase.

Both SNOPT and filterSQP employ sophisticated update strategies in their active-set quadratic programming solvers to carry information from one active-set to the next. Global convergence results have been established for variations of the filter mechanism implemented in filterSQP [13], but this is still an ongoing area of research.

3 Testing Environment

We performed tests using 509 problems from the CUTE collection [2] that have been translated into AMPL [16] by Benson and Vanderbei [24]. We selected a subset of the 740 problems available in the Benson-Vanderbei test set as of December, 2000, according to the following criteria. All linear programming (LP) and quadratic programming (QP) were excluded since we feel that these problems can be treated more efficiently by specifically designed methods and software. We also excluded all feasibility problems, i.e., problems with no objective function. Finally, to have a more manageable set of general constrained problems, we excluded all problems whose only constraints were simple bounds. The dimensions of the problems were the default values specified in the Benson-Vanderbei models.

Table 1 in section 6 gives the number of problems in each category; see www.ece.northwestern.edu/~nocedal/knitro. Since CPU times can be dominated by set-up costs for small problems, we will only report CPU times for problems with more than 100 variables. Function evaluations are reported for all problems.

It was impossible to apply a uniform stopping test on all codes, since their form cannot be changed by the user. Nevertheless, we tried to make them as similar as possible, and in fact we modified KNITRO's stopping test to be almost identical to that of SNOPT. Some differences in final precision were noticed, but these were not significant.

All the tests were performed on a Sun Ultra 5, with 385Mb of memory running SunOS 5.7. The four codes were run in double precision using all their default settings. A limit of 30 minutes of CPU time, and 1000 major iterations, were imposed for each problem; if one of these limits was reached the code was considered to have failed. The stopping tolerance was set at 10^{-6} on all solvers.

The fact that SNOPT uses only first-derivative information[1] and approximates second derivatives by quasi-Newton updating introduces an undesirable disparity with the other three codes. We will nevertheless take advantage of this difference to assess the effectiveness of quasi-Newton approximations.

4 Numerical Results

We will now proceed with the examination of results, by problem class. In order to facilitate their interpretation, we highlight the key features of each method when applied to a specific problem class.

4.1 Unconstrained Problems

When applied to an unconstrained problem, the algorithms implemented in LOQO, KNITRO, and filterSQP reduce to Newton methods, whereas SNOPT amounts to a dense quasi-Newton method.

LOQO is an (exact) Newton method with a backtracking line search requiring sufficient decrease in f. The search direction is computed using a sparse LDL^T factorization of the Hessian $\nabla^2 f(x)$. Negative curvature is handled by adding a multiple of the identity to the Hessian matrix, if necessary.

KNITRO is an (inexact) Newton method with a spherical trust region. The step is computed by an unpreconditioned conjugate gradient iteration, which is terminated using Steihaug's rule [23]. The step is accepted if it provides sufficient reduction in f.

SNOPT is a quasi-Newton method with a backtracking line search requiring sufficient reduction in f. But it is not a traditional unconstrained quasi-Newton method, and no special consideration has been given to make the code efficient for unconstrained problems. SNOPT maintains two approximate Hessians: the limited-memory BFGS approximation of the full Hessian, and a dense reduced Hessian. Therefore, for large problems it will be inefficient to keep both a sparse and a dense version of essentially the same $n \times n$ matrix. An evaluation of the limited memory strategy used in SNOPT is given in [19].

[1] Our numerical tests were conducted in the fall of 2000; at that time a second-derivative version of SNOPT was not available.

FilterSQP is an (exact) Newton method using an ℓ_∞ trust region. The filter mechanism reduces, in the unconstrained case, to a standard sufficient decrease condition on f. The search direction is computed by minimizing a quadratic model of the objective subject to the trust region constraints. Indefiniteness is exploited by ensuring that the step is a local minimizer of the quadratic model of f subject to the trust region constraint.

All the results in this paper will be presented using the performance profiles proposed by Dolan and Moré [10]. Let $t_{p,s}$ denote the time to solve problem p by solver s. Then define the ratios

$$r_{p,s} = \frac{t_{p,s}}{t_p^*}, \tag{8}$$

where t_p^* is the lowest time required by any code to solve problem p. If a code does not solve a problem, the ratio $r_{p,s}$ is assigned a large number, M. Then, define the logarithmic performance profile for each code s, as

$$\pi_s(\tau) = \frac{\text{no. of problems s.t.} \log_2(r_{p,s}) \leq \tau}{\text{total no. of problems}}, \qquad \tau \geq 0. \tag{9}$$

This performance profile will also be used to analyze the number of function evaluations required by the four codes.

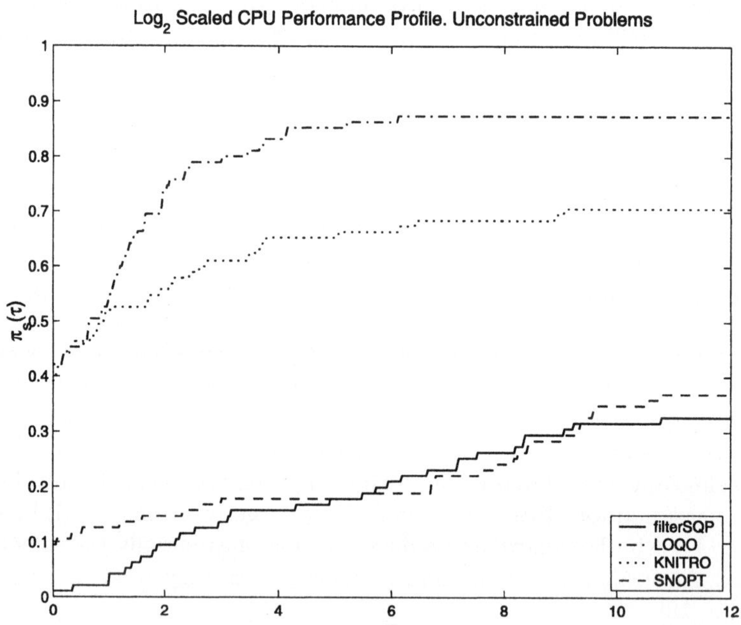

Fig. 1. Performance, in terms of CPU time, on 95 unconstrained problems.

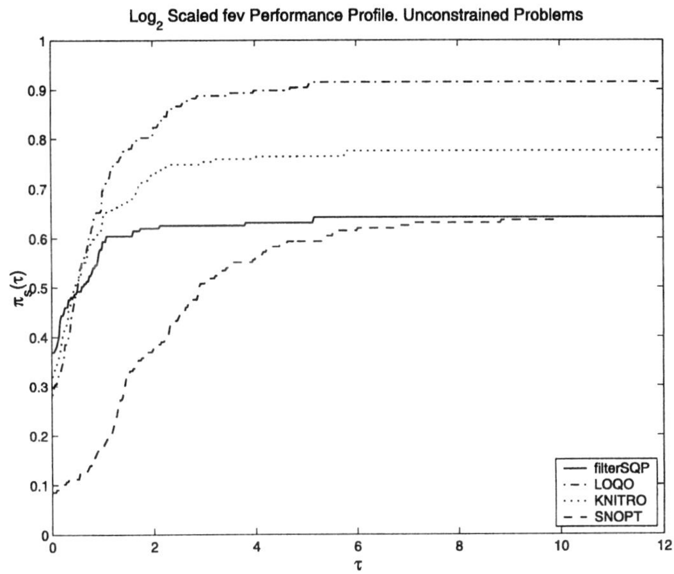

Fig. 2. Peformance, in terms of function evaluations, on 187 unconstrained problems.

In Figures 1 and 2 we plot performance profiles for CPU time and function evaluations, for the unconstrained problems. We recall that CPU profiles are reported only for problems with $n \geq 100$; therefore to study robustness of the codes it is preferable to consider the profiles reporting function evaluations and Table 1 in section 6.

We make the following observations based on these figures, and on a problem by problem examination of the runs. (i) The CPU times for filterSQP are unexpectedly high, and are almost certainly caused by inefficiencies in the factorization algorithm. (ii) The trade-offs between the iterative method used in KNITRO for the step computation, and the direct method in LOQO, can be clearly observed in these test problems (consider e.g. the range $\tau \in [0, 2]$). An examination of KNITRO's output shows that the problems for which it required significantly more time than LOQO are mostly ill-conditioned, demanding a large number of CG iterations. KNITRO's performance would improve on difficult problems with the use of a preconditioner. (iii) SNOPT required consistently more function evaluations than the other codes, as is expected from a quasi-Newton method. Computing times would improve if the dense quasi-Newton approximation were to be replaced by a limited-memory approximation. (iv) LOQO performs well in terms of function evaluations[2]

[2] LOQO 4.05 makes one unnecessary function evaluation at every iteration, something that could easily be corrected. We adjusted for this by discounting this extra evaluation in our results.

and time, but the results must be interpreted with caution because its stopping test for unconstrained problems is often less demanding than that used in the other codes due to scaling differences. (v) KNITRO and filterSQP are very similar in terms of function evaluations, as can be seen by considering the range $\tau \in [0, 2]$. The differences in the shape of the trust region and the inexactness vs. exactness of the Newton iteration do not impact the number of function evaluations in these problems.

4.2 Equality Constrained Problems

When applied to equality constrained problems (1a)-(1b), the algorithms implemented in the four codes belong to the class of sequential quadratic programming (SQP) methods. There are, nevertheless, important algorithmic differences.

LOQO is a line search SQP method. The search direction is obtained by applying Newton's method to the KKT conditions of (1a)-(1b). This gives an iteration of the form

$$\begin{bmatrix} -\nabla^2 L(x,y) & A(x)^T \\ A(x) & 0 \end{bmatrix} \begin{bmatrix} d_x \\ d_y \end{bmatrix} = \begin{bmatrix} \nabla L(x,y) \\ -h(x) \end{bmatrix}, \tag{10}$$

where A denotes the Jacobian of the equality constraints. The linear system in (10) is first reduced, and then solved using a sparse factorization; in case of indefiniteness, a multiple of the identity matrix is added to $\nabla^2 L(x,y)$.

KNITRO is an SQP method with trust regions. The step d is computed by the Byrd-Omojokun approach [3, 21]: d is the sum of a normal component v and a tangential component t, as described in section 2.2. The tangential component is computed using an (unpreconditioned) projected conjugate gradient iteration in which projections are performed using the factors of a matrix of the form (5), but where the slacks are not present. The CG iteration follows the first direction of negative curvature to the boundary of the trust region. The merit function has the form (6) without the barrier term.

SNOPT is a line search, quasi-Newton, SQP method. The search direction is obtained by solving a system of the form (10) in which $\nabla^2 L$ is replaced by a positive definite quasi-Newton matrix. The line search requires sufficient decrease in the augmented Lagrangian merit function. The elastic mode handles problems in which the constraints are ill-conditioned.

In filterSQP, the step is computed by solving an SQP subproblem with trust regions. The step is always a local minimizer of the quadratic model at which no directions of negative curvature exist. Step acceptance is determined by the filter mechanism.

The results on equality constrained problems are given in Figures 3 and 4. They should be taken with caution because the sample of problems is small.

(i) Once more, we observe large CPU times for filterSQP, even though it performs well in terms of function evaluations. The cause, as in the unconstrained case, is likely to be inefficiencies in the factorization. FilterSQP

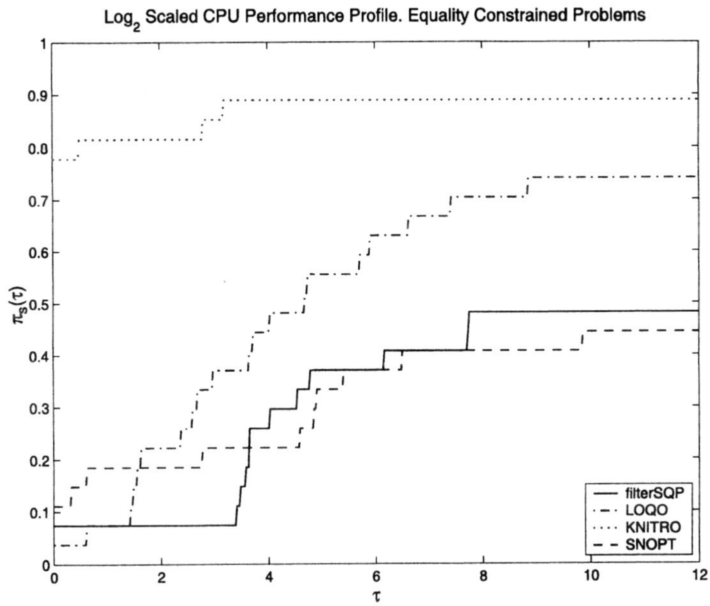

Fig. 3. Comparison, in terms of CPU time, on 27 equality constrained problems.

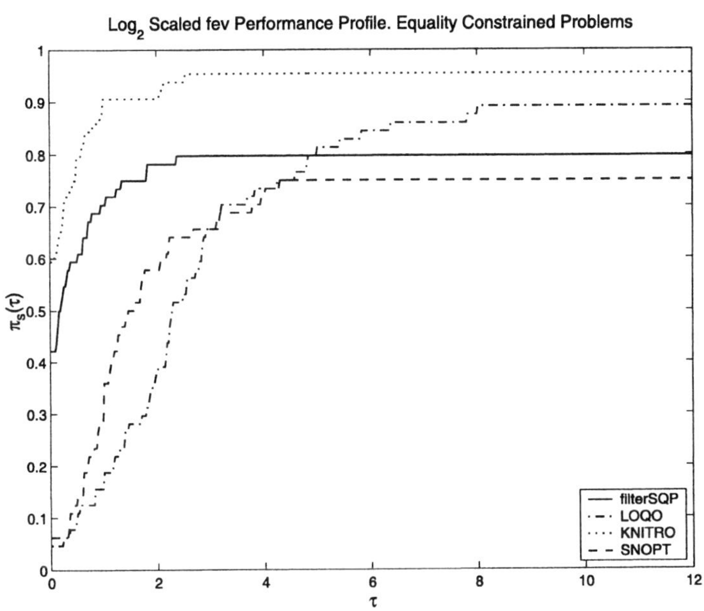

Fig. 4. Comparison, in terms of function evaluations, on 64 equality constrained problems.

applies many second-order corrections on some problems, but we have verified that they are not the cause for the large times. (ii) It is not surprising that SNOPT does not perform well in terms of function evaluations, due to the lack of second-derivative information. (iii) One of the distinctive features of KNITRO, namely its step computation mechanism using the CG method and projections, is fully active in equality constrained problems. The results indicate that it performs efficiently on this test set. (iv) The filter mechanism is also active now, and appears to work well, but a precise evaluation of its benefits is difficult to make: whereas filterSQP is clearly more economical in terms of function evaluations than LOQO and SNOPT, a problem-by-problem comparison with KNITRO's runs suggests that both codes have similar step-acceptance rates. The efficiency of the filter mechanism must, therefore, be further investigated, preferably using a code that provides both filter and merit function options.

5 General Nonlinear Programs

We will now consider problems containing inequality constraints, a setting in which the four methods exhibit major algorithmic differences.

LOQO is a line search, primal-dual interior method for nonlinear programming. The cost of the iteration is dominated by the factorization of the primal-dual matrix (3). The barrier parameter μ is redefined at every iteration using the rule:

$$\mu = 0.1 \min\left(0.05, \frac{1-\xi}{\xi}, 2\right)^3 \frac{s^T y}{q}, \quad \text{with} \quad \xi = \frac{\min_i s_i y_i}{y^T s/q}, \tag{11}$$

where q is the number of inequality constraints. This rule permits increases in the barrier parameter from one iteration to the next. A sufficient reduction in the penalty function (4) is required at each iteration.

KNITRO is a trust region, primal-dual interior method. It can be considered a path-following interior method since the barrier parameter is changed only after each barrier subproblem is solved to a given accuracy, and at that point it is decreased by a (fixed) factor of five. KNITRO uses the non-differentiable merit function (6), and applies second-order correction (SOC) steps to avoid the Maratos effect. The cost of the iteration is spread out, as in the equality constrained case, between three computations:

1. the factorization of the projection matrix (5),
2. $ncg+2$ backsolves using the factors of (5), where ncg denotes the number of CG steps. (One backsolve to compute the normal step v, one to compute Lagrange multipliers y, and ncg backsolves to perform projections during the CG iterations
3. ncg products of the Hessian $\nabla^2 L$ with a vector.

FilterSQP is a trust region SQP method, and SNOPT is a line search SQP method. Their main features have been described in the previous sections. For problems with general constraints, the active-set quadratic programming solvers of both methods are fully active.

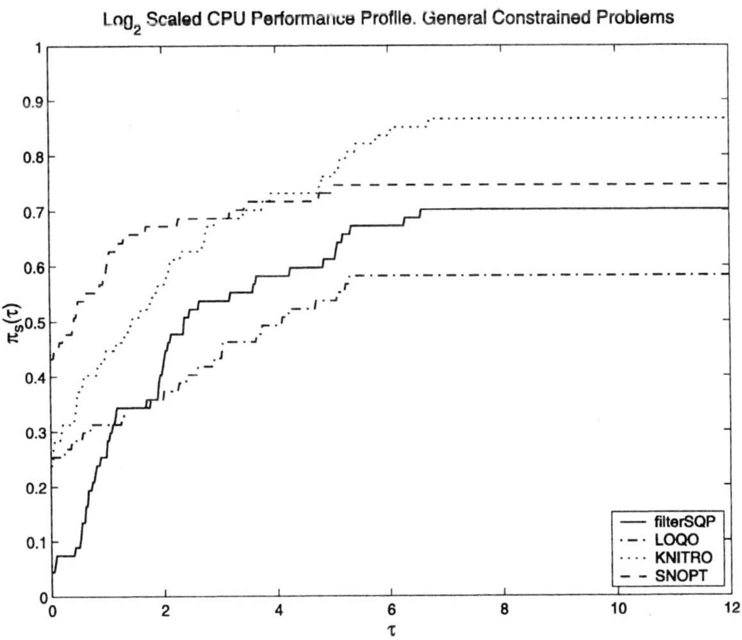

Fig. 5. Comparison, in terms of CPU time, on 67 constrained problems.

The results are presented in Figures 5 and 6.

(i) We observe that SNOPT performs quite well compared to the other three codes despite using only first derivatives. This is remarkable, and contrasts with our observations for unconstrained and equality constrained problems. To test whether this was due to the fact that most of our problems are not very large, we selected 16 problems from the CUTE collection and created three versions of each, with increasing dimensions ranging generally in the thousands. We compared KNITRO and SNOPT, but were not able to discern any clear trend. Whereas in some of the problems SNOPT's CPU time increased much more rapidly than KNITRO's time, on about as many problems SNOPT remained the fastest of the two codes. (ii) The times of filterSQP are high, especially since it requires a small number of iterations compared to the other solvers. On the other hand, filterSQP performs exceptionally well in terms of function evaluations. Whether this can be attributed to efficiencies in the filter mechanism, remains to be established. (iii) SNOP-

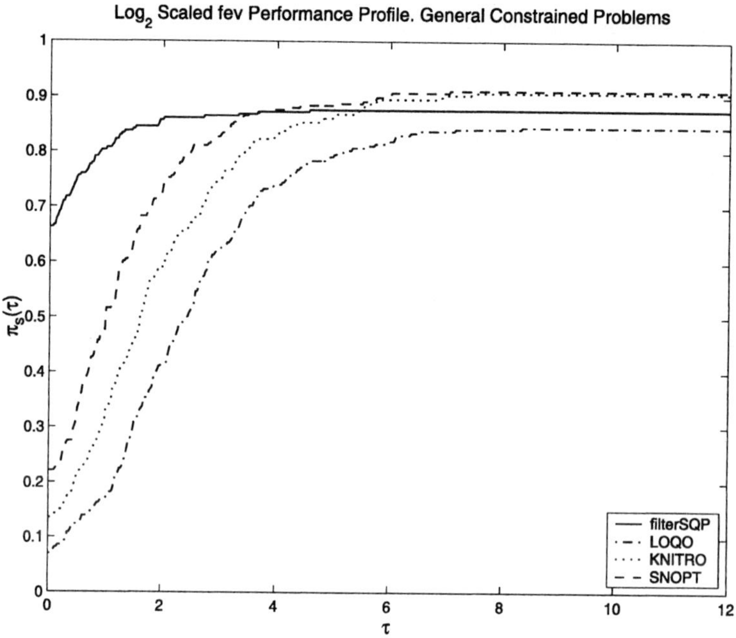

Fig. 6. Comparison, in terms of function evaluations, on 258 constrained problems.

T's performance in terms of function evaluations must be interpreted with caution because the code does not report them in the same manner as the other codes; it may undercount them when the objective function is linear.

We paid close attention to the relative performance of LOQO and KNITRO, to try to evaluate their main algorithmic features. (iv) We observed that KNITRO tends to perform fewer iterations (iterations are not reported in our tables). Considering, for example, all problems with $n \geq 50$ we observed that KNITRO required fewer iterations in about 2/3 of the problems. The difference persisted when looking at problem sets of other dimensions. KNITRO's advantage in this respect could be attributed to the step computation procedure, or more likely, to its path-following approach combined with a conservative rule for decreasing the barrier parameter. (v) LOQO's time per iteration was lower than KNITRO's in a significant percentage of the problems. LOQO's advantage in this respect is attributed to a faster factorization, in some cases, and to a large cost of KNITRO's CG steps, in other cases (usually problems with an ill-conditioned reduced Hessian). (vi) A series of numerical tests indicate that KNITRO's strategy of setting the initial value barrier parameter always as $\mu = 0.1$ can be highly detrimental in some cases; the adaptive choice used by LOQO appears to be more effective in the early iterations.

6 Final Remarks

Our experiments suggest that interior methods constitute a powerful approach for solving medium to moderately large nonlinear programming problems. The two (rather different) interior algorithms implemented in LOQO and KNITRO, appear to be competitive, in terms of robustness and efficiency, with the active-set SQP algorithms implemented in SNOPT and filterSQP. In Table 1 we summarize robustness statistics for the four codes, focusing also on the problems that are not small ($n \geq 100$).

Table 1. Number of problems successfully solved by each code

		FILTER	LOQO	KNITRO	SNOPT	sample size
unconstrained	all	120	171	145	120	187
	$n \geq 100$	33	83	67	35	95
equality constrained	all	51	57	61	48	64
	$n \geq 100$	13	20	24	12	27
general constrained	all	226	218	234	235	258
	$n \geq 100$	47	39	58	50	67

Our numerical experiments indicate that the interior methods also performed efficiently on small problems, as well as in the simpler classes of unconstrained and equality constrained problems. Therefore, LOQO and KNITRO constitute effective general purpose optimization solvers.

It remains to be seen how interior methods perform on larger problems. Our set of inequality constrained problems contained 30 problems with more than 1000 variables, and only 6 problems with more than 10,000 variables. Based on this small sample, the interior methods appear to scale up well compared to the SQP codes. The observations made in the previous sections indicate, however, that a number of algorithmic and implementation features in the interior methods deserve further attention. In particular, KNITRO would benefit from a preconditioning strategy, and from an adaptive choice of the barrier parameter in the initial iterations.

7 Acknowledgments

We are grateful to Sven Leyffer, Philip Gill, and Robert Vanderbei for assistance with their codes, and to David Gay for answering our questions on fine points of AMPL. We thank Hande Benson for translating the CUTE collection into AMPL (something that greatly facilitated our study) and to an anonymous referee who provided many suggestions on how to improve the presentation.

References

1. P. Armand, J.-Ch. Gilbert, and S. Jan-Jégou. A feasible BFGS interior point algorithm for solving strongly convex minimization problems. *SIAM Journal on Optimization*, 11:199–222, 2000.
2. I. Bongartz, A. R. Conn, N. I. M. Gould, and Ph. L. Toint. CUTE: Constrained and Unconstrained Testing Environment. *ACM Transactions on Mathematical Software*, 21(1):123–160, 1995.
3. R. H. Byrd. Robust trust region methods for constrained optimization. Third SIAM Conference on Optimization, Houston, Texas, May 1987.
4. R. H. Byrd, J. Ch. Gilbert, and J. Nocedal. A trust region method based on interior point techniques for nonlinear programming. *Mathematical Programming*, 89(1):149–185, 2000.
5. R. H. Byrd, M. E. Hribar, and J. Nocedal. An interior point algorithm for large scale nonlinear programming. *SIAM Journal on Optimization*, 9(4):877–900, 2000.
6. R. H. Byrd, G. Liu, and J. Nocedal. On the local behavior of an interior point method for nonlinear programming. In D. F. Griffiths and D. J. Higham, editors, *Numerical Analysis 1997*, pages 37–56. Addison Wesley Longman, 1997.
7. A. R. Conn, N. I. M. Gould, D. Orban, and Ph. L. Toint. A primal-dual trust-region algorithm for non-convex nonlinear programming. *Mathematical Programming*, 87(2):215–249, 2000.
8. A. R. Conn, N. I. M. Gould, and Ph. L. Toint. LANCELOT*: a Fortran package for Large-scale Nonlinear Optimization (Release A).* Springer Series in Computational Mathematics. Springer Verlag, Heidelberg, Berlin, New York, 1992.
9. J. Czyzyk, M. Mesnier, and J. J. Moré. The NEOS server. *IEEE Journal on Computational Science and Engineering*, 5:68–75, 1998.
10. E. D. Dolan and J. J. Moré. Benchmarking optimization software with performance profiles. Mathematics and Computer Science Technical Report ANL/MCS-P861-1200, Argonne National Laboratory, Argonne, Illinois, USA, 2001.
11. A. Drud. CONOPT – a large scale GRG code. *ORSA Journal on Computing*, 6:207–216, 1994.
12. A. S. El-Bakry, R. A. Tapia, T. Tsuchiya, and Y. Zhang. On the formulation and theory of the Newton interior-point method for nonlinear programming. *Journal of Optimization Theory and Applications*, 89(3):507–541, June 1996.
13. R. Fletcher, N. I. M. Gould, S. Leyffer, and Ph. L. Toint. Global convergence of trust-region SQP-filter algorithms for nonlinear programming. Technical Report 99/03, Department of Mathematics, University of Namur, Namur, Belgium, 1999.
14. R. Fletcher and S. Leyffer. Nonlinear programming without a penalty function. Numerical Analysis Report NA/171, Department of Mathematics, University of Dundee, Dundee, Scotland, 1997.
15. A. Forsgren and P. E. Gill. Primal-dual interior methods for nonconvex nonlinear programming. *SIAM Journal on Optimization*, 8(4):1132–1152, 1998.
16. R. Fourer, D. M. Gay, and B. W. Kernighan. *AMPL: A Modeling Language for Mathematical Programming.* Scientific Press, 1993.
17. D. M. Gay, M. L. Overton, and M. H. Wright. A primal-dual interior method for nonconvex nonlinear programming. In Y. Yuan, editor, *Advances in Nonlinear Programming*, pages 31–56, Dordrecht, The Netherlands, 1998. Kluwer Academic Publishers.

18. P. E. Gill, W. Murray, and M. A. Saunders. SNOPT: An SQP algorithm for large-scale constrained optimization. Technical Report 97-2, Dept. of Mathematics, University of California, San Diego, USA, 1997.
19. J. L. Morales. A numerical study of limited memory BFGS methods, 2001. to appear in Applied Mathematics Letters.
20. B. A. Murtagh and M. A. Saunders. MINOS 5.4 user's guide. Technical report, SOL 83-20R, Systems Optimization Laboratory, Stanford University, 1983. Revised 1995.
21. E. O. Omojokun. *Trust region algorithms for optimization with nonlinear equality and inequality constraints.* PhD thesis, University of Colorado, Boulder, Colorado, USA, 1989.
22. D. Shanno and R. Vanderbei. Interior-point methods for nonconvex nonlinear programming: Orderings and higher-order methods. Technical Report SOR-99-05, Statistics and Operations Research, Princeton University, 1999.
23. T. Steihaug. The conjugate gradient method and trust regions in large scale optimization. *SIAM Journal on Numerical Analysis*, 20(3):626–637, 1983.
24. R. J. Vanderbei. AMPL models. http://www.sor.princeton.edu/~rvdb/ampl/nlmodels.
25. R. J. Vanderbei and D. F. Shanno. An interior point algorithm for nonconvex nonlinear programming. *Computational Optimization and Applications*, 13:231–252, 1999.
26. H. Yamashita, H. Yabe, and T. Tanabe. A globally and superlinearly convergent primal-dual point trust region method for large scale constrained optimization. Technical report, Mathematical Systems, Inc., Sinjuku-ku, Tokyo, Japan, 1997.

An Interior-Point Algorithm for Large Scale Optimization

John T. Betts[1], Samuel K. Eldersveld[2], Paul D. Frank[1], and John G. Lewis[1]

[1] Mathematics and Engineering Analysis,The Boeing Company, P.O. Box 3707, MC 7L-21, Seattle, Washington 98124-2207
[2] Expedia, Inc. 13810 SE Eastgate Way, Suite 400, Bellevue, WA 98005

Abstract. This paper describes an interior-point algorithm for solving large scale nonlinear programming problems. The fundamental step of the algorithm requires solution of a sparse symmetric indefinite linear system. Row and column scaling are used to ensure that the system is well-conditioned. A globalization strategy based on a nonlinear filter is used instead of a merit function. The computational performance of the algorithm is demonstrated on a high index partial differential-algebraic equation application.

1 Introduction

When a physical process modeled by a partial differential equation is discretized, the result is often a large scale optimization problem. Usually inequality constraints are required in order to accurately model the process. When there are many inequality constraints, determining the correct set of active constraints can be an extremely costly computational burden. To address the shortcomings of active set algorithms, there has been considerable interest in another class of methods for nonlinear applications. In these *interior-point* or *barrier* methods [5, 8, 9], some of the combinatorial issues are dealt with by incorporating the inequality constraints into the objective using a barrier transformation. A sequence of these equality constrained problems is solved with the solution of each subsequent subproblem being used to provide an improvement in the estimate of the solution of the original inequality constrained problem. In addition to attempting to avoid the computational complexity of active-set methods, these barrier methods possess a second attribute that makes them attractive for large-scale optimization. Specifically, the form of the underlying sparse linear equations can be exploited by state of the art direct factorization techniques.

The purpose of this paper is to describe a basic interior-point algorithm designed for large-scale applications. First we define the nonlinear program in a format that is convenient for development of barrier methods. We then present the barrier method including descriptions of important operations. We conclude with a section on computational results. Although our current results were obtained using direct linear algebra, it is expected that most of the methods will extend to applications that require iterative solution of the linear systems.

2 The Nonlinear Programming Problem

The nonlinear programming problem can be stated as follows:

Minimize the function

$$f(\mathbf{y}) \equiv f(\mathbf{x}) \tag{1}$$

of the n variables

$$\mathbf{y} \equiv \begin{pmatrix} \mathbf{x} \\ \mathbf{s} \end{pmatrix}, \tag{2}$$

subject to the m_E nonlinear equality constraints

$$\mathbf{c}(\mathbf{y}) \equiv \begin{pmatrix} \mathbf{a}_{\mathcal{E}}(\mathbf{x}) - \underline{\mathbf{a}}_{\mathcal{E}} \\ \mathbf{a}_{\mathcal{I}}(\mathbf{x}) - \mathbf{s} \end{pmatrix} = \mathbf{0} \tag{3}$$

and the m_B linear bounds

$$\mathbf{b}(\mathbf{y}) \equiv \begin{pmatrix} (\mathbf{x} - \underline{\mathbf{x}})_{\mathcal{B}_1} \\ (\overline{\mathbf{x}} - \mathbf{x})_{\mathcal{B}_2} \\ (\mathbf{s} - \underline{\mathbf{a}}_{\mathcal{I}})_{\mathcal{B}_3} \\ (\overline{\mathbf{a}}_{\mathcal{I}} - \mathbf{s})_{\mathcal{B}_4} \end{pmatrix} \geq \mathbf{0}. \tag{4}$$

Observe that this *internal format* explicitly includes user-defined equality constraints of the form $\mathbf{a}_{\mathcal{E}} = \underline{\mathbf{a}}_{\mathcal{E}}$, whereas user inequalities $\mathbf{a}_{\mathcal{I}}$ are treated using the slack variables $\mathbf{s}$. The bounds $\mathbf{b}(\mathbf{y}) \geq \mathbf{0}$ are formed using only the *finite* values of the user specified bounds. For example finite lower bounds for the variables are denoted by $\underline{\mathbf{x}}$, and we use $\mathcal{B}_1$ to denote the corresponding subset of variable indices. Similar notation applies to the other bounds. When transforming the problem to internal format we also eliminate variables that are fixed and constraints that are ignored. A complete description of the *external format* has been omitted for brevity, but can be found in [3].

Note that the equality constraints (3) may be nonlinear functions of the variables $\mathbf{y}$, whereas the inequality constraints (4) are strictly linear in $\mathbf{y}$. This property makes it straightforward to construct an initial guess $\mathbf{y}^{(0)}$ that is feasible with respect to the inequalities $\mathbf{b}(\mathbf{y}^{(0)}) > \mathbf{0}$. Also, our method constructs iterates that maintain feasibility $\mathbf{b}(\mathbf{y}^{(k)}) > \mathbf{0}$ for all k. However, this also implies that the original (external format) inequalities $\mathbf{a}_{\mathcal{I}}$ may not be strictly feasible until a solution is found. Consequently, we refer to the approach as an *infeasible barrier method.*

3 Interior-Point Algorithm

3.1 Definitions

The *Lagrangian* is defined as

$$L(\mathbf{y}, \boldsymbol{\eta}, \boldsymbol{\lambda}) = f(\mathbf{y}) - \boldsymbol{\eta}^{\mathsf{T}}\mathbf{c}(\mathbf{y}) - \boldsymbol{\lambda}^{\mathsf{T}}\mathbf{b}(\mathbf{y}), \tag{5}$$

where $\boldsymbol{\eta}$ is the m_E-vector of Lagrange multipliers corresponding to the equality constraints, and $\boldsymbol{\lambda}$ is the m_B-vector of Lagrange multipliers corresponding to the inequality (bound) constraints. The gradient of the Lagrangian is

$$\nabla_y L(\mathbf{y}, \boldsymbol{\eta}, \boldsymbol{\lambda}) = \mathbf{g} - \mathbf{C}^{\mathsf{T}}\boldsymbol{\eta} - \mathbf{B}^{\mathsf{T}}\boldsymbol{\lambda}, \tag{6}$$

where the gradient of the objective is just

$$\mathbf{g} = \nabla_y f(\mathbf{y}) = \begin{pmatrix} \nabla_x f(\mathbf{x}) \\ \mathbf{0} \end{pmatrix},$$

and the Jacobian matrices are

$$\mathbf{C} \equiv \begin{pmatrix} \mathbf{A}_{\mathcal{E}} & \mathbf{0} \\ \mathbf{A}_{\mathcal{I}} & -\mathbf{I} \end{pmatrix} \qquad \mathbf{B} \equiv \begin{pmatrix} \mathbf{I}_{\mathcal{B}_1} & \mathbf{0} \\ -\mathbf{I}_{\mathcal{B}_2} & \mathbf{0} \\ \mathbf{0} & \mathbf{I}_{\mathcal{B}_3} \\ \mathbf{0} & -\mathbf{I}_{\mathcal{B}_4} \end{pmatrix}.$$

The equality Jacobian $\mathbf{C}$ is $m_E \times n$, and the bound Jacobian $\mathbf{B}$ is $m_B \times n$. The matrices $\mathbf{I}_{\mathcal{B}_1}$, $\mathbf{I}_{\mathcal{B}_2}$, $\mathbf{I}_{\mathcal{B}_3}$ and $\mathbf{I}_{\mathcal{B}_4}$, are rectangular sub-matrices of an identity matrix; each row having a single non-zero entry of 1.

The Hessian of the Lagrangian is given by

$$\begin{aligned} \nabla^2_{yy} L(\mathbf{y}, \boldsymbol{\eta}, \boldsymbol{\lambda}) &= \nabla^2_{yy} f(\mathbf{y}) - \sum_{k=1}^{m_E} \eta_k \nabla^2_{yy} c_k(\mathbf{y}) - \sum_{k=1}^{m_B} \lambda_k \nabla^2_{yy} b_k(\mathbf{y}) \\ &= \nabla^2_{yy} f(\mathbf{y}) - \sum_{k=1}^{m_E} \eta_k \nabla^2_{yy} c_k(\mathbf{y}) = \begin{pmatrix} \mathbf{H}_L & \mathbf{0} \\ \mathbf{0} & \mathbf{0} \end{pmatrix}, \end{aligned} \tag{7}$$

where

$$\mathbf{H}_L \equiv \nabla^2_{xx} f(\mathbf{x}) - \sum_{k=1}^{m_E} \eta_k \nabla^2_{xx} c_k(\mathbf{x}). \tag{8}$$

When the projected Hessian is not positive definite we will consider a modified Hessian matrix that is defined as

$$\mathbf{W} = \begin{pmatrix} \mathbf{H} & \mathbf{0} \\ \mathbf{0} & \mathbf{0} \end{pmatrix} = \begin{pmatrix} \mathbf{H}_L + \tau(|\sigma| + 1)\mathbf{I} & \mathbf{0} \\ \mathbf{0} & \mathbf{0} \end{pmatrix}, \tag{9}$$

where $0 \leq \tau \leq 1$ is a Levenberg parameter and σ is the Gerschgorin bound for the most negative eigenvalue of $\mathbf{H}_L$.

Let us denote a solution by $(\mathbf{y}^*, \boldsymbol{\eta}^*, \boldsymbol{\lambda}^*)$. The solution is characterized by:

$$\mathbf{g} - \mathbf{C}^\mathsf{T}\boldsymbol{\eta}^* - \mathbf{B}^\mathsf{T}\boldsymbol{\lambda}^* = \mathbf{0}, \tag{10}$$

and

$$\lambda_k^* \geq 0 \qquad \lambda_k^* b_k = 0 \qquad\qquad k = 1, \ldots, m_B. \tag{11}$$

A more complete statement of the optimality conditions is given in [3].

3.2 Logarithmic Barrier Function

A convenient way to motivate the interior-point approach is based on the definition of the *logarithmic barrier function*

$$\beta(\mathbf{y}, \mu) = f(\mathbf{y}) - \mu \sum_{k=1}^{m_B} \ln b_k(\mathbf{y}) \tag{12}$$

where μ is called the *barrier parameter.* Fiacco and McCormick [6] showed that the original constrained problem (1)-(4) could be replaced by a sequence of problems for successively smaller values of the barrier parameter μ. A classical *primal-barrier* approach would solve a sequence of equality constrained problems, namely minimize the barrier function (12) subject to the equalities (3). The necessary optimality conditions for the *barrier subproblem* of minimizing (12) subject to (3) can be stated in terms of the *barrier Lagrangian*

$$L_\beta(\mathbf{y}, \boldsymbol{\eta}_\mu, \mu) = \beta(\mathbf{y}, \mu) - \boldsymbol{\eta}_\mu^\mathsf{T}\mathbf{c}(\mathbf{y}). \tag{13}$$

The gradient of the barrier Lagrangian is given by

$$\begin{aligned} \nabla_y L_\beta &= \nabla_y \beta - \mathbf{C}^\mathsf{T}\boldsymbol{\eta}_\mu & (14)\\ &= \mathbf{g} - \mu\mathbf{B}^\mathsf{T}\mathbf{D}_b^{-1}\mathbf{e} - \mathbf{C}^\mathsf{T}\boldsymbol{\eta}_\mu & (15)\\ &= \mathbf{g} - \mathbf{B}^\mathsf{T}\boldsymbol{\pi}_\mu - \mathbf{C}^\mathsf{T}\boldsymbol{\eta}_\mu, & (16) \end{aligned}$$

where the diagonal matrix $\mathbf{D}_b = \mathrm{Diag}(b_1, b_2, \ldots, b_{m_B})$, and $\mathbf{e}$ is a vector of ones. Observe the components of the vector

$$\boldsymbol{\pi}_\mu = \mu\mathbf{D}_b^{-1}\mathbf{e} \tag{17}$$

are just $(\pi_\mu)_k = \mu/b_k$.

The classical approach is to solve a sequence of equality constrained problems (12),(3) for successively smaller barrier parameters μ. Unfortunately when μ is small the approach requires the solution of an ill-conditioned linear system. This difficulty can be reduced by solving a sequence of problems rather than just beginning with a small value for μ. Let us denote the local minimizer by $\mathbf{y}_\mu$. The necessary conditions for the barrier subproblem are

$$\nabla_y L_\beta = \mathbf{g} - \mathbf{B}^\top \boldsymbol{\pi}_\mu - \mathbf{C}^\top \boldsymbol{\eta}_\mu = \mathbf{0}, \tag{18}$$

$$\nabla_\eta L_\beta = \mathbf{c}(\mathbf{y}_\mu) = \mathbf{0}. \tag{19}$$

Furthermore, it can be shown that

$$\lim_{\mu\to 0} \mathbf{y}_\mu = \mathbf{y}^*, \qquad \lim_{\mu\to 0} \boldsymbol{\eta}_\mu = \boldsymbol{\eta}^*, \qquad \lim_{\mu\to 0} \boldsymbol{\pi}_\mu = \boldsymbol{\lambda}^*,$$

where $(\mathbf{y}^*, \boldsymbol{\eta}^*, \boldsymbol{\lambda}^*)$ is the solution of the original inequality constrained problem (1)-(4). This limiting behavior can be exploited to form a modified set of necessary conditions for the unknown quantities $(\mathbf{y}, \boldsymbol{\eta}, \boldsymbol{\lambda})$. In particular if we treat $(\mathbf{y}_\mu, \boldsymbol{\eta}_\mu, \boldsymbol{\pi}_\mu)$ as estimates for the true values $(\mathbf{y}^*, \boldsymbol{\eta}^*, \boldsymbol{\lambda}^*)$, we obtain the *modified primal-dual necessary conditions*

$$\mathbf{F}_\mu \equiv \begin{pmatrix} \mathbf{g} - \mathbf{C}^\top \eta - \mathbf{B}^\top \boldsymbol{\lambda} \\ \mathbf{c} \\ \mathbf{D}_b(\boldsymbol{\lambda} - \boldsymbol{\pi}_\mu) \end{pmatrix} = \mathbf{0}. \tag{20}$$

The conditions in the last row of (20) often written as

$$b_k(\mathbf{y})\lambda_k - \mu = 0 \tag{21}$$

for $k = 1, \ldots, m_B$ are referred to as the "centering" or "approximate complementarity" conditions, since as $\mu \to 0$, the centering condition (21) approaches the complementarity condition $b_k(\mathbf{y})\lambda_k = 0$. Like other primal-dual methods, this condition is explicitly included here in contrast to a classical approach based solely on (18)-(19).

3.3 Computing a Search Direction

The equations (20) form a nonlinear set of equations $\mathbf{F}_\mu = \mathbf{0}$ in the variables $(\mathbf{y}, \boldsymbol{\eta}, \boldsymbol{\lambda})$. If we consider a Taylor series expansion about the current point, we can derive the Newton equations for the modified primal-dual necessary conditions

$$\mathbf{W}\boldsymbol{\Delta}\mathbf{y} - \mathbf{C}^\top \boldsymbol{\Delta}\eta - \mathbf{B}^\top \boldsymbol{\Delta}\boldsymbol{\lambda} = -(\mathbf{g} - \mathbf{C}^\top \eta - \mathbf{B}^\top \boldsymbol{\lambda}); \tag{22}$$

$$\boldsymbol{\nabla}\mathbf{c}_k^\top(\mathbf{y})\boldsymbol{\Delta}\mathbf{y} = -c_k(\mathbf{y}), \qquad k = 1, \ldots, m_E; \tag{23}$$

$$\lambda_k \boldsymbol{\nabla}\mathbf{b}_k^\top(\mathbf{y})\boldsymbol{\Delta}\mathbf{y} + b_k(\mathbf{y})\Delta\lambda_k = -(b_k(\mathbf{y})\lambda_k - \mu), \qquad k = 1, \ldots, m_B, \tag{24}$$

where $\mathbf{W}$ is the Hessian matrix defined by (9). Rewriting these equations we obtain the following *unsymmetric primal-dual KKT system*

$$\begin{pmatrix} \mathbf{W} & \mathbf{C}^\top & \mathbf{B}^\top \\ \mathbf{C} & \mathbf{0} & \mathbf{0} \\ \mathbf{D}_\lambda \mathbf{B} & \mathbf{0} & -\mathbf{D}_b \end{pmatrix} \begin{pmatrix} \boldsymbol{\Delta}\mathbf{y} \\ -\boldsymbol{\Delta}\eta \\ -\boldsymbol{\Delta}\boldsymbol{\lambda} \end{pmatrix} = - \begin{pmatrix} \mathbf{g} - \mathbf{C}^\top \eta - \mathbf{B}^\top \boldsymbol{\lambda} \\ \mathbf{c} \\ \mathbf{D}_b\,(\boldsymbol{\lambda} - \boldsymbol{\pi}_\mu) \end{pmatrix}, \tag{25}$$

where $\mathbf{D}_\lambda = \mathrm{Diag}(\lambda_1, \lambda_2, \ldots, \lambda_{m_B})$.

The KKT system (25) can be scaled so that it is symmetric. Here we propose using both row and column scaling. We apply scaling only to the third block row and the third block column of (25). We scale the rows of the coefficient matrix by the diagonal matrix $\mathbf{D}_r = \text{Diag}(r_1, r_2, \dots, r_{m_B})$. We scale the columns (or the variables) by the diagonal matrix $\mathbf{D}_v = \text{Diag}(v_1, v_2, \dots, v_{m_B})$. Thus application of the scaling matrices to (25) yields the *symmetric primal-dual KKT system*

$$\begin{pmatrix} \mathbf{W} & \mathbf{C}^\mathsf{T} & \mathbf{B}^\mathsf{T}\mathbf{D}_v \\ \mathbf{C} & 0 & 0 \\ \mathbf{D}_r\mathbf{D}_\lambda\mathbf{B} & 0 & -\mathbf{D}_r\mathbf{D}_b\mathbf{D}_v \end{pmatrix} \begin{pmatrix} \boldsymbol{\Delta}\mathbf{y} \\ -\boldsymbol{\Delta}\boldsymbol{\eta} \\ -\mathbf{D}_v^{-1}\boldsymbol{\Delta}\boldsymbol{\lambda} \end{pmatrix} = - \begin{pmatrix} \mathbf{g} - \mathbf{C}^\mathsf{T}\boldsymbol{\eta} - \mathbf{B}^\mathsf{T}\boldsymbol{\lambda} \\ \mathbf{c} \\ \mathbf{D}_r\mathbf{D}_b(\boldsymbol{\lambda} - \boldsymbol{\pi}_\mu) \end{pmatrix}. \quad (26)$$

There are a number of possible ways for choosing the matrices $\mathbf{D}_r$ and $\mathbf{D}_v$ such that the resulting KKT system is symmetric. In order to achieve symmetry of the KKT system we must have $\mathbf{D}_v = \mathbf{D}_r\mathbf{D}_\lambda$ implying $r_k\lambda_k = v_k$, for $k = 1, \dots, m_B$. Consider

$$v_k = r_k^{-1} = \sqrt{\lambda_k}. \quad (27)$$

In this case, the lower-right diagonal block entry is $b_k(\mathbf{y})$, which should be well-behaved. It is for this reason that we have chosen (27) as the scaling technique here.

A Newton iteration requires solving the linear system (26) for the search direction $(\boldsymbol{\Delta}\mathbf{y}, \boldsymbol{\Delta}\boldsymbol{\eta}, \boldsymbol{\Delta}\boldsymbol{\lambda})$. This sparse symmetric indefinite linear system can be solved efficiently using the multi-frontal algorithm [1]. This technique also provides the *inertia* of the system (26) given as a triple defining the number of positive, negative, and zero eigenvalues respectively. It can be shown that if $In(\mathbf{K}) = (n, m, 0)$ where $\mathbf{K}$ is the matrix on the left hand side of (26), then the projected Hessian matrix $\mathbf{Z}^\mathsf{T}\mathbf{W}\mathbf{Z}$ is positive definite as required by the necessary conditions. The Levenberg parameter τ in (9) can be adjusted to obtain the correct inertia using the procedure described in [4], thereby creating a well-defined subproblem. It is shown in [3] that the above inertia requirements on the barrier method KKT matrix are consistent with the optimality requirements for the original problem.

The search direction computed is a linear prediction for the nonlinear behavior, and as such may not satisfy the inequality constraints $\mathbf{b}(\mathbf{y}) \geq \mathbf{0}$ and $\boldsymbol{\lambda} \geq \mathbf{0}$. However, both of these conditions are linear functions of the variables, and consequently we can modify the *length* of the step such that these conditions are met. Specifically, if we make a linear estimate for the boundary, i.e. $\mathbf{b}(\bar{\mathbf{y}}) = \mathbf{0}$ the step length is given by

$$\sigma_y = \min_k \left[\frac{-b_k(\mathbf{y})}{\nabla \mathbf{b}_k^\mathsf{T}(\mathbf{y})\boldsymbol{\Delta}\mathbf{y}} \right], \quad (28)$$

where the minimization is over all constraints in the downhill direction, that is for all k such that $\mathbf{b}_k^\mathsf{T}(\mathbf{y})\boldsymbol{\Delta}\mathbf{y} < 0$. In a similar fashion we must modify the length of the step in the multipliers so that $\boldsymbol{\lambda} \geq \mathbf{0}$. Again, making a linear estimate for the boundary at $\overline{\boldsymbol{\lambda}} = \mathbf{0}$ we find

$$\sigma_\lambda = \min_k \left[\frac{-\lambda_k}{\Delta\lambda_k} \right], \tag{29}$$

for all k such that $\Delta\lambda_k < 0$. Now, a full length step as defined by the scalars σ_y and σ_λ corresponds to a move to the boundary of the feasible region. In order to maintain strict feasibility we must take some fraction of this step. Following the approach in [8] we define a fraction of the full step according to:

$$\varphi = 1 - \min(.01, 100\mu^2). \tag{30}$$

Thus, we have introduced a "pad" that ensures all subproblem iterates remain strictly feasible. Note that because the format of the problem has only *linear* inequalities this computation is particularly simple. In contrast, a formulation that permits nonlinear inequality constraints to appear directly in the barrier function would require a more complex line search procedure to guarantee the nonlinear constraints remain feasible.

In keeping with most primal-dual methods [10], we attempt to use different step lengths in the primal variables $\mathbf{y}$ and the dual variables $\boldsymbol{\eta}$ and $\boldsymbol{\lambda}$. Specifically, consider a correction of the form:

$$\begin{pmatrix} \overline{\mathbf{y}} \\ \overline{\boldsymbol{\eta}} \\ \overline{\boldsymbol{\lambda}} \end{pmatrix} = \begin{pmatrix} \mathbf{y} \\ \boldsymbol{\eta} \\ \boldsymbol{\lambda} \end{pmatrix} + \alpha \begin{pmatrix} \boldsymbol{\Delta}\mathbf{y} \\ \gamma\boldsymbol{\Delta\eta} \\ \gamma\boldsymbol{\Delta\lambda} \end{pmatrix}. \tag{31}$$

Different primal and dual step scaling is defined by setting $\alpha = \hat{\alpha}$ where

$$\hat{\alpha} = \min(1, \varphi\sigma_y), \tag{32}$$

$$\gamma = \frac{1}{\hat{\alpha}} \min(1, \varphi\sigma_\lambda). \tag{33}$$

The primal and dual steps have the same scaling when

$$\alpha = \tilde{\alpha} = \min(1, \varphi\sigma_y, \varphi\sigma_\lambda), \tag{34}$$

with $\gamma = 1$, and in this case the step is just a multiple of the Newton step. Clearly it is desirable to take a Newton step when converging to a solution. Furthermore we know that a scalar multiple of the Newton step must provide local improvement on the KKT conditions. On the other hand computational experience suggests using different primal-dual scaling may improve efficiency. Consequently we first try to use the scaled primal-dual step. If an acceptable point is obtained with $\alpha^{(0)} = \hat{\alpha}$ and $\gamma \neq 1$ we use the scaled step. If the point is not accepted, we force the same scaling by setting $\alpha^{(0)} = \tilde{\alpha}$ with $\gamma = 1$ and then choose subsequent steps so that $\alpha^{(k)} \leq \hat{\alpha}$ based on the globalization strategy.

3.4 Filter Globalization

The role of a globalization strategy within the context of a line-search method is essentially to decide whether to accept or reject a particular step. Presumably, the point $(\overline{\mathbf{y}}, \overline{\boldsymbol{\eta}}, \overline{\boldsymbol{\lambda}})$ defined by (31) should be accepted if something "good" happens, and otherwise rejected. Many approaches have been proposed for the step acceptance criteria. One possible technique is to monitor the error in the KKT conditions, i.e. the right hand side of (25)

$$\|F_\mu\| \equiv \left\| \begin{matrix} \mathbf{g} - \mathbf{C}^\mathsf{T}\eta - \mathbf{B}^\mathsf{T}\boldsymbol{\lambda} \\ \mathbf{c} \\ \mathbf{D}_b(\boldsymbol{\lambda} - \boldsymbol{\pi}_\mu) \end{matrix} \right\|_\infty . \tag{35}$$

Unfortunately, it is well known that reducing $\|F_\mu\|$ does not necessarily correspond to finding a minimizer of $\beta(\mathbf{y}, \mu)$. Recently, Fletcher and Leyffer [7] have proposed the use of a "filter" as an alternative to constructing a merit function within the context of a trust-region SQP method. We apply the nonlinear filter as a globalization strategy for the equality constrained barrier subproblem. In particular, for *fixed* μ we would like to minimize *both* the objective and constraint violations. Denote the values of the objective and constraint violation at $(\mathbf{y}^{(k)}, \boldsymbol{\lambda}^{(k)})$ by:

$$\beta^{(k)} = f(\mathbf{y}^{(k)}) - \mu \sum_{k=1}^{m_B} \ln b_k(\mathbf{y}^{(k)}) \tag{36}$$

$$v^{(k)} = \left\| \begin{matrix} \mathbf{c}(\mathbf{y}^{(k)}) \\ \mathbf{D}_b(\boldsymbol{\lambda}^{(k)} - \boldsymbol{\pi}_\mu) \end{matrix} \right\|_\infty . \tag{37}$$

When comparing the information at two different points $(\mathbf{y}^{(k)}, \boldsymbol{\lambda}^{(k)})$ and $(\mathbf{y}^{(j)}, \boldsymbol{\lambda}^{(j)})$, a pair $\{\beta^{(k)}, v^{(k)}\}$ is said to *dominate* another pair $\{\beta^{(j)}, v^{(j)}\}$ if and only if both $\beta^{(k)} \le \beta^{(j)}$ and $v^{(k)} \le v^{(j)}$. Using this definition we can then define a *filter* as a list of pairs such that no pair dominates any other. A new point $\{\beta^{(\ell)}, v^{(\ell)}\}$ is said to be acceptable for inclusion in the filter if it is not dominated by any point in the filter. Conversely, if a new point $\{\beta^{(\ell)}, v^{(\ell)}\}$ is dominated by any point in the filter it is not acceptable. Thus, if a trial point produces an improvement in *either* the objective *or* the constraint violation over previous iterates it is accepted. In keeping with [7], we also include two special entries in the filter corresponding to the "northwest" and the "southeast" corners. For the northwest corner, the filter includes $\{\beta_{NW}, 0\}$ where β_{NW} is a liberal estimate for the upper bound on the objective function. At the southeast corner, we include an entry $\{-\infty, c_{max}\}$ where c_{max} is an upper bound on the absolute constraint violation. An estimate for c_{max} must be specified by the user, and may be reduced if necessary to insure convergence of the algorithm. It is important to note that the filter globalization strategy is used within a *single* barrier subproblem. When the barrier parameter is changed, we restart the filter. Consequently, we are only comparing iterates with the same value for the barrier parameter.

We would like to choose the step-length $\alpha^{(k)}$ such that the new point given by (31) is "acceptable" to the filter. If the point is accepted, the information is used to augment the filter for subsequent iterations. If the point is not accepted, we utilize a special line search strategy that forces step contraction, i.e. $\alpha^{(k)} < \alpha^{(k-1)}$ in order to find a point that is acceptable for inclusion in the filter. This step contraction is only utilized when the primal-dual step scaling is equal, i.e. $\gamma = 1$ in Eq. (31). In contrast to a traditional line search, a filter has two different quantities that determine acceptance. Consequently, we compute two different estimates for the step length—one chosen to reduce the objective, and one chosen to reduce the constraint error. Details of the line search can be found in [3]. It should be noted that the approximate complementarity equations (21) appear in the filter. The solution of (25) is guaranteed to have the properties of a solution to the barrier KKT system only if the approximate complementarity equations are satisfied. Thus, filter convergence theory can only be relied on when (21) is satisfied.

3.5 Barrier Parameter Update Strategy

An important practical matter when implementing a barrier method is the choice for the the barrier parameter μ. Clearly, μ must converge to zero, in order that $\mathbf{y}_\mu \to \mathbf{y}^*$. Typical early implementations of interior-point methods, as described in Fiacco and McCormick [6], calculate the solution to the barrier subproblem very accurately. Recent computational experience suggests not only that this is unnecessary, but it is computationally expensive. Instead, the preferred technique is to simply get "close" to the central path and then reduce the barrier parameter. Obviously, some quantitative definition of "close" is required, and we have adopted the approach in [8]. Specifically, we will consider a point "close" to the central path if

$$\|F_\mu\| < \min\left[\kappa\mu, \epsilon_c\right] \tag{38}$$

where (35) defines the error in the KKT conditions, and ϵ_c is a user specified central path tolerance. We will use this relation to decide when the barrier parameter should be reduced.

Let us now describe the procedure for updating the barrier parameter estimate. The following test is applied after completing every step and is essentially a modified version of the procedure described in [8] for computing the new barrier parameter $\widehat{\mu}$:

if $(\alpha \geq .1)$ **and** $\|F_\mu\| < \min\left[\kappa\mu, \epsilon_c\right]$ **then**

$$\widehat{\mu} = \begin{cases} 10\mu^2 & \text{if } \mu < 10^{-4}, \\ \mu/10 & \text{if } \mu \geq 10^{-4}, \end{cases} \tag{39}$$

elseif μ unchanged for N_u iterations

$$\widehat{\mu} = .9\mu. \tag{40}$$

Typically, we use $N_u = 10$ and $\epsilon_c = \kappa = 10$. The philosophy of the procedure is to aggressively reduce the barrier parameter when it appears promising to do so, based on the observed behavior of the KKT error $\|F_\mu\|$. Conversely, if it appears that progress is slow and the barrier parameter is unchanged after N_u iterations, a modest reduction is made.

3.6 Initialization

A fundamental property of the barrier algorithm is that the sequence of iterates remains strictly feasible with respect to the bounds (4) and inequalities $\boldsymbol{\lambda} \geq \mathbf{0}$, as required by (11). In this section, we describe how to initialize the interior-point method. The primal variables $\mathbf{y}$ are simply reset to lie strictly within their bounds as a part of the transformation to internal format (2). However, in general, the user cannot provide good values for the variables $(\boldsymbol{\eta}, \boldsymbol{\lambda})$ and the related barrier parameter μ. There are a number of possibilities and we have implemented three options.

Option 1. For a user-specified μ, compute the central path estimate $\boldsymbol{\lambda} = \boldsymbol{\pi}_\mu$, and compute estimates $\widetilde{\boldsymbol{\eta}}$ to minimize the error in the KKT conditions (10).

Option 2. Use the central path estimate for $\boldsymbol{\lambda}$, and compute estimates $\tilde{\mu}$ and $\widetilde{\boldsymbol{\eta}}$ to minimize the error in the KKT conditions (10).

Option 3. Compute estimates $\tilde{\mu}$, $\widetilde{\boldsymbol{\eta}}$ and $\widetilde{\boldsymbol{\lambda}}$ to minimize the error in the KKT conditions (10).

In the second and third option, if the computed solution $\tilde{\mu} < \mu_L$ where $\mu_L = \max(\mu_\ell, \|c\|_\infty/\bar{\kappa})$ for a user-specified minimum value μ_ℓ, then we set $\mu = \mu_L$ and recompute using Option 1. Otherwise, we use $\tilde{\mu}$ and simply truncate the multipliers $\lambda_k = \max(\epsilon_m, \widetilde{\boldsymbol{\lambda}}_k)$ where ϵ_m is machine precision, to insure they are strictly positive. All three initialization options require the solution of a sparse linear least squares problem. The multifrontal algorithm [1] used for solving the KKT system can also be used to construct the minimum norm solution of the linear least squares.

4 Outline of the Primary Algorithm

The preceding sections described the main features of the interior point algorithm. We are now in a position to outline the main algorithmic operations. The primary algorithm proceeds as follows:

1. Initialization: Compute external to internal transformation; (1)-(4), define sparsity of $\mathbf{B}$, $\mathbf{C}$, $\mathbf{W}$, permutations, etc. Evaluate function and gradient information at the initial point.

2. Multiplier and Barrier Parameter Initialization: Compute initial values for $\boldsymbol{\eta}$, $\boldsymbol{\lambda}$, and μ using *Option 1*, *Option 2*, or *Option 3*.
3. Gradient Evaluation and Convergence Tests: Evaluate gradients, then check error in KKT conditions (35). Terminate if $\|F_\mu\| \leq \epsilon$ and $\mu \leq \epsilon$ for a tolerance ϵ (go to step 9).
4. Step Calculations:
   (a) Compute Log-Barrier Function. Initialize Filter if necessary.
   (b) Iteration Print. Hessian Evaluation.
   (c) Levenberg modification.
   (d) Barrier parameter update using (39) or (40). Restart filter when μ changes.
   (e) Multiplier reset. If $\|\boldsymbol{\eta}\|_\infty$ is too large, then recompute multipliers using *Option 2*, and if this fails using *Option 1*. If they are still too large, terminate (go to step 9).
5. Search Direction: Solve the KKT system (26) and modify the Hessian (9) by adjusting the Levenberg parameter if necessary.
6. Step Calculations: Compute the step scaling (28), (29), and (30)
7. Line Search:
   (a) compute predicted point (31), then evaluate functions and Log-barrier function
   (b) check point against filter—if acceptable then update filter and go on, otherwise reduce α and repeat.
8. Update Information: Set $\mathbf{y} \leftarrow \overline{\mathbf{y}}$, $\boldsymbol{\eta} \leftarrow \overline{\boldsymbol{\eta}}$, and $\boldsymbol{\lambda} \leftarrow \overline{\boldsymbol{\lambda}}$, etc. Return to step 3.
9. Barrier Algorithm Termination.

5 High Index Partial Differential-Algebraic Equation

An example (proposed by Steve Campbell) illustrates a system of differential-algebraic equations (DAE) of very high index. Heat transfer can be described by the partial differential equation

$$\frac{\partial y}{\partial t} = \frac{\partial^2 y}{\partial x^2} \tag{41}$$

where the spatial domain is $0 \leq x \leq \pi$ and the time domain is $0 \leq t \leq 5$. Conditions are imposed on three boundaries of this domain.

$$y(x,0) = y_0(x) = 0 \tag{42}$$
$$y(0,t) = u_0(t) \tag{43}$$
$$y(\pi,t) = u_\pi(t). \tag{44}$$

The input temperatures $u_0(t)$ and $u_\pi(t)$ are viewed as (algebraic) control variables. In addition the temperature over the domain is bounded below according to

$$y(x,t) \geq g(x,t) = c\left[\sin x \sin\left(\frac{\pi t}{5}\right) - a\right] - b \tag{45}$$

where $g(x,t)$ is a prescribed function with $a = .5$, $b = .2$, and $c = 1$. Finally, we would like to choose the controls $u_0(t)$ and $u_\pi(t)$ to minimize

$$\phi = \int_0^5 \int_0^\pi y^2(x,t)dxdt + \int_0^5 \left[q_1 u_0^2(t) + q_2 u_\pi^2(t)\right] dt. \tag{46}$$

For our example we set the constants $q_1 = q_2 = 10^{-3}$.

One way to solve this problem is to introduce a discretization in the spatial direction, i.e. $x_k = k\delta = k\frac{\pi}{N}$ for $k = 0, \ldots, N$. If we denote $y(x_k,t) = y_k(t)$, then we can approximate the partial differential equation (41) by the following system of ordinary differential equations:

$$\dot{y}_1 = \frac{1}{\delta^2}(y_2 - 2y_1 + u_0) \tag{47}$$

$$\dot{y}_k = \frac{1}{\delta^2}(y_{k+1} - 2y_k + y_{k-1}) \qquad k = 2, \ldots, N-2 \tag{48}$$

$$\dot{y}_{N-1} = \frac{1}{\delta^2}(u_\pi - 2y_{N-1} + y_{N-2}). \tag{49}$$

This (method of lines) approximation is obtained by using a central difference approximation to $\frac{\partial^2 y}{\partial x^2}$ and incorporating the boundary conditions (43) and (44) to replace $y_0 = u_0$ and $y_N = u_\pi$. As a consequence of the discretization the single constraint (45) is replaced by the set of constraints

$$y_k \geq g(x_k, t) \qquad k = 0, \ldots, N. \tag{50}$$

The boundary conditions (42), (43), and (44) are imposed by setting $u_0(0) = u_\pi(0) = y_k(0) = 0$. Furthermore if we use a trapezoidal approximation to the integral in the x-direction the objective function (46) becomes

$$\phi = \int_0^5 \left[\frac{1}{2}\delta + q_1\right] u_0^2(t)dt + \delta \sum_{k=1}^{N-1} \int_0^5 y_k^2(t)dt + \int_0^5 \left[\frac{1}{2}\delta + q_2\right] u_\pi^2(t)dt. \tag{51}$$

This problem becomes particularly challenging because when one of the inequality constraints (50) is active, the differential-algebraic equation system has a high index. For our example we set $N = 20$ leading to an index ten DAE. While many interesting issues are introduced by "high index" DAEs when this application is discretized the resulting NLP is impacted in three significant ways. First, the number of NLP variables will become large, because a very large number of grid points must be introduced in order to accurately approximate the continuous problem. Second, a large number of inequality constraints are introduced, making determination of the active constraints very challenging. Finally, the "high" DAE index implies some type of degeneracy in the problem specification. The solution is illustrated in the figure.

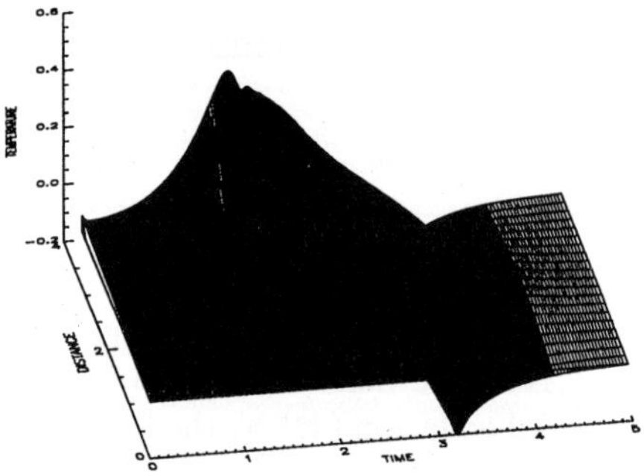

(a) $y(x,t)$

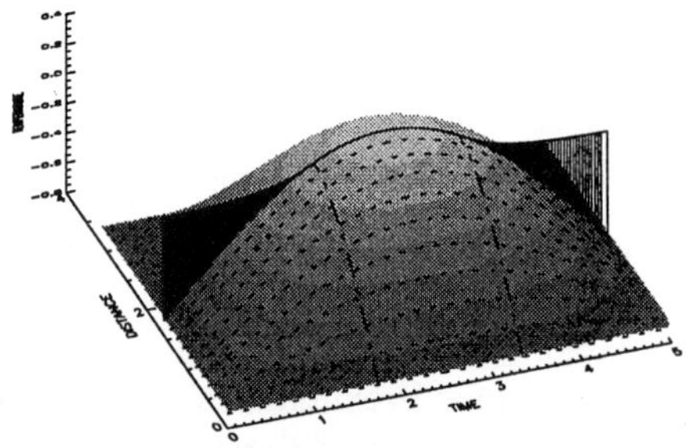

(b) Inequality $y(x_{10},t) \geq g(x_{10},t)$

Barrier Algorithm			SQP Algorithm
CPU Time (sec)	2029.98	598.94	CPU Time (sec)
Linear Alg. Time (sec)	1338.62	498.00	QP Subproblem Time (sec)
Number of KKT Sol.	161	4364	QP Iterations

A comparison of the barrier and SQP algorithms in the optimal control setting is difficult primarily because the NLP is "inside" a more complex numerical process. Nevertheless, it is worthwhile to compare the behavior when the algorithms are used in a "functionally equivalent" manner. In this context "functionally equivalent" implies we are using the same mesh refinement procedure as implemented in SOCS [2] to compute the optimal control solution to a specified relative accuracy of 10^{-7}. In one case we will solve the NLP using the barrier algorithm, and in the other the SQP method. A summary of key performance parameters on this problem is presented in the table. The left side of the table describes the barrier algorithm. The overall optimal control solution was obtained in 2030 seconds, and required 161 solutions of the KKT system (26). The total time required for all of these solutions is 1339 seconds. In comparison, results on the right side of the table for the SQP method required 599 seconds. However, the QP algorithm required 4364 changes to the active set during the course of the solution. Each solution with a new active set requires solution of the KKT system or modification of a previous solution. In contrast the barrier solution required only 161 solutions of the KKT system. In short, the barrier algorithm required far fewer solutions of the KKT system than the active set SQP method. This property is critical when considering large applications that require iterative linear algebra and cannot exploit the efficiencies of a direct matrix factorization. Thus a comparison of CPU times on this example is deceptive, because we do not expect a direct matrix factorization will be possible for very large problems.

Although the "bottom line" in this comparison is very encouraging, there are a number of factors which make a "clean" comparison difficult. When the barrier algorithm was used 11 mesh refinement iterations were performed and the largest problem had 1292 grid points. In contrast, when the SQP algorithm was employed, 11 mesh refinement iterations were executed with a final grid size of 1296 points. Presumably this behavior can be attributed to minor discretization accuracy differences at intermediate NLP solutions. A second factor also obscures the comparison of the SQP and barrier algorithm when used in this mesh refinement context. During mesh refinement the initial guess for the NLP variables are constructed by interpolating the previous (coarse) grid. Thus as the mesh refinement proceeds the initial guess becomes better and better and typically only one QP subproblem is required to compute a solution when the SQP method is employed. In contrast, the barrier method must begin at a feasible point. Thus the "good" guess supplied by interpolation of the coarse grid is essentially ignored. The net affect is that the barrier algorithm required approximately the same number of iterations for coarse and fine grids! In essence, the barrier method cannot exploit a good guess. Finally, it is worth noting that the size of the linear systems that are solved by the respective algorithms can be substantially different. In particular for the SQP method, the KKT system is at most order $2n$ where n is the number of variables. In contrast, for the barrier algorithm the order of the KKT system is $n + m_E + m_B$ which may be considerably larger than the SQP KKT system. Nevertheless, no obvious ill-conditioning was observed in the numerical tests. Similar experience (not reported here) has been observed when the algorithm has been tested on a suite of large sparse nonlinear programs derived from discretization of differential equations.

6 Summary and Conclusions

This paper describes a primal-dual interior point algorithm for solving large sparse nonlinear programming problems. The search direction in the primal and dual variables is computed by direct solution of the KKT system, which is made symmetric by using a technique for dynamically scaling the rows and columns to achieve good numerical conditioning. Numerical experience seems to suggest that the linear systems are not ill-conditioned as the barrier parameter is reduced. A globalization technique based on a nonlinear filter is utilized in lieu of a merit function, with very positive computational experience.

References

1. C. Ashcraft, R. G. Grimes, and J. G. Lewis, *Accurate Symmetric Indefinite Linear Equation Solvers*, SIAM Journal of Matrix Analysis, 20 (1999), pp. 513–561.
2. J. T. Betts, *Practical Methods for Optimal Control using Nonlinear Programming*, Society for Industrial and Applied Mathematics, Philadelphia, PA., 2001.
3. J. T. Betts, S. K. Eldersveld, P. D. Frank, and J. G. Lewis, *An Interior-Point Nonlinear Programming Algorithm for Very Large Scale Optimization*, Mathematics and Computing Technology Report M&CT-TECH-003, Mathematics and Computing Technology, The Boeing Company, PO Box 3707, Seattle, WA 98124-2207, Apr. 2000.
4. J. T. Betts and P. D. Frank, *A Sparse Nonlinear Optimization Algorithm*, Journal of Optimization Theory and Applications, 82 (1994), pp. 519–541.
5. R. H. Byrd, M. E. Hribar, and J. Nocedal, *An Interior Point Algorithm for Large Scale Nonlinear Programming*, SIAM Journal on Optimization, 9 (2000), pp. 877–900.
6. A. V. Fiacco and G. P. McCormick, *Nonlinear Programming: Sequential Unconstrained Minimization Techniques*, John Wiley and Sons,, New York, New York, 1968.
7. R. Fletcher and S. Leyffer, *Nonlinear Programming Without a Penalty Function*, University of Dundee Numerical Analysis Report NA/171, University of Dundee, Department of Mathematics, Dundee, DD1 4HN, Scotland, U.K., 1997.
8. D. M. Gay, M. L. Overton, and M. H. Wright, *A Primal-Dual Interior Method for Nonconvex Nonlinear Programming*, in Advances in Nonlinear Programming, Y. Yuan, ed., Kluwer Academic Publishers, Dordrecht, The Netherlands, 1998, pp. 31–56.
9. R. J. Vanderbei and D. F. Shanno, *An Interior-Point Method for Nonconvex Nonlinear Programming*, Computational Optimization and Applications, 13 (1999), pp. 231–252.
10. S. J. Wright, *Primal-Dual Interior-Point Methods*, Society for Industrial and Applied Mathematics, Philadelphia, 1997.

SQP SAND Strategies that Link to Existing Modeling Systems

Lorenz T. Biegler and Andreas Wächter

Chemical Engineering Department, Carnegie Mellon University, Pittsburgh, PA 15213

Abstract. We consider variants of Successive Quadratic Programming (SQP) applied to Simultaneous Analysis and Design (SAND) implementations for existing engineering modeling systems. Such programs include large scale PDE solvers, chemical process simulators and programs for electrical circuit analysis. Several aspects of the SQP algorithm will be considered. In particular, options need to be selected that are compatible with the information available from the modeling system. Here the following three areas are considered for SAND optimization algorithms: efficient and reliable strategies for global convergence based on a newly developed filter line search method, improved second order information for reduced Hessian methods, and a barrier approach for inequality constraints. The benefits of each of these options will be supported by numerical results and the relevance of these options to PDE models will be explored.

1 Introduction

Engineering modeling systems form the cornerstone of analysis and design in a broad range of disciplines. In computational mechanics and the analysis of transport systems, a wide variety of PDE modeling systems and products are available. In addition to solving discretized partial differential equations, most of these also have capabilities for mesh generation, many options for construction of finite element bases and a wide variety of linear iterative solvers and preconditioners.

These engineering systems represent dozens of man-years of software development. However, virtually all of them were developed for analysis and not for design optimization. As a result, there remain some interesting challenges in leveraging this software investment in order to adapt these systems for optimization. In principle, such systems require smooth functions and first (and possibly second) derivatives to be accessed from the modeling equations by the nonlinear programming (NLP) algorithm. Moreover, in the solution of nonlinear systems, many modeling systems apply a Newton-based approach to solve the finite element equations. In some implementations, an exact Jacobian can be constructed for the Newton step. More often, however, the Jacobian matrix is simplified so that many nonlinear terms are neglected. Since modeling systems aim to solve the nonlinear equations efficiently with suitable preconditioners for the iterative solution of linear systems that comprise Newton steps, an inexact Jacobian is often satisfactory for engineering analysis. However, the absence of accurate Jacobian information presents a major hurdle for the implementation of efficient gradient-based NLP algorithms. Finally, even if accurate Jacobian information is computed, the Jacobian matrix is rarely constructed explicitly. Hence, it may be difficult to make its elements available to the NLP algorithm.

This paper explores the conceptual requirements of interfacing existing PDE modeling systems with reduced Hessian SQP (rSQP) NLP solvers. We consider some recent advances in rSQP algorithms and discuss the suitability of these advances for these implementations. In particular, we consider methods for improving global convergence, providing second order information and handling inequalities. Many of the options in these areas require a strong interaction between the modeling system and the NLP algorithm.

To motivate this discussion, we consider four levels of implementation to NLP algorithms. At the most basic level, we consider a *black-box* or Nested Analysis and Design (NAND) implementation. This interface requires very little interaction of the NLP solver with the PDE model. Gradients for the NLP solver are obtained by finite difference, the NLP is fairly small for design optimization and the modeling system, which is solved at every NLP iteration, requires little modification for the interface. On the other hand, the black-box implementation can suffer from repeated and time-consuming solution of the PDE model and the possibility of intermediate convergence failures of the model.

At the other extreme we consider an *open implementation* that allows complete access to the discretization of the PDE model and its first and second derivatives. This approach leads to full flexibility for optimization along with a SAND strategy. With this implementation one can apply fast NLP algorithms, but to a much larger problem that includes all of the discretized model and decision variables. In the research and development of NLP algorithms, full information from an open implementation is always assumed to be available. In practice, however, a fully open implementation would require many engineering modeling systems to be rewritten, with little reuse of their existing components.

In this study, we consider two intermediate approaches with implementations that are *tailored* to the engineering modeling system. These two approaches are distinguished by whether linear systems involving the transpose of the accurate Jacobian can be solved. Here we assume that a reduced Hessian SQP approach is used and, for simplicity, we emphasize a line search option for global convergence. Nevertheless, trust region approaches can also be considered and are discussed in other papers at this workshop. Both tailored approaches considered here are true SAND optimization strategies and are labeled as follows:

- In the *direct tailored approach* reduced gradients and search directions for rSQP are calculated directly using Newton steps from the PDE modeling system and the "sensitivity" of Newton steps to the decision variables.

- In the *adjoint tailored approach* reduced gradients and search directions for rSQP are calculated using *both* the model Jacobian and its transpose. One advantage of this approach is the availability of multiplier estimates.

This paper outlines improvements that can be made to the rSQP algorithm for more efficient and reliable performance. In the next section we sketch the rSQP approach for the two intermediate levels of implementation. Section 3 deals with a better globalization approach through a line search implementation of the filter method, while Section 4 discusses the incorporation of better second-order information for these two intermediate strategies. Section 5 deals with the treatment of inequality constraints and the introduction of a barrier method that also takes advantage of the advances in Sections 3 and 4. Section 6 provides a numerical case study to illustrate the benefits of these advances. Finally, Section 7 closes with some key points and summarizes the paper.

2 Tailored SAND Optimization

We consider optimization problems that take the form:

$$\min_{x \in \mathbf{R}^n} \; f(x) \tag{1}$$

$$\text{s.t.} \;\; c(x) = 0, \;\; x \in [x^L, x^U], \tag{2}$$

where $f : \mathbf{R}^n \to \mathbf{R}$ and $c : \mathbf{R}^n \to \mathbf{R}^m$ are generally assumed to be smooth functions. Here we also assume $n, m \gg n - m$. While the vector $c(x)$ is generally straightforward to construct for PDE models, in many modeling systems these equations are not constructed explicitly and their functional form is generally not available through an NLP interface. Similarly, gradients of f and c may not be available to the interface, although we may assume that exact derivatives are generated and used internally in the modeling system. Also, for this study we assume that second derivatives are not available from the modeling system.

For PDE optimization, the NLP method of choice is usually some variant of Successive Quadratic Programming (SQP), but this method is applied differently for each implementation. At an iterate x_k, SQP generates a search direction d_k by solving

$$\min_{d \in \mathbf{R}^n} \; g(x_k)^T d + \frac{1}{2} d^T W(x_k) d \tag{3}$$

$$\text{s.t.} \;\; c(x_k) + A(x_k)^T d = 0, \;\; x_k + d \in [x^L, x^U] \tag{4}$$

where g denotes the gradient of f, W denotes the Hessian of the Lagrangian function $L(x, \lambda) = f(x) + \lambda^T c(x)$ and A denotes the $n \times m$ matrix of constraint gradients

$$A(x) = [\nabla c_1(x), ..., \nabla c_m(x)].$$

A new iterate is then computed as

$$x_{k+1} = x_k + \alpha_k d_k,$$

where α_k is a step length parameter chosen in a line search procedure that tries to ensure global convergence. For example, this can be done by enforcing sufficient reduction in a merit function, e.g., an exact penalty function

$$\phi(x) = f(x) + \gamma \| c(x) \|, \tag{5}$$

where γ is a penalty parameter.

For fully open implementations, a large NLP is considered directly by the SQP algorithm. All equations and variables are accessible and gradient calculations are straightforward. Nevertheless, care must be taken to solve these problems with a large-scale implementation of SQP. For this purpose, both sparse full-space methods and reduced Hessian variations of SQP have been proposed and demonstrated for PDE applications (as seen in several papers in this volume).

For black box implementations, SQP is applied only to a subset of the problem variables, as the model equations and variables, a subset of x, are eliminated implicitly. As a result, the size of the optimization problem remains small (typically less than 100 variables) and no large-scale extensions are required for SQP. On the other hand, because the model is solved repeatedly and finite difference derivatives are required, black box optimization is usually more costly than optimization based an open implementation.

With *tailored* implementations we combine the best features of both open and black box approaches. Here we assume that the model variables and equations are solved with Newton's method within the PDE solve and at each SQP iteration we require only one iteration of Newton's method for these equations. Hence the model equations are not solved repeatedly as in the black box approach. To motivate this implementation, we consider a partition of all the variables x into m state variables y and $n-m$ decision variables u with $A(x)^T = [\nabla_y c(x)^T \;\; \nabla_u c(x)^T] = [C \;\; N]$. We assume that C is nonsingular. We further define $Z(x) \in \mathbf{R}^{n \times (n-m)}$, a basis for the tangent space of the equality constraints, by:

$$Z = \begin{bmatrix} -C^{-1}N \\ I \end{bmatrix}. \tag{6}$$

We denote the reduced gradient as $Z(x)^T g(x)$ and note that $Z(x)^T A(x) = 0$.

We first consider the *direct tailored* approach where we assume that only operations with C and C^{-1} can be performed. This precludes calculation of the multipliers and reduced gradient by:

$$\begin{aligned} \lambda &= C^{-T} \nabla_y f \\ Z^T g &= -N^T \lambda + \nabla_u f. \end{aligned} \tag{7}$$

Instead we need to calculate the quantity $C^{-1}N$ with direct or iterative solvers. In fact, even if the exact C is not calculated, a preconditioned Krylov (PK) method with differenced matrix vector products can be applied to solve systems of the form $Cz = b$. The reduced gradient $Z^T g$ is then obtained using $C^{-1}N$ and g directly. While this direct approach is more expensive than in (7), it is easy to implement for Newton-based models and is also easily parallelized.

A key point in the direct method is that the multiplier estimates λ need not be calculated explicitly. This property is useful for the tailored approach, where direct access to the constraint normals (i.e., the Jacobians of the PDE models) may be difficult. On the other hand, for the adjoint tailored approach, transposed systems are solved and the evaluation of $C^{-T}\nabla_y f$ provides multiplier information for better performance.

A sketch of both tailored approaches follows. By neglecting the variable bounds in (1)–(2), the solution of the quadratic program (3)–(4) can be written in a simple reduced form. We consider a nonsingular matrix of dimension n, which we write as $[Y_k\ Z_k]$ where $Y_k \in \mathbf{R}^{n\times m}$ and $Z_k \in \mathbf{R}^{n\times(n-m)}$. (From now on we denote $A(x_k)$ as A_k, $g(x_k)$ as g_k, etc.) We can now express d_k, the solution to (3)–(4), as

$$d_k = Y_k p_Y + Z_k p_Z, \tag{8}$$

for some vectors $p_Y \in \mathbf{R}^m$ and $p_Z \in \mathbf{R}^{n-m}$. Because $A_k^T Z_k = 0$, the linear equality constraints in (4) become:

$$c_k + A_k^T Y_k p_Y = 0. \tag{9}$$

Since $A_k^T Y_k$ is nonsingular, p_Y is determined by (9):

$$p_Y = -(A_k^T Y_k)^{-1} c_k. \tag{10}$$

Two popular choices for Y_k are the coordinate decomposition, $Y_k^T = [I\ \ 0]$ and an orthogonal decomposition, $Y_k^T = [I\ \ C_k^{-1}N_k]$. For the development in this paper we will use the simpler coordinate decomposition, although we note that an orthogonal decomposition is preferred if $n - m$ is small. In particular, the orthogonal decomposition leads to the smallest step for $Y_k p_Y$, is insensitive to the partitioning of A_k to form C_k and N_k (as long as C_k is nonsingular), often performs better on highly nonlinear problems and its computational cost is $O(m(n-m)^2)$).

The quadratic programming (QP) problem is now solved exclusively in terms of the variables p_Z. Substituting (8) into (3) with $Y_k p_Y$ determined, we obtain the unconstrained QP:

$$\min_{p_Z \in \mathbf{R}^{n-m}} (Z_k^T g_k + \zeta_k Z_k^T W_k Y_k p_Y)^T p_Z + \frac{1}{2} p_Z^T (Z_k^T W_k Z_k) p_Z. \tag{11}$$

Here ζ_k is a damping parameter that attempts to preserve a descent property for p_Z in early iterations. In [3] it was shown that ζ_k can always be set to one in the neighborhood of the solution. If $Z_k^T W_k Z_k$ or its quasi-Newton approximation is positive definite, then the solution of (11) is:

$$p_Z = -(Z_k^T W_k Z_k)^{-1}[Z_k^T g_k + \zeta_k Z_k^T W_k Y_k p_Y]. \tag{12}$$

In our previous studies [3, 4], the reduced Hessian, $Z_k^T W_k Z_k$, was approximated by the BFGS update, say as B_k, and the cross term $[Z_k^T W_k Y_k]p_Y$ was approximated by a vector w_k given by:

$$w_k = Z(x_k + Y_k p_Y)^T g(x_k + Y_k p_Y) - Z_k^T g_k$$

for the direct tailored approach, or by:

$$w_k = Z_k^T(\nabla L(x_k + Y_k p_Y, \lambda_k) - g_k)$$

by the adjoint tailored approach. With these changes, the QP step is given by:

$$p_Z = -B_k^{-1}[Z_k^T g_k + \zeta_k w_k]. \tag{13}$$

Also, for the exact penalty line search, we choose γ so that for some $\rho > 0$:

$$\begin{aligned}\gamma\|c(x)\| &\geq |\lambda(x)^T c(x)| + 2\rho\|c(x)\| \\ &= |g(x)^T Y(x) p_Y| + 2\rho\|c(x)\|.\end{aligned}$$

This multiplier-free approach is required for the direct tailored strategy. Alternately, for the adjoint tailored approach, well-known updating rules that satisfy the relation $\gamma_k > \|\lambda_k\|_D$ can be used, where $\|\cdot\|_D$ is the dual norm of $\|\cdot\|$.

Finally, to handle the bound constrained problem (1)–(2) we consider the decomposition in (8) and (9), and compute the null space step p_Z through the solution of a low dimensional quadratic programming problem. Because $n-m$ is small, solution of the QP problem is usually inexpensive and the modification is relatively easy to implement. For this problem we have adapted a version of the Goldfarb and Idnani QP algorithm [12]. A complete description of the QP algorithm and its integration with the reduced Hessian SQP method is given in [16].

To solve for p_Z the reduced dimension QP problem is given by:

$$\begin{aligned}\min_{\xi, p_Z \in \mathbf{R}^{n-m}} \quad & (Z_k^T g_k + \zeta_k w_k)^T p_Z + \frac{1}{2} p_Z^T B_k p_Z + M(\xi + \xi^2/2) \\ \text{s.t.} \quad & x_k + (1-\xi) Y_k p_Y + Z_k p_Z \in [x^L, x^U],\ \xi \geq 0\end{aligned} \tag{14}$$

Here, the scalar variable ξ is added to ensure that the QP always has a solution and that the resulting search direction d_k keeps x_{k+1} between its bounds. The quadratic term in ξ is required because the Goldfarb and Idnani algorithm applies only to problems with strictly positive definite Hessian matrices. Provided that M is sufficiently large, the ξ variable remains zero unless inconsistent constraint linearizations are encountered.

The QP (14) is substituted for the calculation of p_Z in (13) and the search direction

$$d_k = (1-\xi)Y_k p_Y + Z_k p_Z \tag{15}$$

is substituted in (8).

For the direct tailored approach, the solution of the inequality constrained QP also leads to a minor modification in the calculation of the penalty parameter γ_k by including the bound multipliers, ν_k, calculated from the bound constraints in the QP, i.e.,

$$\gamma_k = \begin{cases} \gamma_{k-1} & \text{if } \gamma_{k-1}\|c_k\| \geq |(1-\xi)(g_k+\nu_k)^T Y_k p_Y| + 2\rho\|c_k\| \\ \frac{|(1-\xi)(g_k+\nu_k)^T Y_k p_Y|}{\|c_k\|} & +3\rho \text{ otherwise.} \end{cases}$$

The direct tailored approach has been applied to a number of optimization problems where existing modeling systems are used. In process engineering applications [17], this approach was demonstrated on a hydrodealkylation process with detailed distillation models. The tailored approach was also applied [1, 18] to COLDAE [2], a two-point boundary-value solver. In these studies, Newton's method was used to solve the equations, but the Jacobian matrices were never constructed explicitly and the linear subproblems were solved with specialized block banded solvers. As a result, the elements of the Jacobian matrix were difficult to extract for the optimization algorithm. More recently, direct tailored implementations have been made to MPSalsa, a finite element code at Sandia National Lab. In this case, both FORTRAN and C++ versions of the rSQP algorithm were linked to this system for the design optimization of CVD reactor models [14, 19]. Again, Newton's method was used to solve the finite element equations, and preconditioned Krylov solvers were used for the arising linear systems. This approach was an order of magnitude faster than the black box approach.

Using the rSQP algorithm in our previous work, particularly with the option of multipliers-free quasi-Newton updates, one might expect that multiplier estimates are unnecessary. Moreover, if variable bounds are included in the reduced space quadratic program (14), then the Z_k matrix and $C_k^{-1}N_k$ are required explicitly. As a result, there would seem to be no advantage to constructing the reduced gradient in (7) by solving linear systems that contain C^T. However, in some cases the usage of multipliers in the quasi-Newton approximation leads to more accurate and efficient estimates for the reduced Hessian. Certainly, if linear systems with C^T can be solved, then the availability of multiplier estimates leads to less work to construct reduced gradients via (7), especially with iterative linear solvers. In addition, it allows a wider variety of merit functions (e.g., augmented Lagrangian) to be used and more options for second order information.

3 The Filter Line Search

Having solved the QP (11) or (14) to obtain the search direction d_k, we need to choose a step length $\alpha_k \in (0,1]$ to obtain the next iterate

$$x_{k+1} = x_k + \alpha_k d_k.$$

In our previous work [4] we performed an Armijo line search using the exact penalty function (5). In this study, we also apply a line search strategy based on the filter approach recently proposed by Fletcher and Leyffer [11]. This leads to larger step sizes and more reliable performance. A detailed derivation and convergence analysis of this method can be found in [20].

The filter is based on a bicriterion minimization with $f(x)$ and $\theta(x) = \|c(x)\|$ as competing objectives. Instead of decreasing a linear combination of $\theta(x)$ and $f(x)$, we require sufficient decrease in only *one* of those objectives. Details of the filter line search method are given in [20], but the basic concepts of this method can be stated as follows. At x_k, if a certain descent criterion for $f(x)$ is not satisfied or the feasibility measure $\theta(x_k)$ is above a specified threshold, then the conditions for sufficient decrease are given by:

$$f(x_k + \alpha d_k) \le f(x_k) - \gamma_f[\theta(x_k)] \qquad (16)$$

$$\theta(x_k + \alpha d_k) \le \theta(x_k) - \gamma_\theta[\theta(x_k)], \qquad (17)$$

where γ_f and γ_θ are small positive constants. Since we are no longer concerned with a descent property for a merit function, we set $\zeta = 1$ in (12) or (13).

In the other case, when $\theta(x_k)$ is small or the descent criterion on $f(x)$ is satisfied, we require only an Armijo-type decrease in $f(x)$, i.e.:

$$f(x_k + \alpha d_k) \le f(x_k) + \alpha \eta g_k^T d_k \qquad (18)$$

with some constant $\eta \in (0, \frac{1}{2})$ instead of (16)–(17). Further, if both the descent condition on $f(x)$ and (18) do not hold, then we add $(\theta(x_k), f(x_k))$ to a set of previous $(\theta(x), f(x))$-pairs (called the *filter*) and force future iterates to improve on points in this set (in the sense of (16)–(17)). Future iterates that fall in this category are *acceptable to the filter*. Finally, if no admissible step size α can be found that satisfies the above conditions for sufficient decrease, the method switches to a *feasibility restoration phase* in which $\theta(x)$ is minimized directly in order to find a less infeasible point which is eventually acceptable to the filter.

Since this approach does not require multiplier estimates, it shares the advantage with the multiplier-free merit function option. Consequently, it is suitable for the direct tailored approach, and the numerical results in Section 6.1 indicate efficient performance. On the other hand, the augmented Lagrangian merit function, which requires multiplier estimates, cannot be used with the direct tailored approach. In Section 6 we will consider performance issues with all of these line search strategies.

4 Incorporating Second Order Information

In [4, 7] the reduced Hessian in (12) was approximated by means of a quasi-Newton method, i.e., $B_k \approx Z_k^T W_k Z_k$ using either BFGS and SR1 quasi-Newton updates for B_k. On the other hand, direct solution of (12) with an exact reduced Hessian $[Z_k^T W_k Z_k]$ provides a faster convergence rate and could greatly reduce the number of NLP iterations. This can be done in two ways.

First, the exact reduced Hessian can be employed directly. This can be done by calculating matrix-vector products of the Hessian of the Lagrangian, $[Z_k^T W_k Z_k]e_i$, for all coordinate vectors e_i. For this task, one may use some automatic differentiation package such as ADOL-C [13] applied to the model equations. Second, the linear system (13) can be solved with a preconditioned conjugate gradient (PCG) method. These iterative methods also rely on matrix-vector products (e.g., $[Z_k^T W_k Z_k]q$). In both cases, these products can be substituted by divided differences of first derivatives. For instance the first term in these products can be approximated by:

$$Z_k^T W_k Z_k q \approx \frac{Z_k^T(\nabla L(x_k + tZ_k q) - \nabla L(x_k))}{t},$$

where t is a small scalar. Then second derivatives need not be supplied at all.

For efficient performance of the conjugate gradient method, it is essential to supply an effective preconditioner. Here the preconditioning matrix P_k needs to approximate $[Z_k^T W_k Z_k]^{-1}$. For this task, Morales and Nocedal [15] apply a BFGS update for $P_k \approx [Z_k^T W_k Z_k]^{-1}$ based on matrix-vector products, $[Z_k^T W_k Z_k]w$, supplied from the PCG iterations. At the beginning of a PCG procedure, the current approximation is copied and used as (fixed) preconditioner. In order to prepare the preconditioner for the PCG procedure in the next IP iteration $k+1$, the BFGS update to P_k is applied at *every PCG iteration*, hence collecting as much information as possible.

Since the reduced Hessian can be indefinite, we terminate the PCG procedure prematurely if a direction of negative curvature is encountered. Otherwise, we note that this approach can achieve efficient performance because very good approximations of the reduced Hessians are provided. However, it is clear that a practical implementation of this approach requires multiplier estimates and this is not possible with a direct tailored approach. As a result, the benefits of faster convergence can only be obtained if adjoint systems involving C_k^T can be solved.

For equality constrained problems, we will see that significant performance improvements can be achieved with this preconditioned CG approach over quasi-Newton options. However, if inequality constraints are present, this approach needs to be considered more carefully. Certainly, this approach would be impractical in an active set strategy (used to solve (14)) because many linear systems need to be solved at each iteration as active sets are updated. However, for the conjugate gradient method, there are no cheap active set updates and therefore solving multiple linear systems from scratch would be extremely expensive. As a result, we need to consider a different approach to handle inequalities, and for this we turn to the barrier strategy in the next section.

5 Handling Inequalities

In order to simplify the presentation of our algorithm we consider, without loss of generality, the following NLP with all variables having only lower bounds of zero:

$$\begin{aligned} \min\ & f(x) \\ \text{s.t.}\ & c(x) = 0,\ x \geq 0. \end{aligned} \tag{19}$$

The algorithm uses a barrier approach, where the bound constraints are replaced by a logarithmic barrier term. This term is added to the objective function to give

$$\min\ \varphi_\mu(x) = f(x) - \mu \sum_{i=1}^{n} \ln(x^{(i)}) \tag{20}$$

$$\text{s.t.}\ c(x) = 0 \tag{21}$$

with a barrier parameter $\mu > 0$. Here, $x^{(i)}$ denotes the ith component of the vector x. Since the objective function of this barrier problem becomes arbitrarily large as x approaches the boundary of the nonnegative orthant $\{x \mid x \geq 0\}$, it is clear that a local solution $x_*(\mu)$ of this problem lies in the interior of this set, i.e., $x_*(\mu) > 0$. The degree of influence of the barrier is determined by the size of μ, and under certain conditions [10] $x_*(\mu)$ converges to a local solution x_* of the original problem (19) as $\mu \to 0$. Consequently, a strategy for solving the original NLP is to solve a sequence of barrier problems (20)–(21) for decreasing barrier parameters μ_l, where l is the counter for the sequence of subproblems. Since the exact solution $x_*(\mu_l)$ is not of interest for large μ_l, the corresponding barrier problem is solved only to a relaxed accuracy ϵ_l, and the approximate solution is then used as a starting point for the next barrier problem with $\lim_{l\to\infty} \epsilon_l = 0$.

To solve the barrier problem for a fixed value of μ_l we follow a primal-dual approach (see e.g. [9]), which generates search directions for primal variables $x > 0$ as well as for *dual* variables $v > 0$. These dual variables correspond to the Lagrange multipliers for the bound constraints in (19) as $\mu_l \to 0$.

After defining the dual variables by $v = \mu X^{-1} e$, the optimality conditions of (20)–(21) can be written as:

$$\begin{aligned} \nabla f(x) + A(x)\lambda - v &= 0 \\ XVe - \mu e &= 0 \\ c(x) &= 0, \end{aligned}$$

where the components of λ are the Lagrange multipliers for the equality constraints (21). Throughout this section, e denotes the vector of appropriate dimension of all ones, and a capital letter of a vector name (e.g. X) denotes the diagonal matrix with the vector elements on the diagonal. Obtaining a Newton step for this nonlinear system of equations is equivalent to solving the following quadratic program at (x_k, λ_k, v_k):

$$\begin{aligned} \min_{d \in \mathbf{R}^n} \quad & \nabla \varphi_\mu(x_k)^T d + \frac{1}{2} d^T \left(W_k + \Sigma_k\right) d \\ \text{s.t.} \quad & A_k^T d + c_k = 0 \end{aligned}$$

if the matrix $W_k + \Sigma_k$ is positive definite in the null space of A_k^T. Here $\Sigma_k = X_k^{-1} V_k$ and the solution d_k is the search direction for x. The similarity of this QP to (3)–(4) allows us to employ the decomposition presented in Section 2. As before, we partition the overall primal step into two components, $d_k = Z_k p_Z + Y_k p_Y$. The step for the second component can be obtained using (10), while p_Z can directly be computed as

$$p_Z = -[Z_k^T (W_k + \Sigma_k) Z_k]^{-1} \left(Z_k^T \nabla \varphi_\mu(x_k) + w_k\right) \tag{22}$$

with

$$w_k = Z_k^T (W_k + \Sigma_k) Y_k p_Y .$$

The barrier approach avoids the bound constrained quadratic programming subproblem (14) and leads to a conceptually simpler implementation. Solution of (22) can be done in a number of ways. First, we consider an approach to calculate the Hessian terms exactly as they are stated above. Second, we solve (22) using a PCG method. Here the preconditioner needs to be considered more carefully. For this task we consider two preconditioners. The first is due to Morales and Nocedal [15] and applies a BFGS update for $P_k \approx [Z_k^T (W_k + \Sigma_k) Z_k]^{-1}$ based on matrix-vector products, $[Z_k^T (W_k + \Sigma_k) Z_k] w$, supplied from the PCG iterations. At the beginning of a PCG procedure, the current approximation is copied and used as a (fixed) preconditioner. In order to prepare the preconditioner for the PCG procedure in the next IP iteration $k+1$, the BFGS update to P_k is applied at *every PCG iteration*, hence collecting as much information as possible. The second preconditioner separates the two terms; here the BFGS update is applied only to the first term $B_k \approx [Z_k^T W_k Z_k]$ and the second $Z_k^T \Sigma_k Z_k$ is calculated directly. Once updated, we have $P_k = (B_k + Z_k^T \Sigma_k Z_k)^{-1}$. Again the BFGS update is applied to B_k at *every PCG iteration* but the resulting preconditioner itself is kept constant during the solution of the linear system at x_k.

Finally, we also consider quasi-Newton methods in (22); we replace $Z_k^T W_k Z_k$ by an SR1 or BFGS update matrix B_k and add $Z_k^T \Sigma_k Z_k$. Also, we simplify w_k to $Z_k^T \Sigma_k Y_k p_Y$. Note that if SR1 is used, the eigenvalues of $Z_k^T (W_k + \Sigma_k) Z_k$ are monitored and modified if necessary, to ensure positive definiteness of the reduced Hessian. In the next section we provide a numerical study that demonstrates the effectiveness of each of these approaches.

6 Numerical Case Studies

In this section we compare a number of rSQP algorithms on two sets of test problems. In particular, we are interested in assessing the effect of providing multiplier estimates for the tailored approach. Multiplier estimates are not needed with multiplier-free quasi-Newton updates or for multiplier-free and filter line searches. They are, however, essential for the augmented Lagrangian line search and for the exact and PCG options in solving (22) or (12).

For the first set of test problems, we consider NLPs from the COPS test set [6, 8] where $n - m$ is small to moderate and only equality constraints are present. These problems demonstrate the effects of improved line searches and second-order information. In the second set of test problems we consider a distillation optimization using the barrier approach. These distillation problems share much of the same structure of dynamic PDE optimization problems. Here a number of second order strategies are considered with a reduced space barrier method.

6.1 Equality Constrained Problems

For these equality constrained problems we consider the effect of three different line search approaches (filter, exact penalty ℓ_2 merit function and augmented Lagrangian merit function) and four strategies for second-order information (BFGS and SR1 updates, exact Hessians and PCG solutions). Here we choose four scalable problems from the COPS collection [6, 8]. All of these examples are derived from parameter estimation problems with discretizations of nonlinear differential equations. They range from 503 to 4814 variables with $n - m$ ranging from 3 to 22. Moreover, the original formulations include inequality constraints, which we replace by active equality constraints. The iteration counts for the different line search options and treatments of the reduced Hessian are shown in Tables 1, 2 and 3 for four differently sized formulations of each of the test problems. We used the multiplier-free options for the BFGS and SR1 quasi-Newton updates, as well as for updating the penalty parameter for the exact ℓ_2-penalty function in Table 1. On the other hand, in Table 2 we consider the same line search options but with multipliers used in the quasi-Newton and penalty parameter updates. Finally, in Table 3 we show the results obtained using exact second derivative information, either within the PCG framework, or by computing the exact reduced Hessian explicitly.

Note that we chose the ℓ_2-penalty function for two reasons. First, it is differentiable everywhere except when $c(x) = 0$ and, second, the dogleg step for $Y_k p_Y$ is a descent direction for the ℓ_2 norm (but not the ℓ_1 norm). While this option was not needed here, incorporating the dogleg step for $Y_k p_Y$ has been useful on ill-conditioned, nonlinearly constrained problems.

From Table 1 we see that quasi-Newton methods can require many iterations for convergence and exhibit failure for several problems. The primary

Table 1. Computational results for COPS problems using quasi-Newton updates without multipliers. *Numbers in last six columns list iteration counts. Numbers followed by a '*' indicate that a higher final objective value was obtained for this run. The following failures occurred: "—[1]" Maximal number of 3000 Iterations exceeded; "—[2]" Line search failure; "—[3]" Failure probably due to ill-conditioning*

Name	Size		BFGS w/o λ			SR1 w/o λ		
	n	m	Filter	ℓ_2	aug.Lag.	Filter	ℓ_2	aug.Lag.
gasoil_eq_50	503	500	—[1]	60	69	36	33	93
gasoil_eq_100	1003	1000	—[1]	29	76	65	36	87
gasoil_eq_200	2003	2000	—[1]	93	66	36	43	60
gasoil_eq_400	4003	4000	81*	108	45	36	46	73
marine_eq_25	614	592	264	—[1]	—[2]	190	—[3]	234
marine_eq_50	1214	1192	257	306	257	145	—[3]	653
marine_eq_100	2414	2392	267	—[1]	258	174	—[3]	153
marine_eq_200	4814	4792	263	—[1]	263	282	—[3]	148
methanol_eq_50	604	600	21	21	18	28	26	27
methanol_eq_100	1204	1200	21	19	19	25	17	16
methanol_eq_200	2404	2400	20	19	20	18	15	16
methanol_eq_400	4804	4800	20	20	20	19	15	18
pinene_eq_25	505	500	44	—[1]	43	59	—[3]	59
pinene_eq_50	1005	1000	43	—[2]	43	55	—[3]	54
pinene_eq_100	2005	2000	46	—[1]	42	50	—[3]	47
pinene_eq_200	4005	4000	44	—[1]	42	44	—[3]	43

reason for failure is that the 3000 iteration limit is exceeded. Failure because of ill-conditioning of the updates was also encountered, particularly for the exact penalty line search. Nevertheless, for the BFGS update the filter method was able to solve 13 out of 16 problems successfully, while the exact penalty function solved 9 and the augmented Lagrangian solved 15. For the solved problems the filter method was similar to the exact penalty function and the augmented Lagrangian merit function. For the SR1 update the filter and augmented Lagrangian methods solved all 16 problems successfully, while the exact penalty function line search solved only 8. For the SR1 update the results for the filter method appear to be better than for the other two options. This provides an indication that filter line searches should be considered for tailored implementations with quasi-Newton updates.

Better performance can be seen if multiplier estimates are used in the quasi-Newton updates. From Table 2 we again see that quasi-Newton methods require many iterations for convergence and exhibit failure for several problems. Nevertheless, for the BFGS update the filter and augmented Lagrangian methods were able to solve all 16 problems, while the exact penalty function solved only 9. Similarly, for the SR1 update the filter method solved 14 out of 16 problems successfully, while the exact penalty function solved only 8 and the augmented Lagrangian solved all 16. On the other hand, for the solved problems the filter method has better overall performance than the augmented Lagrangian line search.

Table 2. Computational results for COPS problems using quasi-Newton updates with multipliers. *Numbers in last six columns list iteration counts. Numbers followed by a '*' indicate that a higher final objective value was obtained for this run. The following failures occurred: "—[1]" Maximal number of 3000 Iterations exceeded; "—[2]" Line search failure; "—[3]" Failure probably due to ill-conditioning*

Name	Size		BFGS w/ λ			SR1 w/ λ		
	n	m	Filter	ℓ_2	aug.Lag.	Filter	ℓ_2	aug.Lag.
gasoil_eq_50	503	500	94*	69	212	23	35	138
gasoil_eq_100	1003	1000	37	238	311	32	31	159
gasoil_eq_200	2003	2000	100	30	246	33	38	130
gasoil_eq_400	4003	4000	66	40	381	32	27	197
marine_eq_25	614	592	277	—[1]	353	176	—[3]	224
marine_eq_50	1214	1192	266	—[1]	284	211	—[3]	147
marine_eq_100	2414	2392	270	—[1]	263	215	—[3]	250
marine_eq_200	4814	4792	273	—[1]	256	161	—[3]	173
methanol_eq_50	604	600	19	152	16	25	22	26
methanol_eq_100	1204	1200	20	17	19	18	644	18
methanol_eq_200	2404	2400	24	18	20	22	30	15
methanol_eq_400	4804	4800	18	18	20	18	18	15
pinene_eq_25	505	500	38	200	39	—[3]	—[1]	42
pinene_eq_50	1005	1000	38	—[2]	38	59	—[1]	50
pinene_eq_100	2005	2000	38	—[1]	44	53	—[3]	53
pinene_eq_200	4005	4000	38	—[1]	47	—[3]	—[1]	39

Finally, with the exact and PCG options in Table 3, all 16 problems with all line search options are now solved with far fewer iterations than with either quasi-Newton method. Here the results are very similar for the PCG and exact options. Also the iteration counts are comparable among all three methods, although the augmented Lagrangian method requires more iterations on the first four problems.

When quasi-Newton updates are used for the reduced Hessian, the results of these experiments show a clear advantage for the filter approach over the exact penalty function line search. Compared with augmented Lagrangian line searches, the results are mixed; the filter method is less reliable but generally requires fewer iterations. Therefore the filter method is preferred for the *direct tailored approach* where multiplier information is not available. With more accurate second order information, iteration counts are reduced significantly in all cases; there is little difference among the first two line search options, and the augmented Lagrangian line search is somewhat slower. Here we see even more strongly the influence of accurate reduced Hessian information in these algorithms, along with the benefits of multiplier estimates and the ability to solve linear systems that include C^T. On the other hand, while the exact and PCG options lead to a reduction in the number of NLP iterations, the cost of each iteration can be considerably higher than with quasi-Newton methods. This difference can play an important role if function evaluations are cheap and when we extend the NLP to include inequality constraints. In

Table 3. Computational results for COPS problems using reduced Hessian information obtained by finite difference

Name	Size		PCG			exact		
	n	m	Filter	ℓ_2	aug.Lag.	Filter	ℓ_2	aug.Lag.
gasoil_eq_50	503	500	13	13	19	13	13	19
gasoil_eq_100	1003	1000	11	11	16	11	11	16
gasoil_eq_200	2003	2000	13	13	22	13	13	22
gasoil_eq_400	4003	4000	21	21	35	21	21	35
marine_eq_25	614	592	6	6	6	5	5	5
marine_eq_50	1214	1192	6	6	6	5	5	5
marine_eq_100	2414	2392	7	7	7	6	6	6
marine_eq_200	4814	4792	7	7	7	6	6	6
methanol_eq_50	604	600	4	4	4	4	4	4
methanol_eq_100	1204	1200	4	4	4	4	4	4
methanol_eq_200	2404	2400	4	4	4	4	4	4
methanol_eq_400	4804	4800	4	4	4	4	4	4
pinene_eq_25	505	500	6	6	6	6	6	6
pinene_eq_50	1005	1000	7	6	6	7	6	6
pinene_eq_100	2005	2000	7	6	6	7	6	6
pinene_eq_200	4005	4000	6	6	6	6	6	6

the next subsection, we consider another set of numerical experiments that explore these issues as well.

6.2 Distillation Optimization

This example shares many characteristics with the optimization of distributed dynamic PDE systems. The differential algebraic equation (DAE) model for both cases consists of dynamic MESH (Mass, Equilibrium, Summation and Heat) equations as described in detail in [5]. In this example we simulate a simple air separation using a continuous distillation column with 15 trays without the reaction terms. The feed is assumed to have a composition of 78.1% Nitrogen, 21% Oxygen, and 0.9% Argon. The purity of Nitrogen taken out at the top of the column is 99.8%. The complete model consists of 70 differential equations and 356 algebraic equations. Here, we simulate a change of set point in the distillate flow rate from $D(0) = 301.8\ mol/h$ to $D^{set} = 256.0\ mol/h$. The objective is to minimize the offset produced during the change from one steady state to another by controlling the feed flowrate F, and the NLP is given by:

$$\begin{aligned} \min \quad & \int_0^{t_f} (D - D^{set})^2 dt \\ \text{s.t.} \quad & \text{DAE model} \\ & 0\ kmol/h \leq F \leq 2\ kmol/h. \end{aligned}$$

Here we compare the following cases for solution of this problem:

- The *QN-BFGS* option uses a BFGS update to approximate $Z_k^T W_k Z_k$ as in our previous study [5]. In order to obtain the overall reduced Hessian in (22), the term $Z_k^T \Sigma_k Z_k$ is computed explicitly.
- Similarly, the *QN-SR1* option uses the SR1 update to approximate $Z_k^T W_k Z_k$ as in our previous study [5]. Here the overall reduced Hessian is modified if the update is not positive definite.
- The *exact* option uses finite difference approximations to calculate $Z_k^T W_k Z_k$.
- The *PCG1* option due to [15] constructs a preconditioner from a BFGS update. Here $P_k = B_k \approx \left[Z_k^T (W_k + \Sigma_k) Z_k\right]^{-1}$.
- The *PCG2* option applies a damped BFGS for $B_k \approx Z_k^T W_k Z_k$. The preconditioner is constructed from $P_k = \left[B_k + Z_k^T \Sigma_k Z_k\right]^{-1}$.

We note that for this example only control variable bounds are imposed on this problem. As a result $Z_k^T \Sigma_k Z_k = \Sigma_u$, which is a diagonal matrix corresponding to the control bounds and their multipliers, is much cheaper to evaluate. This accounts for the fast performance in options *PCG2, QN-BFGS* and *QN-SR1*. To present more realistic performance behavior (here for the *PCG2* case) for general control problems we also provide the calculation for a complete $Z_k^T \Sigma_k Z_k$, even though it is not used. This option is labeled as *PCG2**.

To assess the computational cost, all of these cases were run on an 800 MHz Pentium III Intel processor; they were initialized to a feasible solution corresponding to the steady state at $t = 0$. Table 4 shows the computational results for this example using three collocation points and for different numbers of finite elements.

It should be mentioned that these results are about five times faster than with the direct tailored rSQP approach, for smaller air separation examples described and compared in [5]. Also, our current implementation of the previous tailored rSQP approach was unable to handle problems of this size. This is largely because the matrix $C_k^{-1} N_k$ no longer needs to be computed

Table 4. Computational Results for Air Separation Optimization

Elements	n/m	QN-BFGS [Iter/CPU]	QN-SR1 [Iter/CPU]	Exact [Iter/CPU]
50	67620/67470	206/5.0	82/1.93	10/7.35
100	135170/134870	238/11.35	124/6.03	9/25.6
300	405370/404470	288/52.47	151/27.65	11/269.98
600	810670/808870	336/210.52	218/133.83	11/1041.36

Elements	n/m	PCG1 [Iter/CPU]	PCG2 [Iter/CPU]	PCG2* [Iter/CPU]
50	67620/67470	15/3.78	12/1.45	12/5.98
100	135170/134870	15/13.03	12/3.43	12/19.6
300	405370/404470	17/99.93	16/17.93	16/209.98
600	810670/808870	20/280.6	18/66.22	18/949.62

Table 5. Number for Hessian-Vector products for Air Separation Optimization

Elements	Exact	PCG1	PCG2
50	1500	742	190
100	2700	1359	237
300	9900	3440	341
600	19800	4579	480

and instead the factors of C are now stored, thus eliminating a significant storage requirement.

In addition, from the results in this table several features are worth noting. First it is clear that the Newton-based options (*exact, PCG1, PCG2* and *PCG2**) require far fewer interior point iterations than the quasi-Newton methods. In most cases this factor exceeds an order of magnitude. However, the Newton methods clearly require more work per iteration. In particular, the exact approach incurs a large cost due to $n - m$ of Hessian vector products per IP iteration. The numbers of products are tabulated in Table 5 for the exact and *PCG* options. Here an interesting feature can be seen with the two *PCG* options. The first preconditioner is four to six times slower than the second one. This can be explained because the first preconditioner requires up to an order of magnitude more *PCG* iterations and Hessian vector products. *PCG2* requires less work because its preconditioner has separated terms which are updated more accurately, particularly when μ changes. On the other hand, *PCG2* (as well as *QN-BFGS* and *QN-SR1*) is greatly aided by the simple form of $Z_k^T \Sigma_k Z_k$ for this particular example. For more generally constrained problems, $Z_k^T \Sigma_k Z_k$ is more costly to update. This can be seen with option *PCG2**, which requires the same low number of *PCG* iterations but is now about four times as expensive as *PCG1*.

Nevertheless, we see that the improved IP algorithm is quite attractive for solving very large nonlinear programming problems. As can be seen from the table, even the largest problem with over 800,000 variables and 1800 degrees of freedom could be solved in less than 67 CPU minutes.

7 Conclusions

We consider the implementation of SAND optimization strategies for existing engineering modeling systems, particularly for finite element applications. In order to reuse most of the components of these systems, we consider reduced Hessian SQP strategies that can be tailored to the discretized PDE model and interfaces. Here we consider two variants of the tailored approach: a direct approach where the Jacobian is not accessible but Newton steps of the model equations can be extracted, and an adjoint approach where linear systems can be solved that involve the transpose of the model Jacobian. Our previous studies with rSQP and quasi-Newton updates showed that both tailored approaches would have similar convergence behavior.

Here we consider recent improvements to rSQP: a filter line search strategy, better second-order information and a barrier method to handle variable bounds. Taken together, they lead to significant improvements over the previous rSQP strategy. On the other hand, except for the filter line-search option, these improvements require multiplier estimates which cannot be obtained from the direct tailored approach. Consequently, algorithmic implementers should consider the performance benefits of an adjoint tailored approach and allow for transposed linear systems to be solved in a straightforward manner within these modeling systems.

References

1. Alkaya, D., S. Vasantharajan and L. T. Biegler, "Generalization of a Tailored Approach for Process Optimization" I & EC Research , 39, 6, pp. 1731-1742 (2000)
2. Bader, G. and U. Ascher, A New Basis Implementation for Mixed Order Boundary Value ODE Solver, *SIAM J. Sci. Comput.* **8** (1987) 483–500.
3. Biegler, L. T., J. Nocedal and C. Schmid, "A Reduced Hessian Method for Large Scale Constrained Optimization," SIAM J. Optimization, 5, 2, p. 314 (1995)
4. Biegler, L. T., Claudia Schmid, and David Ternet, "A Multiplier-Free, Reduced Hessian Method For Process Optimization," Large-Scale Optimization with Applications, Part II: Optimal Design and Control, L. T. Biegler, T. F. Coleman, A. R. Conn and F. N. Santosa (eds.), p. 101, IMA Volumes in Mathematics and Applications, Springer Verlag (1997)
5. Biegler, L. T., A. Cervantes and A. Wächter, "Advances in Simultaneous Strategies for Dynamic Process Optimization," *Chemical Engineering Science*, to appear (2002)
6. A. S. Bondarenko, D. M. Bortz, and J. J. Moré, "COPS: Large-Scale Nonlinearly Constrained Optimization Problems," Technical Memorandum ANL/MCS-TM-237, September 1998, revised October 1999.
7. Cervantes, A., A. Waechter, R. Tutuncu and L. T. Biegler, "A reduced space interior point strategy for optimization of differential algebraic systems," *Comp. Chem. Engr.*, **24**, p. 39-51 (2000)
8. E. D. Dolan and J. J. Moré, "Benchmarking Optimization Software with COPS," Technical Memorandum ANL/MCS-TM-246, November 2000.
9. A. S. El-Bakry, R. A. Tapia, T. Tsuchiya, and Y. Zhang, On the Formulation and Theory of the Newton Interior-Point Method for Nonlinear Programming, *J. Optim. Theory App.* **89** (1995) 507–541.
10. A. V. Fiacco and G. P. McCormick. *Nonlinear Programming: Sequential Unconstrained Minimization Techniques.* John Wiley, New York, USA, 1968. Reprinted by SIAM Publications, 1990.
11. Fletcher, R. and S. Leyffer, "Nonlinear programming without a penalty function," Technical Report NA/171, University of Dundee Numerical Analysis Report, Dundee, Scotland, UK (1998)
12. Goldfarb, D. and A. Idnani, "A Numerically Stable Dual Method for Solving Strictly Convex Quadratic Programs," Mathematical Programming, 27 (1983) 1-33.

13. Griewank A., D. Juedes, and J. Utke, ADOL-C: A Package for the Automatic Differentiation of Algorithms Written in C/C++, *TOMS* 22(2) (1996), pp. 131-167
14. G. Itle, A. Salinger, R. Pawlowski, D. Alkaya and L. Biegler "Constraint Aggregation for Large-scale NLP Algorithms," presented at Annual AIChE meeting, Los Angeles, CA, November, 2000
15. Morales, J. L. and J. Nocedal, "Automatic Preconditioning by Limited Memory Quasi-Newton Updating," *SIAM J. Optimization*, 10,4, pp. 1079-1096 (2000)
16. Schmid, C. and L. T. Biegler, "Quadratic Programming Methods for Tailored Reduced Hessian SQP," Computers and Chemical Engineering, 18, 9, p. 817 (1994)
17. Schmid, C. and L. T. Biegler, "A Simultaneous Approach for Process Optimization with Existing Modeling Procedures," Proc. European Symposium on Computer Aided Process Engineering, I Chemical Engineering Symposium Series No. 133, p. 231 (1994). (Also in Chemical Engineering Research and Design, Part A, 72, p. 382 (1994))
18. Tanartkit, P., and L. T. Biegler, "Reformulating Ill-Conditioned DAE Optimization Problems," I & EC Research, 35, 6, p. 1853 (1996)
19. van Bloemen Waanders, B., A. Salinger, R. Pawlowski, L. Biegler and R. Bartlett, "SAND Optimization of Massively Parallel Simulation Codes Using and Object Oriented Framework," presented at SIAM Meeting on Parallel Computing, Norfolk, VA, March (2001)
20. Wächter, A. and L. T. Biegler. "Global and local convergence of line search filter methods for nonlinear programming," Technical Report CAPD B-01-09, Department of Chemical Engineering, Carnegie Mellon University, Pittsburgh, PA, USA (2001)

Interior Methods For a Class of Elliptic Variational Inequalities

Randolph E. Bank, Philip E. Gill, and Roummel F. Marcia*

Department of Mathematics, University of California, San Diego, La Jolla, CA

Abstract. We consider the application of primal-dual interior methods to the optimization of systems arising in the finite-element discretization of a class of elliptic variational inequalities. These problems lead to very large (possibly non-convex) optimization problems with upper and lower bound constraints.

When interior methods are applied to the discretized problem, the resulting linear systems have the same zero/nonzero structure as the finite-element equations solved for the unconstrained case. This crucial property allows the interior method to exploit existing efficient, robust and scalable multilevel algorithms for the solution of partial differential equations (PDEs).

We illustrate some of these ideas in the context of the elliptic PDE package PLTMG.

1 Introduction

We consider a class of elliptic partial differential equation (PDEs) for which the solution is required to satisfy certain inequality constraints (for example, so-called obstacle problems in the class of elliptic variational inequalities [18–22]). The proposed method combines an adaptive finite-element method with a finite-dimensional primal-dual interior method for optimization.

As background for the problem to be discussed, we introduce the self-adjoint, positive-definite elliptic boundary value problem:

$$\begin{aligned} -\nabla \cdot (a\nabla u) + bu &= f \text{ for } x \in \Omega, \\ (a\nabla u) \cdot n &= g \text{ for } x \in \partial\Omega_1, \\ u &= 0 \text{ for } x \in \partial\Omega_2 \equiv \partial\Omega - \partial\Omega_1. \end{aligned} \tag{1}$$

Here, $a(x)$ and $b(x)$ are smooth functions with $a(x) > 0$ and $b(x) \geq 0$, but more generally we could let $a(x)$ and $b(x)$ be symmetric positive-definite and positive semidefinite matrix functions. It is assumed throughout that a unique solution exists (which will follow, for example, if $b > 0$ or $\partial\Omega_2$ is nonempty). The methods to be considered exploit the so-called weak formulation of problem (1) and hence require the definition of certain function spaces. Let

$$H \equiv \mathcal{H}^1_E(\Omega) = \Big\{u : \int_\Omega |\nabla u|^2\, dx + \int_\Omega u^2\, dx < \infty \text{ and } u = 0 \text{ on } \partial\Omega_2\Big\}.$$

* Research supported by National Science Foundation grants DMS-9973276 and ACI-0082100.

For $u \in H$, the energy inner-product norm is $\|u\|^2 = a(u,u)$, where $a(u,v)$ denotes the operator

$$a(u,v) = \int_\Omega a(x)\nabla u \cdot \nabla v \, dx + \int_\Omega b(x) uv \, dx, \quad u, v \in H.$$

For u, $v \in \mathcal{L}^2(\Omega)$, the standard inner-product norm is $\|u\|^2 = (u,u)$, where $(u,v) = \int_\Omega uv\,dx$. Similarly, on the boundary we use the inner product $\langle u, v \rangle = \int_{\partial\Omega_1} uv\,ds$ for u, $v \in \mathcal{L}^2(\partial\Omega_1)$.

The Ritz variational formulation of (1) involves the minimization problem

$$\underset{u \in H}{\text{minimize}} \;\; q(u) = a(u,u) - 2\{(f,u) + \langle g, u \rangle\}. \tag{2}$$

Many nonlinear PDE's have a similar variational formulation. However, unlike (2), the objective function $q(u)$ is not necessarily quadratic in u.

Let $\mathcal{S} \subset H$ be the n dimensional space of continuous piecewise-linear polynomials corresponding to a triangulation $\mathcal{T}$ of Ω. The finite-element approximation $u_h \in \mathcal{S}$ to $u \in H$ solves the minimization problem:

$$\underset{u_h \in \mathcal{S}}{\text{minimize}} \;\; q(u_h). \tag{3}$$

Let $\{\phi_i\}_{i=1}^n$ denote the usual nodal basis for $\mathcal{S}$. Any $u_h \in \mathcal{S}$ can be written in the form $u_h = \sum_{i=1}^n U_i \phi_i$, which implies that (3) can be formulated as the finite-dimensional minimization problem

$$\underset{U \in \mathbb{R}^n}{\text{minimize}} \;\; Q(U).$$

This problem is usually solved using Newton's method. Given an approximate solution U, a direction of improvement ΔU is computed from the Newton equations $\nabla^2 Q(U)\Delta U = -\nabla Q(U)$, where $\nabla Q \in \mathbb{R}^n$ is the gradient and $\nabla^2 Q \in \mathbb{R}^{n\times n}$ is the Hessian of Q. The new estimate is then $U + \alpha\Delta U$, where α is a step length used to enforce convergence. In the case of the linear PDE (1), the objective is quadratic, with

$$Q(U) = U^T AU - 2U^T F,$$

where A is the sparse symmetric positive-definite stiffness matrix with $A_{ij} = a(\phi_j, \phi_i)$, and $F_i = (f, \phi_i) + \langle g, \phi_i \rangle$. In this case, the optimal U can be computed from the single sparse symmetric system $AU = F$. (Matrices with similar nonzero structure arise in the nonlinear case, but the entries of the stiffness matrix generally depend upon the current value of U.) Such linear systems are solved using iterative methods. In particular, the finite-element code PLTMG [1] uses the conjugate-gradient method with a preconditioner based on the multilevel multigraph technique [2]. Multigraph preconditioners are related to hierarchical basis multigrid preconditioners, which have proved to be very robust for problems posed on a sequence of nonuniform, adaptively refined meshes. However, multigrid methods rely on the refinement structure generated through the adaptive refinement process. This limits their applicability when geometrically complex domains require many elements just for the definition of the domain, or when adaptivity is derived from moving mesh points rather than refinement. In these situations, it is possible to have fine, highly nonuniform meshes with no refinement history available to create a hierarchical basis. Multigraph methods overcome this limitation by creating a hierarchical basis when no natural geometry is present. The reader is referred to [2] for a detailed description of the multigraph preconditioner used in PLTMG. It suffices here to note that the preconditioner considered in this paper involves incomplete sparse factorizations of A and its restriction to a sequence of subspaces of decreasing dimension.

The purpose of this paper is to extend the multigraph finite-element method to the case where u is subject to inequality constraints. In this case, the continuous problem is

$$\underset{u\in H}{\text{minimize}} \;\; q(u), \;\; \text{subject to} \;\; \underline{b} \le u \le \bar{b}. \tag{4}$$

If $\mathcal{I}$ denotes the interpolation operator, then the finite-element formulation of this problem can be written as

$$\underset{u_h\in \mathcal{S}}{\text{minimize}} \;\; q(u_h), \;\; \text{subject to} \;\; \mathcal{I}(\underline{b}) \le u_h \le \mathcal{I}(\bar{b}).$$

This leads to the finite-dimensional optimization problem

$$\underset{U\in \mathbb{R}^n}{\text{minimize}} \;\; Q(U), \; \text{subject to} \; \underline{B} \le U \le \bar{B}, \tag{5}$$

where $\underline{B}$ and $\bar{B}$ define the expansions $\mathcal{I}(\underline{b}) = \sum_{i=1}^n \underline{B}_i\phi_i$, and $\mathcal{I}(\bar{b}) = \sum_{i=1}^n \bar{B}_i\phi_i$. Many methods have been proposed for solving problems in this form. These methods may be broadly categorized as active-set methods (see, e.g., [3, 4, 8, 23]) and *interior methods* (see, e.g., [7, 9, 17, 27]). As in the unconstrained case, the quadratic form of the objective can be exploited (see, e.g., [11, 24]).

2 Solving the Finite-Dimensional Problem

In this section we focus on the solution of the finite-dimensional bound-constrained problem (5). Accordingly, we change our notation to that used in the optimization literature. In particular, we write the problem (5) in the form

$$\underset{x\in \mathbb{R}^n}{\text{minimize}} \;\; f(x) \;\; \text{subject to} \;\; x \ge 0, \tag{6}$$

where $f(x)$ is a general smooth nonlinear function of n variables x_1, x_2, ..., x_n. For simplicity, we start with a problem with only nonnegativity constraints. Throughout, we will assume that f is twice-continuously differentiable, with gradient $\nabla f(x)$ and Hessian $\nabla^2 f(x)$ denoted by $g(x)$ and $H(x)$ respectively. Since the original continuous problems are usually convex, we will assume that $f(x)$ is a convex function. However, we emphasize that all the methods to be discussed can be extended to the nonconvex case.

The first-order necessary conditions for a solution of (6) are that there exist nonnegative x_i and z_i (Lagrange multipliers) such that $z_i = g_i(x)$ and $x_i z_i = 0$. In this section we adopt a convention common to the literature of interior-point methods of using upper-case letters to denote diagonal matrices whose diagonal consists of the components of the vector represented by the corresponding lower-case letter. With this convention, we can write the first-order conditions in the compact form

$$\begin{aligned} z &= g(x), & z &\geq 0, \\ Xz &= 0, & x &\geq 0. \end{aligned} \tag{7}$$

The pair (x, z) satisfies the property of strict complementarity if $x+z > 0$, i.e., one of the x_i and z_i is nonzero for each i. If both x_i and z_i are zero, a solution is said to be degenerate, or more precisely, dual degenerate. Second-order sufficient conditions are that (7) hold; x and z are strictly complementary and $XHX + Z$ is positive definite. At any point satisfying the first-order conditions, an active nonnegativity constraint (i.e., a variable on its lower bound of zero) causes the corresponding row and column of XHX to be zero.

Primal-dual interior methods are based on solving a system of $2n$ nonlinear equations that represent the first-order optimality conditions (7) with each condition $x_i z_i = 0$ perturbed by a scalar μ (known as the *barrier parameter*). For a given μ, the equations are written as $G^\mu(x, z) = 0$, where

$$G^\mu(x, z) = \begin{pmatrix} g(x) - z \\ Xz - \mu e \end{pmatrix}, \tag{8}$$

where e is the vector of ones. Let (x, z) be an *interior point*, i.e., $x > 0$ and $z > 0$. For a given μ, the equations $G^\mu(x, z) = 0$ are solved using a form of Newton's method in which x and z are maintained to be interior. Linearizing the perturbed conditions (8) at an interior point (x, z) gives

$$\begin{aligned} \Delta z &= g(x) + H(x)\Delta x - z \\ X\Delta z + Z\Delta x &= \mu e - Xz, \end{aligned}$$

where $Z = \operatorname{diag}(z_1, z_2, \ldots, z_n)$. This yields the linear system

$$\begin{pmatrix} H & -I \\ Z & X \end{pmatrix} \begin{pmatrix} \Delta x \\ \Delta z \end{pmatrix} = - \begin{pmatrix} g - z \\ X(z - \pi) \end{pmatrix}, \tag{9}$$

where π is the vector of *primal multipliers* such that $\pi = \mu X^{-1} e$. (The dependencies on x, z and μ have been suppressed for clarity.) If H is positive definite this system is nonsingular at all interior points (x, z). If v denotes a combined $2n$ vector of unknowns (x, z) and $G^\mu(v)$ denotes the function $G^\mu(x, z)$, then (9) are the equations $G^\mu(v)'\Delta v = -G^\mu(v)$ for the Newton direction $\Delta v = (\Delta x, \Delta z)$. Applying block elimination to (9) gives Δx and Δz as

$$(H + X^{-1}Z)\Delta x = -(g - \pi) \quad \text{and} \quad \Delta z = g + H\Delta x - z.$$

Scaling this system with $X^{1/2}$ gives the solution

$$\Delta x = X^{1/2}\Delta\bar{x}, \quad \text{where } \Delta\bar{x} \text{ solves } \; (X^{1/2}HX^{1/2} + Z)\Delta\bar{x} = -X^{1/2}(g - \pi). \tag{10}$$

The crucial feature of this system is that $X^{1/2}HX^{1/2} + Z$ has the same dimension and sparsity pattern as H. If the second-order sufficient conditions hold, then for μ sufficiently small, a differentiable *trajectory* of solutions $(x(\mu), z(\mu))$ exists such that $(x(\mu), z(\mu)) \to (x^*, z^*)$ as $\mu \to 0^+$. Primal-dual interior methods attempt to follow this trajectory by finding an approximate solution of $G^\mu(x, z) = 0$ for a decreasing sequence of μ-values such that $\mu \to 0^+$. As the solution is approached, the scaled matrix $X^{1/2}HX^{1/2}$ converges to a row and column-scaled version of H in which the zero rows and columns correspond to the active bounds. It follows that if the problem is nondegenerate, the rows and columns of $X^{1/2}HX^{1/2} + Z$ corresponding to the active bounds are diagonal.

The step length is chosen using a standard interior-point backtracking line search (for more details see, e.g., [28]). Consider the calculation of new iterates $(x_{k+1}, z_{k+1}) = (x_k + \alpha_k\Delta x_k, z_k + \alpha_k\Delta z_k)$ at the kth iteration. First, an upper bound on the step is computed such that $\alpha_M = \min\{1, .99\tau\}$, where τ is the largest positive α such that $(x_k + \alpha\Delta x_k, z_k + \alpha\Delta z_k)$ is feasible. The step α_k is then the first member of the sequence $\{\gamma_c^j\alpha_M\}_{j=0}^\infty$ such that

$$\|G^\mu(x_k + \alpha_k\Delta x_k, z_k + \alpha_k\Delta z_k)\|_2 \le (1 - \alpha_k\eta_s)\|G^\mu(x_k, z_k)\|_2,$$

for fixed scalars η_s $(0 < \eta_s < \frac{1}{2})$ and γ_c $(0 < \gamma_c < 1)$ with typical values $\eta_s = \frac{1}{4}$ and $\gamma_c = \frac{1}{2}$. (Throughout this section, $\|\cdot\|_2$ refers to the usual vector two-norm.)

In the general case where f is not convex, a more sophisticated strategy must be used to guarantee convergence to a point satisfying the second-order necessary conditions for problem (6). The description of such strategies is beyond the scope of this paper. However, it is relatively straightforward to formulate methods that are not only provably convergent to second-order points, but are also able to exploit the properties of the multilevel iterative solver. One approach is to define line search or trust region methods that minimize the function

$$M^\mu(x, z) = f(x) - \mu\sum_{j=1}^{n}\left(\ln x_j + \ln\left(\frac{x_j z_j}{\mu}\right) + \left(\frac{\mu - x_j z_j}{\mu}\right)\right)$$

(see [13, 14]). This function is well defined for all (x, z) such that $x_j > 0$ and $z_j > 0$ and has a local minimizer at a point (x, z) such that $G^\mu(x, z) = 0$. It can be arranged that algorithms for minimizing $M^\mu(x, z)$ solve a system that has identical structure to that of (10). However, the multilevel iterative solver must satisfy two requirements: the multigraph algorithm must always implicitly generate a positive-definite preconditioner, and the conjugate-gradient method must be modified to detect indefiniteness in the matrix of system (10) (see, e.g., [16, 25]).

2.1 Treatment of upper and lower bounds

Now consider problem (6) with finite upper and lower bounds $b_l \le x \le b_u$. If $L = \mathrm{diag}(b_l)$ and $U = \mathrm{diag}(b_u)$, let $X_1 = X - L$ and $X_2 = X - U$. Assume that x is interior, so that $X_1 > 0$ and $X_2 < 0$. Let z_1 and z_2 denote estimates of the multipliers associated with the constraints $x \ge b_l$ and $x \le b_u$ respectively. The perturbed optimality conditions are

$$\begin{aligned} g(x) - (z_1 + z_2) &= 0 \\ X_1(z_1 - \pi_1) &= 0 \\ X_2(z_2 - \pi_2) &= 0, \end{aligned}$$

with $\pi_1 = \mu X_1^{-1} e$ and $\pi_2 = \mu X_2^{-1} e$. Linearizing these conditions at a point (x, z_1, z_2) such that $b_l < x < b_u$, $z_1 > 0$ and $z_2 < 0$ gives the following system analogous to (9)

$$\begin{pmatrix} H & -I \\ \widehat{Z} & \widehat{X} \end{pmatrix} \begin{pmatrix} \Delta x \\ \Delta z \end{pmatrix} = - \begin{pmatrix} g - z \\ \widehat{X}(z - \pi) \end{pmatrix},$$

where $\widehat{X} = -X_1 X_2$, $\widehat{Z} = -(X_2 Z_1 + X_1 Z_2)$, $z = z_1 + z_2$ and $\pi = \pi_1 + \pi_2$.

It follows that Δx can be calculated as $\Delta x = \widehat{X}^{1/2} \Delta \bar{x}$, where $\Delta \bar{x}$ satisfies the system

$$(\widehat{X}^{1/2} H \widehat{X}^{1/2} + \widehat{Z}) \Delta \bar{x} = -\widehat{X}^{1/2}(g - \pi). \tag{11}$$

Again, the relevant matrix has the same dimension and sparsity pattern as H.

3 Interior Methods for the Variational Problem

Now we consider the application of the primal-dual interior method to a *sequence* of finite-dimensional optimization problems (5) defined in a finite-element discretization with adaptive mesh refinement.

3.1 Scaling the finite-element discretization

It is well known that problem scaling has a substantial effect on the efficiency of optimization methods. Scaling is particularly relevant when the problem is defined by an adaptive finite-element method. For a given mesh, as the solution is approached, the magnitude of each objective gradient element g_i depends on the support of its constituent nodal basis functions. This implies that the gradients (and hence the Lagrange multipliers) can vary widely in magnitude depending on the degree of refinement in a particular region.

Here we use a scaling that balances the magnitudes of the Lagrange multipliers and nonlinear equations as the mesh is refined. Let D be the positive diagonal matrix with entries $d_i = 2\|\phi_i\|^2$, where $\{\phi_i\}$ are the nodal basis functions. The $\{d_i\}$ form the diagonal elements of the so-called *lumped mass matrix* M, with entries $M_{ij} = (\phi_i, \phi_j)$. Note that M has the same sparsity pattern as the stiffness matrix A. The matrix D is used to replace the first-order optimality condition $g - z = 0$ in (7) by the scaled condition $g - Dz = 0$. Proceeding as in the previous section, we obtain Δx as

$$\Delta x = \widehat{X}^{1/2}\Delta\bar{x},$$

$$\text{where } \Delta\bar{x} \text{ satisfies } (\widehat{X}^{1/2}H\widehat{X}^{1/2} + D\widehat{Z})\Delta\bar{x} = -\widehat{X}^{1/2}(g - D\pi). \quad (12)$$

In the line search, a norm of the balanced vectors $g - Dz$ and $DX(z - \pi)$ is used to measure the proximity to the solution. In particular, the two-norm of G^μ (8) is replaced by the elliptic norm

$$\begin{aligned}\|G^\mu\|^2_{D^{-1}} &\equiv \|g - Dz\|^2_{D^{-1}} + \|DX(z-\pi)\|^2_{D^{-1}} \\ &= (g - Dz)^T D^{-1}(g - Dz) + (z - \pi)^T X^T DX(z - \pi).\end{aligned}$$

Similarly, the variables are balanced using $\|v\|^2_D \equiv \|x\|^2_D + \|z\|^2_D = x^T Dx + z^T Dz$.

3.2 Mesh refinement and the choice of barrier parameter

The proposed algorithm has two levels of iteration: the multilevel computations associated with a given finite-element mesh, and the Newton iterations associated with the given barrier parameter μ. Here we link the choice of barrier parameter to the mesh size, the main idea being that it is appropriate to accurately solve the optimization problem (i.e., choose μ small) only in the final stages of the mesh refinement.

Consider the approximate solution of a sequence of bound-constrained optimization problems BC^j, each associated with a given level of mesh refinement. We associate a barrier parameter $\mu^{(j)}$ with each problem, where $\mu^{(j-1)} > \mu^{(j)}$. The initial point for problem BC^j is obtained by interpolating the continuous finite-element solution associated with the final point from BC^{j-1}. In this situation, the value of $\mu^{(j)}$ is crucial, since a poor choice may result in the interpolated finite-element solution from BC^{j-1} being far from the trajectory for BC^j.

From standard theory, the finite-element approximation $u_h \in \mathcal{S}$ has the error estimate $\|u_h - u^*\| = O(n^{-1/2})$, where u^* is the optimal solution. It follows that if n_j is the number of variables associated with BC^j, then the error in the finite-element approximation is reduced by a factor of $O\big((n_{j-1}/n_j)^{1/2}\big)$ at each refinement. In practice, n is usually increased geometrically. In the numerical examples of Section 4, the n for adaptive refinement is increased by a factor of four, which implies that the error is reduced by a factor of two after each refinement.

Next we estimate the error in any point $v(\mu)$ on the primal-dual trajectory as an estimate of the optimal point v^*. If $dv(\mu)/d\mu$ exists and is bounded, then for μ sufficiently small, $\|v(\mu) - v^*\| = O(\mu)$ (see [12, Theorem 15], and [29, p. 8]). It follows that if $u_h(\mu) \in \mathcal{S}$ denotes the solution associated with barrier parameter μ, we would expect the total error in $u_h(\mu)$ to be $O(\mu)+O(n^{-1/2})$. This bound suggests the choice of initial barrier parameter $\mu^{(0)} = O(n_0^{-1/2})$, with subsequent values reduced at the same rate as the finite-element error estimate, i.e., $\mu^{(j)} = (n_{j-1}/n_j)^{1/2}\mu^{(j-1)}$.

3.3 Choosing the starting point for BC^j

The initial values of the x-variables for BC^j are found by interpolating the final values for BC^{j-1}. The initial z-variables may be defined in several ways. First, as in the primal case, the z values can be interpolated from the final iteration of problem BC^{j-1}. This choice will tend to give small initial residuals for the first set of equations $g - Dz = 0$ of the system $G^\mu(v) = 0$. However, the second set of equations $Xz - \mu e = 0$ will have a large residual since the refined z values are generally increased by the scaling with D. Another option is to use the complementarity conditions to define new dual variables from the interpolated primal values. In this case, $(z_1)_i = (\pi_1)_i$ and $(z_2)_i = (\pi_2)_i$, where π_1 and π_2 are the initial primal multipliers $\pi_1 = \mu X_1^{-1}e$ and $\pi_2 = \mu X_2^{-1}e$. Another possibility is to define z so that the residual for $g - Dz$ is small. For example, if $g_i > 0$, then we can set $(z_1)_i = g_i/d_i$ and $(z_2)_i = (\pi_2)_i/d_i$, where π_2 are the initial primal multipliers $\pi_2 = \mu X_2^{-1}e$. In this case, the residual for the equation $g_i - d_i z_i = 0$ is $g_i - d_i(z_1 + z_2)_i = (\pi_2)_i$.

Any one of these alternatives is reasonable when the "continuous" Lagrange multipliers are smooth. However, when the PDE solution is not differentiable, the continuous multipliers can resemble delta functions (see, e.g., the minimal surface problem of Section 4) and naive refinement techniques can lead to poor initial estimates of the multipliers. To deal with this problem, the multipliers are smoothed using the inverse Laplacian before being interpolated. The interpolated multipliers are then mapped back to the original space to initialize the optimization. Further details of the smoothing procedure are given in Section 3.4.

Adaptive mesh refinement introduces some algorithmic issues that are unique to interior methods. Adaptive refinement involves not only the introduction of new elements where needed, but can also be combined with mesh regularization (i.e., the movement of elements) to further improve the quality of the mesh. Either of these procedures can cause the upper and lower bounds of the discretized problem to change, which implies that the approximate solution of BC^{j-1} may not be a good estimate of the solution for BC^{j}. In particular, the solution of BC^{j-1} may not even lie in the interior of the feasible region for problem BC^{j}. In this situation, it is necessary to redefine the initial point for BC^{j} so that it is interior. In the numerical results of Section 4 the initial point is forced to lie at a distance of at least $\nu\mu$ from their nearest bound, where $0 < \nu \leq 1$.

3.4 Lagrange multiplier smoothing

Let ζ denote the Lagrange multipliers associated with the continuous bound-constrained problem (4). In general, the natural space for ζ is $\mathcal{H}^{-1}(\Omega)$, the class of functions that are integrable against functions in $\mathcal{H}^{1}(\Omega)$ (for the precise definition, see, e.g., [6]). As $\mathcal{H}^{-1}(\Omega)$ functions need not be continuous and can exhibit δ-function-like behavior, interpolating in a region that includes a jump discontinuity can not fully capture the behavior of ζ. This lack of smoothness is problematic when ζ is approximated by a piecewise-linear function derived by linear interpolation from a coarse mesh. If ζ is poorly behaved, then the interpolant will invariably be a poor initial guess for the problem on the finer mesh.

An alternative approach is to smooth the coarse mesh multipliers before interpolation. For any $\zeta \in \mathcal{H}^{-1}(\Omega)$, the solution w of

$$-\Delta w = \zeta \quad \text{in } \Omega \tag{13}$$

is in $\mathcal{H}^{1}(\Omega)$ and it follows that the inverse Laplacian Δ^{-1} can be used as a "smoothing" map for the approximate multipliers. Given a multiplier function $\zeta = \zeta_c$ on the coarse mesh, the Poisson equation (13) is solved for $w = w_c$. The linear interpolant w_f of w_c is computed for the refined mesh, and the new set of Lagrange multipliers ζ_f is defined as $\zeta_f = -\Delta w_f$. Thus, the interpolation is actually applied to $\Delta^{-1}\zeta_c$, and not on ζ_c itself. The discontinuous behavior of ζ is better reflected in this interpolation.

4 Numerical Results

In this section we discuss the results from applying PLTMG to three problems. Each problem was solved using adaptive mesh refinement, starting from (unless otherwise indicated) a uniform 5×5 mesh. At each refinement the number of unknowns was increased by a factor of k_{ref}, where $k_{ref} = 4$. Two variants of PLTMG were used to solve each problem.

Method A. This method was designed to illustrate the influence of the initial interpolated solution and the change of barrier parameter. For each level of refinement, the finite-dimensional problem was first solved using a so-called *refinement step* in which the value of μ is inherited from the previous mesh. In this case, the initial point for the subproblem was the final solution on the previous mesh. The subproblem was then solved again in a *continuation step*, starting from the fine mesh solution, but with μ reduced by a factor of two. All times reported are the cumulative times to each point in the calculation. For the refinement steps, the reported figures include error estimation and mesh refinement times as well as time spent setting up and solving the variational inequality. In each case we report the number `Iter` of major Newton iterations to solve the problem at the given step.

Method B. In this method, the refinement and μ-reduction were performed simultaneously. This is the strategy recommended for practical computation, and is the basis of Algorithm 4.1 given below.

In all cases, the Newton iteration was stopped when the relative increment $\|\Delta v\|_D/\|v\|_D$ was reduced by a factor $\eta = 10^{-2}$ compared to its initial value. This test implies that at least two Newton steps are performed for each subproblem. A more stringent convergence criterion would result in a more accurate solution of the discrete variational inequality, but no improvement in the accuracy of the underlying continuous problem. Indeed, given the typical second-order convergence of the the piecewise linear discretization, 10^{-2} could be considered conservative.

For each problem, the initial point for the first subproblem was $\frac{1}{2}(b_l + b_u)$. The initial values for the multipliers z_1 and z_2 were the primal multipliers $\mu_0/(x - b_l)$ and $\mu_0/(x - b_u)$. The scale factor C_1 for the initial barrier parameter was chosen so that $\mu_0 = 1$. The parameter for shifting variables away from their bounds was $\nu = 1$. The line search parameters were $\gamma_c = \frac{1}{2}$ and $\eta_s = \frac{1}{4}$.

4.1 The algorithm

ALGORITHM 4.1. PRIMAL-DUAL METHOD.

Define $n_{\max}$; η, γ_c $(0 < \eta, \gamma_c < 1)$; η_s $(0 < \eta_s < \frac{1}{2})$; C_1 $(C_1 > 0)$;
Choose n $(0 < n \le n_{\max})$;
$\mu = C_1 n^{-1/2}$; $v = (x, z_1, z_2)$;
while $n \le n_{\max}$ **do**
  tol $= 0$; *first* $=$ **true**; $\Delta v = v$;
  while $\|G^\mu(v)\|_{D^{-1}} > 0$ **and** $\|\Delta v\|_D/\|v\|_D > tol$ **do**
    Compute $\Delta v = (\Delta x, \Delta z_1, \Delta z_2)$ from (12);
    $\tau = \max\{\alpha : \alpha > 0,\ v + \alpha\Delta v$ is feasible $\}$;
    $\alpha = \min\{1, .99\tau\}$;
    while $\|G^\mu(v + \alpha\Delta v)\|_{D^{-1}} > (1 - \alpha\eta_s)\|G^\mu(v)\|_{D^{-1}}$ **do**

$\alpha \leftarrow \gamma_c \alpha$;
end do
$v \leftarrow v + \alpha \Delta v$;
if *first* **then**
$tol = \eta \|\Delta v\|_D / \|v\|_D$; *first* = **false**;
end if
end do
$n \leftarrow k_{ref} n$; Refine the mesh;
$\mu \leftarrow \mu / \sqrt{k_{ref}}$;
end do

All runs were made on an SGI Octane workstation with 512MB of RAM and two 250MHz R10000 processors (only one being used for each problem solution). The f90 compiler was used with `-n32 -O` options specifying 32-bit mode and full code optimization.

Example 1: *Elliptic variational inequality.* Consider the variational problem

$$\min_{u \in K} \; q(u) = \int_{\Omega} |\nabla u(x)|^2 - 2c(x)u(x)\, dx,$$

where $\Omega = [0,1]\times[0,1]$, $K = \{u \in \mathcal{H}_0^1(\Omega) : |u| \le \frac{1}{4} - \frac{1}{10}\sin(\pi x_1)\sin(\pi x_2) \text{ for } x \in \Omega\}$, and $c(x)$ is chosen such that $-\Delta w = c$, with $w = \sin(3\pi x_1)\sin(3\pi x_2)$.

The results are summarized in Table 1. Columns 3 and 4 give the times and iterations for the separate refinement and continuations steps. Columns 5 and 6 give the details for Algorithm 4.1.

Table 1. Elliptic variational inequality.

		Method A		Method B	
n	μ	Iter	cpu secs	Iter	cpu secs
25	1	8	0.01	8	0.01
64	1	6	0.04		
64	2^{-1}	3	0.05	6	0.04
250	2^{-1}	8	0.19		
250	2^{-2}	3	0.25	9	0.19
1000	2^{-2}	12	1.05		
1000	2^{-3}	3	1.34	14	1.08
4000	2^{-3}	10	4.84		
4000	2^{-4}	3	6.34	15	5.74
16000	2^{-4}	10	24.37		
16000	2^{-5}	3	32.23	12	26.49
64000	2^{-5}	9	136.10		
64000	2^{-6}	3	189.75	11	144.47

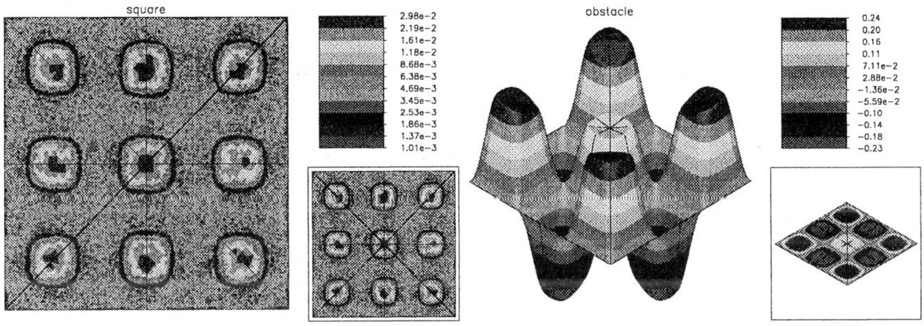

Fig. 1. Elliptic variational inequality. On the left is an plot of the element size in the adaptive mesh with $n = 64000$ vertices. On the right is the corresponding solution (see Plate 4 on page 347).

Fig. 1 gives the final adaptive mesh and corresponding approximate solution. The results in Table 1 indicate that the refinement steps generally require more Newton iterations than the continuation steps. This mainly reflects the quality of the initial guesses for these problems, since the underlying discrete variational inequalities associated with each mesh are quite similar. To some extent, the additional iterations are needed because the adaptive refinement tends to introduce new grid points in areas where the solution is least accurate. This is shown quite clearly in Fig. 1, where we observe that the smallest elements resolve the interface between contact and non-contact zones, and the largest elements appear near the centers of the contact zones, where the solution is well-defined by the obstacle. The initial guess for refinement steps is interpolated from the previous mesh, and is clearly not as good as the fine-grid solution that serves as initial guess for the continuation steps.

Example 2: *Elastic-plastic torsion.* This problem concerns an infinitely long cylindrical bar made up of an isotropic elastic perfectly plastic material. Starting from a zero-stress initial state, an increasing torsion moment is applied to the bar. The elastic-plastic torsion problem [15, p. 41-46] can be formulated in terms of the cross-section Ω of the cylinder and the torsion angle constant c per unit length. The stress field is the solution of the variational problem

$$\min_{u \in K} \; q(u) = \int_{\Omega} |\nabla u(x)|^2 - 2cu(x)\, dx,$$

where $\Omega = [0,1] \times [0,1]$, $K = \{ u \in \mathcal{H}_0^1(\Omega) \; : \; |u| \leq \operatorname{dist}(x, \partial\Omega) \text{ for } x \in \Omega \}$, and $\operatorname{dist}(x, \partial\Omega)$ denotes the distance from x to the boundary of Ω. The symbol $\mathcal{H}_0^1(\Omega)$ denotes the space of functions with gradients in $L^2(\Omega)$ that vanish on the boundary of Ω. If there is no constraint, then u will satisfy the

Table 2. Elastic-plastic torsion problem.

		Method A		Method B	
n	μ	Iter	cpu secs	Iter	cpu secs
25	1	5	0.01	5	0.01
64	1	4	0.03		
64	2^{-1}	3	0.04	4	0.03
250	2^{-1}	4	0.12		
250	2^{-2}	3	0.17	4	0.11
1000	2^{-2}	5	0.60		
1000	2^{-3}	3	0.85	6	0.57
4000	2^{-3}	5	2.93		
4000	2^{-4}	3	4.31	6	2.92
16000	2^{-4}	5	15.43		
16000	2^{-5}	3	23.34	5	14.45
64000	2^{-5}	5	95.48		
64000	2^{-6}	4	161.47	5	89.72

boundary-value problem

$$-\Delta u = c \quad \text{for} \quad x \in \Omega, \quad \text{and} \quad u = 0 \quad \text{for} \quad x \in \partial\Omega.$$

In all the numerical results, the constant $c = 5$ was chosen to be consistent with the elastic-plastic torsion problem in the COPS test set [5]. We repeated the same experiments as in the first example, beginning with an identical 5×5 mesh. Although subsequent meshes have the same numbers of mesh points, adaptive refinement made the meshes quite different. The results of the calculation are summarized in Table 2, and the final adaptive mesh and solution are shown in Fig. 2.

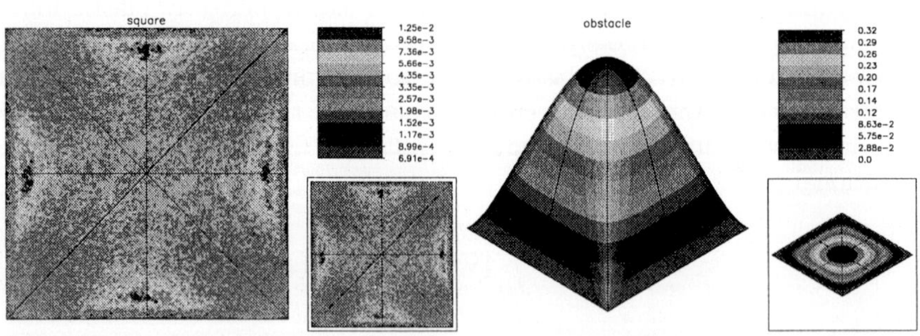

Fig. 2. Elastic-plastic torsion problem. On the left is an plot of the element size in the adaptive mesh with $n = 64000$ vertices. On the right is the corresponding solution (see Plate 5 on page 347).

Generally, this was an easier problem with a smoother solution than our first example. This is reflected in a smaller numbers of Newton iterations needed for the refinement steps. The final adaptive mesh is much more uniform, although, once again, the largest elements appear in the interior of the contact zone where the solution is well-defined by the obstacle.

Example 3: ***Minimal surface with an obstacle.*** This problem involves finding the least amount of total energy to maintain, and thus enclose a given area/volume with as little perimeter/surface area as possible. This formulation is identical to the one in the COPS test set [10]. The aim is to minimize the area function $q\colon K \to \mathbb{R}$, such that

$$q(u) = \int_{\Omega} \left(1 + |\nabla u(x)|^2\right)^{1/2} dx, \tag{14}$$

where $\Omega = [0,1] \times [0,1]$, $K = \{u \in \mathcal{H}^1(\Omega) : u(x) = u_\Omega(x) \text{ for } x \in \partial\Omega,\ u(x) \geq u_L(x) \text{ for } x \in \Omega\}$. The function $u_\Omega\colon \partial\Omega \to \mathbb{R}$ defines the boundary data, and the function $u_L\colon \Omega \to \mathbb{R}$ defines the obstacle. It is assumed that $u_L \leq u_\Omega$ on the boundary. If no obstacle is present, the minimal surface satisfies the boundary-value problem

$$\operatorname{div}\left((1 + |\nabla u|^2)^{-1/2} \nabla u\right) = 0 \text{ for } x \in \Omega, \quad \text{and} \quad u = u_\Omega \text{ for } x \in \partial\Omega$$

(see, e.g., [26, p. 174]). In the case considered here, the boundary constraint u_Ω is defined such that

$$u_\Omega = \begin{cases} 1 - (2x_1 - 1)^2, & \text{if } x_2 = 0, 1; \\ 0, & \text{otherwise.} \end{cases}$$

The obstacle is

$$u_L = \begin{cases} 1, & \text{if } |x_1 - \frac{1}{2}| \leq \frac{1}{4} \text{ and } |x_2 - \frac{1}{2}| \leq \frac{1}{4}; \\ 0, & \text{otherwise.} \end{cases}$$

To make u_L continuous, we added a small band abound the center box where u_L varied linearly from 0 to 1. This is illustrated in Fig. 3.

Table 3 shows the results of the computation. Because the initial mesh had $n = 685$ vertices, the computation was started with $\mu = .25$ in order to ensure that the refined meshes correspond to those of the previous examples. However, this problem differs from the previous ones in some other significant respects. The definition of the initial coarse mesh provides a convenient framework in which to define u_L (see Fig. 3). However, the use of this mesh also means that the boundary of the central square where u_L is nearly discontinuous is automatically and exactly incorporated into each of the refined meshes. The fact that this (anticipated) interface is described exactly in the mesh is, of course, beneficial to the adaptive meshing procedure. The solution and refined mesh are shown in Fig. 4.

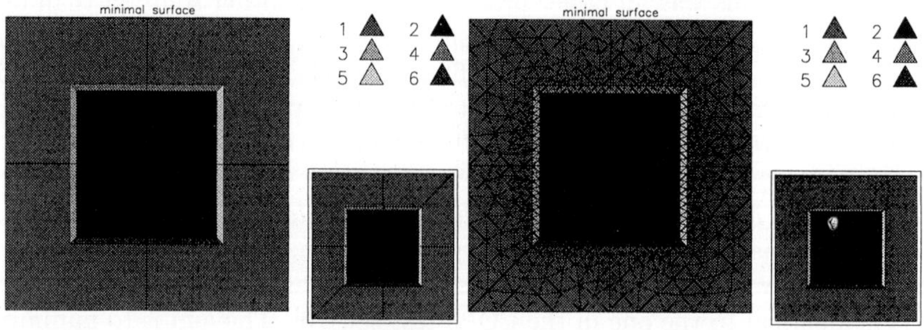

Fig. 3. Minimal surface problem: On the left, we illustrate the skeleton corresponding to the definition of u_L. On the right, we show the corresponding initial mesh with 685 mesh points (see Plate 6 on page 347).

Table 3. Minimal surface problem.

		Method A		Method B	
n	μ	Iter	cpu secs	Iter	cpu secs
685	2^{-2}	11	0.52	11	0.52
1000	2^{-2}	7	1.09		
1000	2^{-3}	3	1.41	5	1.00
4000	2^{-3}	14	6.10		
4000	2^{-4}	3	7.70	9	4.31
16000	2^{-4}	7	21.86		
16000	2^{-5}	6	36.27	5	16.27
64000	2^{-5}	5	109.61		
64000	2^{-6}	7	199.06	5	92.11

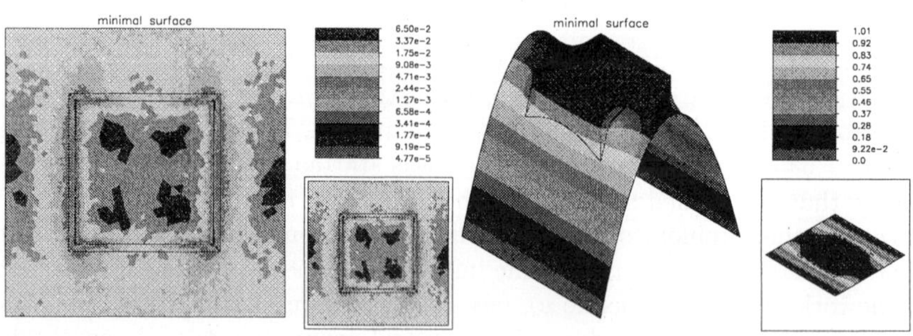

Fig. 4. Minimal surface problem. On the left is an plot of the element size in the adaptive mesh with $n = 64000$ vertices. On the right is the corresponding solution (see Plate 7 on page 348).

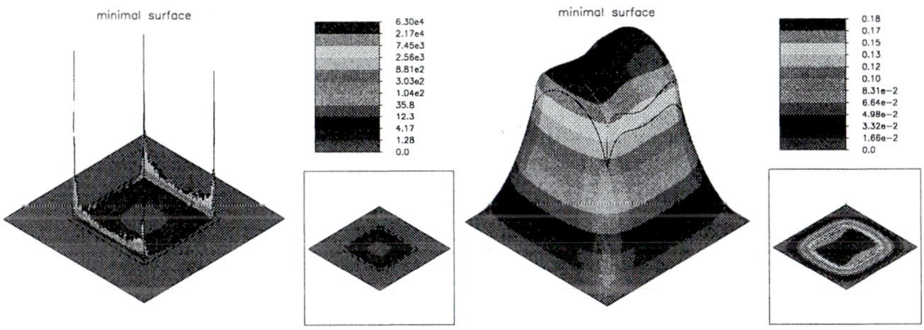

Fig. 5. Minimal surface problem. On the left is a plot of the Lagrange multipliers for the mesh with $n = 64000$ vertices. On the right is a plot of the smoothed Lagrange multipliers (see Plate 8 on page 348).

Another difference we observed in this problem is that the Lagrange multipliers are quite rough. In Fig. 5, we show both the Lagrange multipliers and the smoothed Lagrange multipliers used for interpolation. It is clear that the multipliers exhibit a δ-function-like behavior on the boundary of the contact zone, particularly at the corners of the square. The smoothed multipliers are visibly much nicer. We also note that the number of Newton iterations for the refinement and continuation steps is roughly comparable for this problem. As before, this behavior mostly reflects the quality of the initial guesses. For this problem, the roughness of the Lagrange multipliers is probably responsible for the additional Newton iterations needed for the continuation steps, in comparison with previous examples.

Finally, Table 4 gives the cpu seconds required for some of the principal routines of the algorithm. All initialization and preprocessing, including the computation of the sparsity pattern of H and the setup for the data structures for each level, is done in subroutine `mginit`. The multigraph incomplete LU factorization is computed in subroutine `mgilu`. These factors are used in

Table 4. Minimal surface problem. Breakdown of the computation time. The first set of times are for explicit refinement and continuation steps.

Task	cpu secs	cpu secs
Mesh generation	19.19	15.17
Subroutine mginit	20.40	10.35
Subroutine mgilu	31.16	14.01
Subroutine mg	26.77	10.07
Subroutine linsys	50.19	17.89
Subroutine tpick	15.43	5.94
Subroutine smlm	31.41	16.29
Other	4.51	2.39

subroutine `mg` to solve the main system (11). Subroutine `tpick` performs the line search and computes required directional derivatives. Subroutine `linsys` computes the finite element Jacobian and Newton residual. The computation of the smoothed Lagrange multipliers is done in subroutine `smlm`.

Acknowledgements

We thank Michael Saunders for many helpful comments on the performance and implementation of iterative solvers for interior methods.

References

1. R. E. BANK, *PLTMG: A Software Package for Solving Elliptic Partial Differential Equations, Users' Guide 8.0*, Software, Environments and Tools, Vol. 5, SIAM, Philadelphia, 1998.
2. R. E. BANK AND K. R. SMITH, *The incomplete factorization multigraph algorithm*, SIAM J. Sci. Comput., 20 (1999), pp. 1349–1364 (electronic).
3. D. P. BERTSEKAS, *On the Goldstein-Levitin-Polyak gradient projection method*, IEEE Trans. Automatic Control, AC-21 (1976), pp. 174–184.
4. ———, *Projected Newton methods for optimization problems with simple constraints*, SIAM J. Control Optim., 20 (1982), pp. 221–246.
5. A. BONDARENKO, D. BORTZ, AND J. J. MORÉ, *COPS: Large-scale nonlinearly constrained optimization problems*, Technical Report ANL/MCS-TM-237, Mathematics and Computer Science division, Argonne National Laboratory, Argonne, IL, 1998. Revised October 1999.
6. D. BRAESS, *Finite elements*, Cambridge University Press, Cambridge, second ed., 2001. Theory, fast solvers, and applications in solid mechanics, Translated from the 1992 German edition by Larry L. Schumaker.
7. M. A. BRANCH, T. F. COLEMAN, AND Y. LI, *A subspace, interior, and conjugate gradient method for large-scale bound-constrained minimization problems*, SIAM J. Sci. Comput., 21 (1999), pp. 1–23 (electronic).
8. T. D. CHOI AND C. T. KELLEY, *Estimates for the Nash-Sofer preconditioner for the reduced Hessian for some elliptic variational inequalities*, SIAM J. Optim., 9 (1999), pp. 327–341.
9. T. F. COLEMAN AND Y. LI, *An interior trust region approach for nonlinear minimization subject to bounds*, SIAM J. Optim., 6 (1996), pp. 418–445.
10. E. D. DOLAN AND J. J. MORÉ, *Benchmarking optimization software with performance profiles*, Math. Program., 91 (2002), pp. 201–213.
11. Z. DOSTÁL, *Box constrained quadratic programming with proportioning and projections*, SIAM J. Optim., 7 (1997), pp. 871–887.
12. A. V. FIACCO AND G. P. MCCORMICK, *Nonlinear Programming: Sequential Unconstrained Minimization Techniques*, Classics in Applied Mathematics, SIAM, Philadelphia, 1990. ISBN 0-89871-254-8.
13. A. FORSGREN AND P. E. GILL, *Primal-dual interior methods for nonconvex nonlinear programming*, SIAM J. Optim., 8 (1998), pp. 1132–1152. (Electronic).
14. E. M. GERTZ AND P. E. GILL, *A primal-dual trust region algorithm for nonlinear programming*, Report to appear, Department of Mathematics, University of California, San Diego, 2002.

15. R. GLOWINSKI, *Numerical methods for nonlinear variational problems*, Springer-Verlag, New York, 1984. ISBN 0-387-12434-9.
16. N. I. M. GOULD, S. LUCIDI, M. ROMA, AND P. L. TOINT, *Solving the trust-region subproblem using the Lanczos method*, SIAM J. Optim., 9 (1999), pp. 504–525 (electronic).
17. M. HEINKENSCHLOSS, M. ULBRICH, AND S. ULBRICH, *Superlinear and quadratic convergence of affine-scaling interior-point Newton methods for problems with simple bounds without strict complementarity assumption*, Math. Program., 86 (1999), pp. 615–635.
18. R. H. W. HOPPE AND R. KORNHUBER, *Adaptive multilevel methods for obstacle problems*, SIAM J. Numer. Anal., 31 (1994), pp. 301–323.
19. D. KINDERLEHRER AND G. STAMPACCHIA, *An introduction to variational inequalities and their applications*, Academic Press Inc. [Harcourt Brace Jovanovich Publishers], New York, 1980.
20. ———, *An introduction to variational inequalities and their applications*, Society for Industrial and Applied Mathematics (SIAM), Philadelphia, PA, 2000. Reprint of the 1980 original.
21. R. KORNHUBER, *Monotone multigrid methods for elliptic variational inequalities. I*, Numer. Math., 69 (1994), pp. 167–184.
22. ———, *Monotone multigrid methods for elliptic variational inequalities. II*, Numer. Math., 72 (1996), pp. 481–499.
23. C.-J. LIN AND J. J. MORÉ, *Newton's method for large bound-constrained optimization problems*, SIAM J. Optim., 9 (1999), pp. 1100–1127 (electronic). Dedicated to John E. Dennis, Jr., on his 60th birthday.
24. J. J. MORÉ AND G. TORALDO, *On the solution of large quadratic programming problems with bound constraints*, SIAM J. Optim., 1 (1991), pp. 93–113.
25. S. G. NASH, *Newton-type minimization via the Lánczos method*, SIAM J. Numer. Anal., 21 (1984), pp. 770–788.
26. J. RAUCH, *Partial differential equations*, Springer-Verlag, New York, 1991.
27. M. ULBRICH, S. ULBRICH, AND M. HEINKENSCHLOSS, *Global convergence of trust-region interior-point algorithms for infinite-dimensional nonconvex minimization subject to pointwise bounds*, SIAM J. Control Optim., 37 (1999), pp. 731–764 (electronic).
28. S. J. WRIGHT, *Primal-dual interior-point methods*, Society for Industrial and Applied Mathematics (SIAM), Philadelphia, PA, 1997. ISBN 0-89871-382-X.
29. ———, *On the convergence of the Newton/log-barrier method*, Technical report MCS-P681-0897, Mathematics and Computer Science Division, Argonne National Laboratory, Argonne, IL, 1997, Revised April, 2000.

Hierarchical Control of a Linear Diffusion Equation

Anthony J. Kearsley[1], Jon W. Tolle[2], and Paul T. Boggs *[3]

[1] Department of Mathematical Science, Carnegie Mellon University, Pittsburgh, PA 15213-3890 (anthonyk@andrew.cmu.edu)
[2] Departments of Mathematics and Operations Research, University of North Carolina, Chapel Hill, NC, 27599-3180 (tolle@email.unc.edu)
[3] Computational Sciences and Mathematics Research Department, Sandia National Laboratories, Livermore, CA 94550 (ptboggs@sandia.gov)

Abstract. As part of our study of techniques for solving PDE-constrained optimization problems, we investigate a pertinent class of hierarchical control problems. In a first step towards addressing the difficulties encountered in simulating reservoirs for optimal well placement and other multi-objective control problems, we consider the approximate pointwise control of a linear parabolic system with multiple targets. Assuming a hierarchy among the objectives, we derive optimality conditions and obtain numerical results for a particular test problem.

1 Introduction

Optimal control problems constitute an interesting case of PDE-based optimization problems. There is a rich history of work in this area, beginning with the development of the calculus of variations and work in the control of ordinary differential equations. More recently, researchers have begun to investigate the control of PDEs (see the survey papers [9] and [10]). Instances of these types of problems abound in applications. Thus the development of efficient numerical methods for the solution of these problems has also been the subject of significant recent research.

In this paper we examine a particular instance of optimal control problems where multiple controls seek to force behavior close to multiple "targets" simultaneously. These problems belong to the class of problems called *multi-criteria optimization.* No unique mathematical formulation of these types of problems exists. In practice these problems are often solved by minimizing a weighted sum of the deviations from the targets with the weights corresponding to the preferences (see [11]). Another formulation, sometimes referred to as *goal programming,* insists that the preferred targets must be satisfied to within certain tolerances and the others are reduced as much as possible within these constraints (see [12]). Both of these approaches involve the

* Sandia is a multiprogram laboratory operated by Sandia Corporation, a Lockheed Martin Company, for the United States Department of Energy under contract DE-AC04-94-AL85000.

choice of a set of weights or tolerances for which there can be little theoretical guidance. Yet another approach, called *bilevel optimization,* nests optimization problems, optimizing with respect to one target and using the solutions as parameters in a larger problem (see [15]).

Motivated by specific applications, especially those arising in optimal well placement in reservoir engineering, we investigate means of formulating a special class of optimal control problems in which there is a partition of the targets into two categories, one of which is clearly more important than the other. This approach, based on the work of von Stackelberg [16] in an economic context, requires that a certain level of target satisfaction be achieved in the more important group, called the "leader" set, as a priority over reducing the error in the second target group, called the "follower" set. This type of optimal control problem has been termed "hierarchical control". One way of formulating this problem is in terms of a nested optimization structure in which the follower targets are minimized subject to fixed values of certain of the control variables, the "inner minimization", and then an "outer minimization" is performed over the remaining control variables to obtain optimal leader target satisfaction. The resulting accuracy on the follower targets in therefore determined by and is subordinate to the optimization over the leader targets. This type of bilevel optimization problem has been the object of a great deal of research (see [15] for an exhaustive bibliography) in finite-dimensional optimization and was given a theoretical grounding in the work of Lions [14] in the context of linear hyperbolic state equations. In this paper, we carry out the analysis for a specific parabolic system and obtain preliminary numerical results that we believe illustrate the promise of this type of approach.

The paper is organized as follows. In section 2 we survey various formulations of the hierarchical control problem, with special emphasis on the method of Lions. In section 3, we derive optimality conditions for a special problem and show that these give rise to an interesting formulation. We consider the numerical solution of this example in section 4. Finally, in section 5, we provide some added discussion and consider future extensions to this approach.

2 Model Formulations

Our model problem in hierarchical control is based on a standard linear parabolic equation in two space dimensions with pointwise controls:

$$y_t - \mathbf{A}\, y = f(x,t) + \sum_{j=1}^{k} v_j(t)\, \delta(x - a_j(t)), \quad (x,t) \in \Omega \times (0,T), \tag{1}$$

$$y(x,0) = b_0(x), \quad x \in \Omega, \tag{2}$$

$$y(x,t) = b_1(x,t), \quad (x,t) \in \Gamma_1 \times (0,T), \tag{3}$$

$$y_\eta(x,t) = b_2(x,t), \quad (x,t) \in \Gamma_2 \times (0,T), \tag{4}$$

where Ω is a bounded open subset of $\Re^2$, $T > 0$ is finite, and $\Gamma_1 \cup \Gamma_2$ is a partition of the boundary of Ω. We assume that the functions in the model are well-behaved, i.e., $f(x,t) \in L^2(0,T;\Omega), b_0(x) \in L^2(\Omega)$, and $b_j(x,t) \in L^2(0,T;\Gamma_j), j = 1,2$. Here $\mathbf{A}$ is an elliptic operator and y_η represents the normal derivative of y. For this system we seek to determine controls $v_j(t), j = 1,\ldots,k$ and locations $a_j(t),\ j = 1,\ldots,k$ that will cause the resulting temperature distribution at time T, $y(x,T)$, to be "close" to multiple targets, $y_1(x),\ldots,y_k(x)$. As this is a multi-objective optimization problem, choosing an adequate formulation of the problem that reflects practical considerations, has a firm theoretical foundation, and is amenable to numerical solution poses significant challenges. A further complication induced in applications of this problem is introduced by the addition of restrictions on the locations, $a_j(t)$. These restrictions are typically expressed as nonlinear, possibly nonconvex, inequality constraints, e.g., the sites may not get too near to each other or too close to fixed regions either inside or on the boundary.

There are a number of possible formulations for this problem that depend upon how the various targets are prioritized. A common approach is to weight the targets according to some preassigned idea of their relative importance. For example, we could choose to solve the problem

$$\begin{aligned} \min_{v,a,y} \ & \tfrac{1}{2}\sum_{j=1}^{k} \left\{ \int_0^T v_j^2(t)\,dt + \gamma_j \, \| \, y(x,T) - y_j(x) \, \|_{L^2(\Omega)}^2 \right\} \\ \text{subject to:} \quad & y \text{ satisfies } (1) - (4), \\ & g(a_1(t), a_2(t), \ldots, a_k(t)) \leq 0, \end{aligned}$$

for some fixed constants $\gamma_j,\ j = 1,\ldots,k$. The inequality constraint involving $g(a_1(t),\ldots,a_k(t))$ represents all of the restrictions on the locations. In this problem, and most other formulations that we consider, the terms in the objective function involving the minimization of the controls are required to guarantee the well-posedness of the problems; although from a practical point of view these terms represent a cost functional that could be more general than that given here.

This formulation essentially treats the targets as equal modulo the choice of the constants, γ_j; there is no structural preference for any particular set of targets. Another approach is to put restrictions on the error allowed in the targets, i.e., to solve:

$$\begin{aligned} \min_{v,a,y} \ & \tfrac{1}{2}\int_0^T \sum_{j=1}^{k} v_j^2(t)\,dt \\ \text{subject to:} \quad & y \text{ satisfies } (1) - (4), \\ & \| \, y(x,T) - y_j(x) \, \|_{L^2(\Omega)}^2 \leq \beta_j, \quad j = 1,\ldots,k, \\ & g(a_1(t), a_2(t), \ldots, a_k(t)) \leq 0. \end{aligned}$$

This problem requires an *a priori* choice of the constants β_j that may give rise to an infeasible problem and again does not establish a structural preference among the targets.

In contrast to the above we might consider approaches that reflect the priorities of the targets in terms of the structure of the optimization problem. To do this we first partition the targets into two sets, L and F, with the targets indexed by L being of the highest priority and those by F being of lesser priority. The labels L and F stand for the *leader* and *follower* targets and the idea is to satisfy the follower target goals only subject to the satisfaction of those target conditions of the leader. We also identify each pair, $(v_j,\ a_j)$, with the corresponding target, $y_j(x)$, thus leading to the partitions v_L, v_F and a_L, a_F.

One formulation of this type of problem might be:

$$\begin{aligned}
&\min_{v,a} \tfrac{1}{2}\sum_{j=1}^{k}\int_0^T v_j^2(t)\,dt + \tfrac{1}{2}\sum_{j\in F}\gamma_j \,||\, y(x,T) - y_j(x)\,||^2_{L^2(\Omega)} \\
&\text{subject to:} \qquad y \text{ satisfies } (1) - (4), \\
&\qquad\qquad ||\, y(x,T) - y_j(x)\,||^2_{L^2(\Omega)} \le \beta_j, \quad j \in L, \\
&\qquad\qquad g(a_1(t), a_2(t), \dots, a_k(t)) \le 0.
\end{aligned}$$

In this formulation the conditions on the targets of lower priority are satisfied after those of higher priority are satisfied to within a given tolerance.

Here we consider a different model, called the *hierarchical control* model, motivated by the theoretical results of Lions [14]. This formulation uses an inner optimization problem over a subset of the follower variables. First we fix the variables $a_j, j = 1, \dots, k$ and v_L and consider the minimization over the follower controls v_F, and y:

$$\begin{aligned}
&\min_{y,v_F} \tfrac{1}{2}\int_0^T \sum_{j\in F} v_j^2(t)\,dt + \tfrac{1}{2}\sum_{j\in L}\gamma_j \,||\, y(x,T) - y_j(x)\,||^2_{L^2(\Omega)} \\
&\text{subject to: } y_t - \mathbf{A}\, y = f(x,t) + \sum_{j=1}^{k} v_j(t)\,\delta(x - a_j(t)), \\
&\qquad\qquad\qquad\qquad (x,t) \in \Omega \times (0,T), \\
&\qquad y(x,0) = b_0(x), \quad x \in \Omega, \\
&\qquad y(x,t) = b_1(x,t), \quad (x,t) \in \Gamma_1 \times (0,T), \\
&\qquad y_\eta(x,t) = b_2(x,t), \quad (x,t) \in \Gamma_2 \times (0,T).
\end{aligned} \tag{IP}$$

In this theoretical formulation, the inner problem (IP) is not solved explicitly but the necessary conditions for its solution are obtained and used as constraints, together with the constraints on the leader control, to formulate an outer problem,

$$\begin{array}{lll} \min\limits_{v,a,y} \frac{1}{2}\int_0^T \sum\limits_{j=1}^k v_j^2(t)\,dt & & \\ \text{subject to:} & g(a_1(t), a_2(t), \ldots, a_k(t)) \leq 0, & (OP) \\ & \|\, y(x,T) - y_j(x) \,\|_{L^2(\Omega)}^2 \leq \beta_j,\ j \in L & \\ & y, v_F \text{ satisfy the necessary conditions for } (IP). & \end{array}$$

For a problem with one leader and one follower target and with corresponding *boundary controls*, Lions showed that this sequence of problems can be solved with the restriction on the deviation from the leader target (β) arbitrarily small, i.e., the problem in the leader variable is approximately controllable. As mentioned in the introduction, the hierarchical formulation of a multi-objective problem was first suggested in the finite-dimensional case by von Stackelberg [16]. This finite-dimensional analog (often called bilevel optimization) has been theoretically analyzed by Chen and Florian ([6]).

This formulation is unsatisfactory from a computational standpoint because very small values of β_j need to be chosen which force the feasible region to be small and because the existence of quadratic inequality constraints in the state variables y complicates the optimization process. Thus, in this paper we consider another formulation of (OP) that has been useful for approximately controllable problems (see [9]); namely, we consider

$$\begin{array}{ll} \min\limits_{v,a,y} \frac{1}{2}\int_0^T \sum\limits_{j=1}^k v_j^2(t)\,dt + \frac{1}{2}\sum\limits_{j\in L} \gamma_j \,\|\, y(x,T) - y_j(x) \,\|_{L^2(\Omega)}^2 & \\ \text{subject to:} & g(a_1(t), a_2(t), \ldots, a_k(t)) \leq 0, \\ & y, v_F \text{ satisfy the necessary conditions for } (IP), \end{array}$$

for some choice of $\gamma_j, j \in L$. In the next section we will provide a more explicit description of this formulation for our test problem.

Several additional comments need to be made concerning this formulation. First, the relative sizes of the constants $\gamma_j, j \in F$ and $\gamma_j, j \in L$ affect how accurately the different targets can be approximated. One of the goals of the numerical study described in Section 4 was to determine how the effect of differing scales of magnitude on the choice of these leader and follower constants affected the optimal solutions. Secondly, it is clear that this formulation of the problem is fundamentally different from other models. As is well documented in the finite-dimensional cases of bilevel programming, an optimal solution to a bilevel optimization problem need not be a *Pareto optimal* solution in the sense of multi-objective optimization and we have not been guided by that requirement. Also, the inclusion of the follower control sites a_F as part of the outer optimization, rather than the inner optimization problem, may seem inconsistent. In formulating the problem in this manner, we were again motivated by an effort to make the problem tractable; complicated (and possibly nonconvex) inequality constraints in the control locations would seriously degrade the ability to express concisely the necessary conditions for the inner problem. All of these points speak to the difficulty in formulating and in solving large scale multi-criteria optimization problems. The results presented here represent an initial effort in this direction.

Finally, we observe that hierarchical control might profitably be used to formulate a multitude of important scientific applications. For example in the area of oil reservoir simulation, one can formulate optimal well placement problems as hierarchical control problems where hard constraints on well locations form mandatory (or leader) targets while revenue or efficiency based targets can be optimized in a secondary (follower) fashion. Problems in optimal airfoil design can be viewed in a similar way with structural constraints being posed as leader objectives and vorticity minimizing goals being follower targets. Remote manipulator systems, like those employed by space-craft, are required to rapidly solve optimal control problems. In some instances, these systems must accomplish a goal while maintaining prescribed distances from other pieces of machinery. One could formulate a class of hierarchical control in which leader targets include primary objectives and follower targets maintain minimal separation from sensitive machinery whenever possible.

3 The Test Problem

In this paper we present results for a relatively simple version of the general formulation described in the preceding section. In particular, we consider the case in which there are two targets and two locations and the underlying PDE is the heat equation in two variables on the unit square with homogeneous Dirichlet boundary conditions given by

$$y_t - \triangle y = v_L(t)\,\delta(x - a_L(t)) + v_F(t)\,\delta(x - a_F(t)), \quad (x,t) \in Q, \tag{5}$$

$$y(x,0) = 0, \quad x \in \Omega, \tag{6}$$

$$y(x,t) = 0, \quad (x,t) \in \partial\Omega \times (0,T). \tag{7}$$

Here $\Omega = (0,1)^2$, $Q = \Omega \times (0,T)$, and $\triangle$ represents the Laplacian operator. We have labelled the two controls $v_L(t)$ and $v_F(t)$ (with the corresponding labelling for the locations) so that each control is identified with a particular target. As above, the labels on the targets stand for the *leader* and *follower* targets.

In our adaptation of the Lions [14] hierarchical model to this problem the control variables $v_L(t)$ and the locations $a_L(t)$ and $a_F(t)$ are fixed and an inner optimization problem is solved to determine the PDE solution, $y(x,t)$, and the follower control, $v_F(t)$, as functions of these variables. Formally, we seek to solve

$$\begin{aligned}
&\min_{(v,a,y)} \frac{1}{2}\int_0^T \{v_L{}^2 + v_F{}^2\}\, dt + \tfrac{1}{2}\gamma_L \|\, y(x,T) - y_L(x)\,\|^2_{L^2(\Omega)} \\
&\text{subject to:} \\
&\qquad g(a_L, a_F) \leq 0, \\
&\qquad \min_{(v_F,y)} \frac{1}{2}\int_0^T v_F^2\, dt + \tfrac{1}{2}\gamma_F \|\, y(x,T) - y_F(x)\,\|^2_{L^2(\Omega)} \\
&\qquad \text{subject to: } y_t - \triangle\, y = v_L(t)\,\delta(x - a_L(t)) \\
&\qquad\qquad\qquad\qquad + v_F(t)\,\delta(x - a_F(t)), \\
&\qquad\qquad\qquad\qquad (x,t) \in Q, \\
&\qquad\qquad y(x,0) = 0, \quad x \in \Omega, \\
&\qquad\qquad y(x,t) = 0, \quad (x,t) \in \partial\Omega \times (0,T).
\end{aligned}$$

In this format the solution of the inner optimization problem can be thought of as a black box, yielding a function, $y(x,t)$, and a control, $v_F(t)$, for given values of v_L, a_L, and a_F.

We proceed by deriving the optimality conditions for the inner problem, which yields the variables y and v_F in terms of the values of the variables v_L, a_L, and a_F. Then we derive the adjoint system for the resulting outer problem obtaining a surprisingly elegant formulation of this problem in an all-at-once format.

For fixed v_L, a_L, and a_F, the variational equation for the inner optimization problem is

$$\int_0^T v_F\, \hat{v}_F\, dt + \gamma_F \int_\Omega (y(x,T) - y_F(x))\, \hat{y}\, dx = 0 \tag{8}$$

for all admissible $\hat{v}_F$ and $\hat{y}$, i.e., $\hat{v}_F$ and $\hat{y}$ satisfying

$$\hat{y}_t - \triangle\, \hat{y} = \hat{v}_F\, \delta(x - a_F), \quad (x,t) \in Q, \tag{9}$$

$$\hat{y}(x,0) = 0, \quad x \in \Omega, \tag{10}$$

$$\hat{y}(x,t) = 0, \quad (x,t) \in \partial\Omega \times (0,T). \tag{11}$$

If we define $p(x,t)$ to be a solution to the adjoint system

$$p_t + \triangle\, p = 0, \quad (x,t) \in Q, \tag{12}$$

$$p(x,t) = 0, \quad (x,t) \in \partial\Omega \times (0,T), \tag{13}$$

$$p(x,T) = -\gamma_F\, (y(x,T) - y_F(x)), \quad x \in \Omega, \tag{14}$$

then multiplying (9) by $p(x,t)$, integrating over Q, and applying Green's theorem yields the variational equation (8) provided that we choose

$$v_F(t) = p(a_F(t), t), \quad t \in (0,T). \tag{15}$$

Our optimization problem can now be written as

$$\min_{v,a,y} \frac{1}{2} \int_0^T \{{v_L}^2 + {v_F}^2\}\, dt + \tfrac{1}{2}\gamma_L \,||\, y(x,T) - y_L(x) \,||^2_{L^2(\Omega)}$$
subject to: (5), (6), (7), (12), (13), (14), (15).

Next we derive a set of necessary conditions for this problem assuming that a_L and a_F are fixed. The variational equation for the objective function in this case is

$$\int_0^T (v_L\, \hat{v}_L + v_F\, \hat{v}_F)\, dt + \gamma_L \int_\Omega (y(x,T) - y_L(x))\, \hat{y}\, dx = 0 \qquad (16)$$

for admissible variations $\hat{v}_L$, $\hat{v}_F$, and $\hat{y}$. In light of equations (14) and (15) we see that the variations depend also on a variation in $p(x,t)$ which we denote by $\hat{p}$(x,t). Examining the variations in (5) – (7), (12) – (14), and (15) yields the following admissibility equations:

$$\hat{y}_t - \triangle\, \hat{y} = \hat{v}_L\, \delta(x - a_L) + \hat{v}_F\, \delta(x - a_F), \quad (x,t) \in Q, \qquad (17)$$
$$\hat{y}(x,0) = 0, \quad x \in \Omega, \qquad (18)$$
$$\hat{y}(x,t) = 0, \quad (x,t) \in \partial\Omega \times (0,T), \qquad (19)$$
$$\hat{p}_t + \triangle\, \hat{p} = 0, \quad (x,t) \in Q, \qquad (20)$$
$$\hat{p}(x,t) = 0, \quad (x,t) \in \partial\Omega \times (0,T), \qquad (21)$$
$$\hat{p}(x,T) = -\gamma_F\, \hat{y}(x,T), \quad x \in \Omega, \qquad (22)$$
$$\hat{v}_F(t) = \hat{p}(a_F(t),t), \; t \in (0,T). \qquad (23)$$

To obtain the appropriate set of adjoint equations from this system we introduce the functions $Y(x,t)$ and $P(x,t)$ that satisfy the following partial differential equations and boundary conditions:

$$Y_t + \triangle\, Y = 0, \quad (x,t) \in Q, \qquad (24)$$
$$Y(x,T) = -\gamma_F\, P(x,T) - \gamma_L\, (y(x,T) - y_L(x)), \quad x \in \Omega, \qquad (25)$$
$$Y(x,t) = 0, \quad (x,t) \in \partial\Omega \times (0,T), \qquad (26)$$
$$P_t - \triangle\, P = -\delta(x - a_F)\, (p(x,t) - Y(x,t), \quad (x,t) \in Q, \qquad (27)$$
$$P(x,0) = 0, \quad x \in \Omega, \qquad (28)$$
$$P(x,t) = 0, \quad (x,t) \in \partial\Omega \times (0,T). \qquad (29)$$

With these definitions, if we multiply (17) by $Y(x,t)$ and (20) by $P(x,t)$ and integrate the resulting equations by parts, then the variational equation (16) is satisfied provided that we choose

$$v_L(t) = Y(a_L(t),t). \qquad (30)$$

The resulting form of the optimization problem after these manipulations has the following compact form:

$$
\begin{array}{ll}
\min\limits_{a_L, a_F} \frac{1}{2} \displaystyle\int_0^T \{ Y(a_L(t),t)^2 + p(a_F(t),t)^2 \} \, dt + \frac{1}{2}\gamma_L || \, y(x,T) - y_L(x) \, ||^2_{L^2(\Omega)} & \\
\text{subject to:} \qquad y \text{ satisfies } (5) - (7), & \\
\qquad p \text{ satisfies } (12) - (14), & \\
\qquad Y \text{ satisfies } (24) - (26), & \\
\qquad P \text{ satisfies } (27) - (29), & \\
\qquad v_F(t) = p(a_F(t),t), \quad t \in (0,T) & \\
\qquad v_L(t) = Y(a_L(t),t), \quad t \in (0,T) & \\
\qquad g(a_L, a_F) \leq 0. & \\
\end{array}
\tag{LF}
$$

For a more detailed description of the derivation of these equations, see [2].

4 Numerical Results

In order to test the efficacy of our problem formulation, (LF), we discretize the domain and solve the resulting nonlinear programming problem. For the state-equation portion of our constraints, we use a combination of finite-difference discretization in time with finite-element discretization in space. For each test problem, the domain is simply the unit square in $\Re^2$, $\Omega = (0,1) \times (0,1)$. While we explain the discretization in this context it is clear that the explanation is valid for any piecewise polygonal domain.

We employ a discretization of second-order accuracy for our state equation and the associated adjoint equations. We will give only an abbreviated description here. For further details and accuracy estimates of this discretization, see [5].

We begin by describing the time discretization. Let N_T be the number of time steps desired, so that $\Delta t = \frac{T}{N_T}$. We will denote the estimate of y at the nth time step by y^n where $n = 1 \ldots N_T$. Denoting the number of spatial steps in the x_1 and in the x_2 directions by N_X, the spatial step is denoted by h with $h = \frac{1}{N_X}$. The discrete approximation to y is

$$y(n\Delta t, ih, jh) \approx y^n_{i,j}.$$

We follow the *two-step implicit scheme* for parabolic problems as outlined in Glowinski [8]. Accordingly, we define

$$\frac{\partial y}{\partial t}((n+1)\,\Delta t) \approx \frac{1}{2\Delta t}\left(3y^{n+1}_{i,j} - 4y^n_{i,j} + y^{n-1}_{i,j}\right).$$

Experience with this time discretization has led us to use it on stiff problems when we need to integrate to large values of T. In such cases, the fact that it assures unconditional stability and produces an accuracy to second order in time amply justifies the storage costs.

At each time step we must solve an elliptic problem to obtain $y_{i,j}^{n+1}$. The domain is so simple that we use the very common finite-element triangulation of Ω consisting of bisected squares. The space of polynomials of degree ≤ 1 is used to form a finite dimensional approximation to $L^2(\Omega)$ and $H^1(\Omega)$. More sophisticated schemes are certainly available for both linear and nonlinear parabolic equations. However, for testing optimization formulations of the control problem here, this simple numerical scheme is both adequate and appropriate. For specific applications, more specialized or hybrid discretizations may be called for (see for example [7]).

Two target states, $y_L(x)$ and $y_F(x)$, are used to test the performance of the formulation of the control problem from section 3. While a myriad of test shapes is possible, we choose one specific pair of test shapes that illustrates behavior seen in most of our numerical experiments. The leader target shape is a smooth function with a peak of approximately 1.3 at the point $x_1 = 1/3$, $x_2 = 1/2$ and the follower is a pyramid with a peak of unity at the point $x_1 = \frac{9}{10}, x_2 = \frac{1}{2}$ (see Figures 1 and 2). Specifically, the target functions are

$$\begin{aligned} y_L(x) &= 35x_1x_2(1-x_2)(1-x_1)^2, \\ y_F(x) &= 2\min\{5x_1-4, 3-5x_2, 5-5x_1, 5x_2-2\}. \end{aligned} \tag{31}$$

These test problems are similar to those used to study hierarchical control with stationary controls ([1]).

Secondly we add the constraint that the location of the pointwise controls (the parameters a_L and a_F) are each constrained to be contained inside disjoint balls. The leader location, a_L is constrained to lie within the circle centered at $(\frac{1}{4}, \frac{1}{2})$ where the follower location is constrained to lie within the circle centered at $(\frac{3}{4}, \frac{1}{2})$. Both constraints have radius 0.15 so that the two circles do not intersect (see Figure 4).

The optimization problem that arose from our formulation was solved using a sequential quadratic programming (SQP) algorithm. The specifics of the algorithm are contained in [4] and a theoretical analysis can be found in [3].

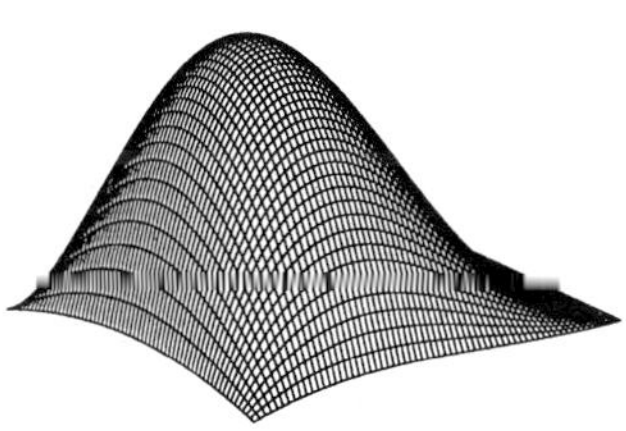

Fig. 1. The leader target

Fig. 2. The follower target

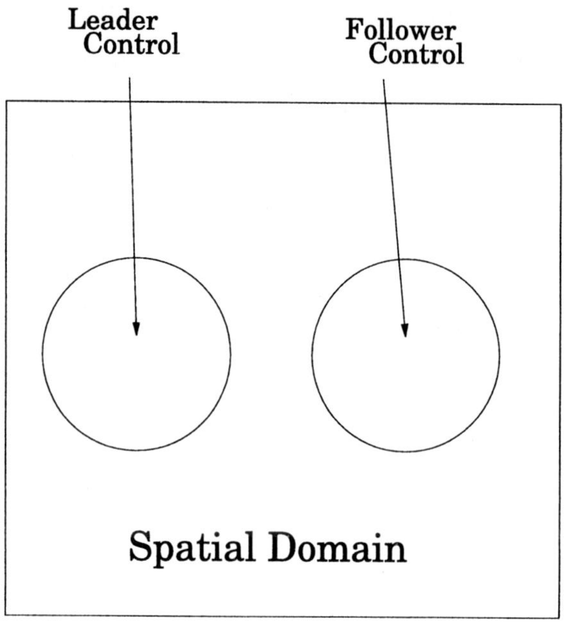

Fig. 3. Geometric constraints separating the controls

The numerical results are summarized in Table 1 together with Figures 4–9. The first two columns of Table 1 give the problem size. The values of γ_L and γ_F are combined in the next column. The next two columns give the discrepancy between state variables and targets in the L^2 norm. The final two columns of the table report on the L^2 norm of the controls.

Table 1. Numerical Performance Summary

N_X	N_T	(γ_F, γ_L)	$\|y_F - y\|_{L^2}/\|y_F\|_{L^2}$	$\|y_L - y\|_{L^2}/\|y_L\|_{L^2}$	$\|v_F\|$	$\|v_L\|$
64	32	(1.e+3,1.e+3)	2.395302	0.4873116	25.774	3.4065
64	32	(1.e+6,1.e+3)	2.181885	0.4978943	151.05	28.420
64	32	(1.e+3,1.e+6)	2.415231	0.2635630	34.279	448.49
128	32	(1.e+3,1.e+3)	2.371236	0.4732074	28.195	3.5591
128	32	(1.e+6,1.e+3)	2.200413	0.5009123	155.89	29.093
128	32	(1.e+3,1.e+6)	2.418927	0.2701232	34.861	449.12

Our problem formulation worked well with our numerical optimization algorithm. In numerical results we were able to solve problems with values of γ as large as $1.e16$ and values approaching machine precision, which indicates the robustness of our formulation with respect to these constants. We present here the results for more reasonable values of γ. In Figures 4, 6 and 8 the dot-

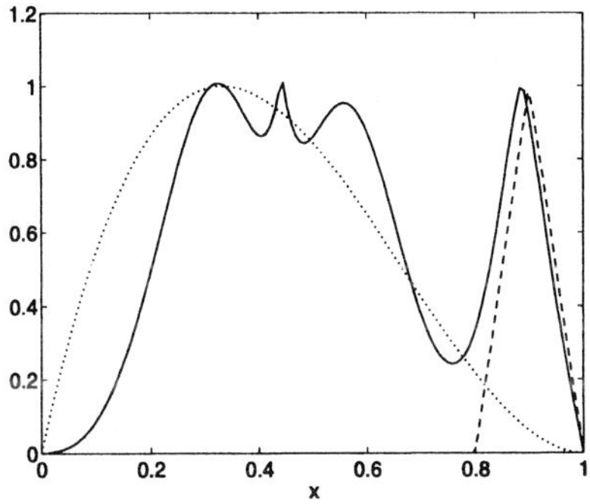

Fig. 4. The state variables restricted to the line $x_2 = \frac{1}{2}$ with $\gamma_L = \gamma_F = 1.e+3$

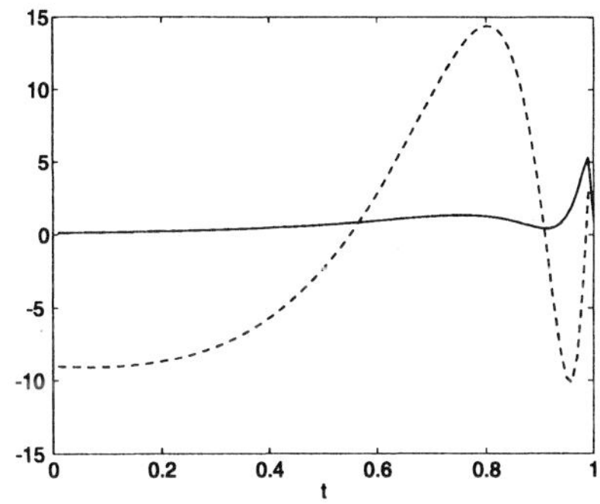

Fig. 5. The control variables with $\gamma_L = \gamma_F = 1.e+3$

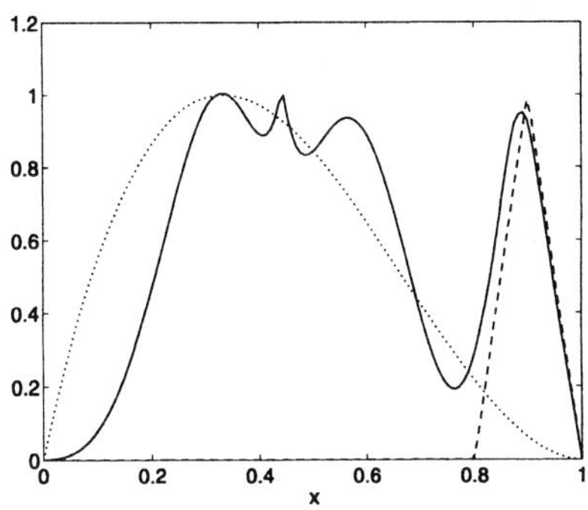

Fig. 6. The state variables restricted to the line $x_2 = \frac{1}{2}$ with $\gamma_L = 1.e+6$ and $\gamma_F = 1.e+3$

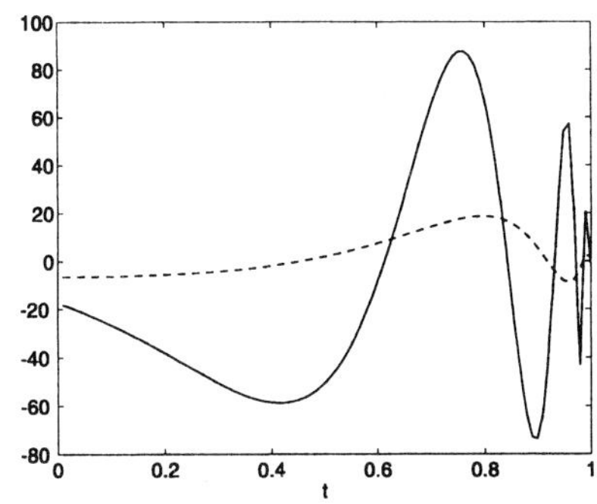

Fig. 7. The control variables with $\gamma_L = 1.e+6$ and $\gamma_F = 1.e+3$

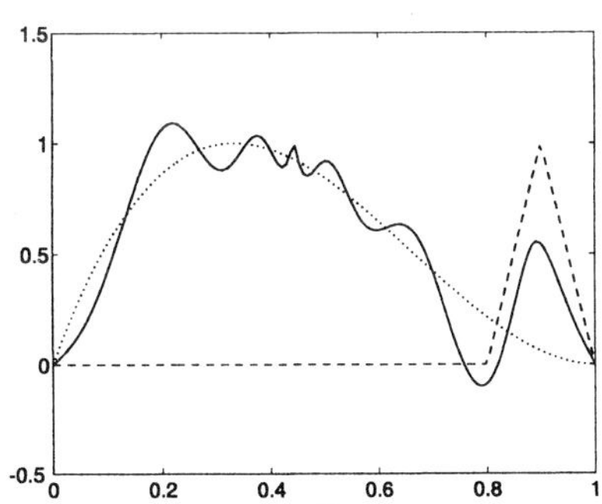

Fig. 8. The state variables restricted to the line $x_2 = \frac{1}{2}$ with $\gamma_L = 1.e+3$ and $\gamma_F = 1.e+6$

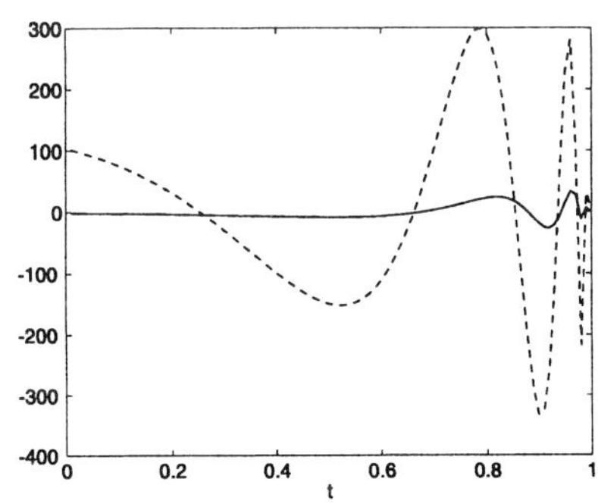

Fig. 9. The control variables with $\gamma_L = 1.e+3$ and $\gamma_F = 1.e+6$

ted and dashed profiles respectively denote leader and follower target profiles along the line $x_2 = \frac{1}{2}$. The solid lines are the state variable y at terminal time $T = 1$ also along the line $x_2 = \frac{1}{2}$. Clearly both the leader and follower targets were approximately attained. In Figures 5, 7 and 9 the leader and follower controls, $v_L(t)$ and $v_F(t)$ are shown with the leader control shown as the solid line. In Figure 5 the total variations in the two controls are comparable while in Figure 9 the value of γ_F is large enough, when compared with γ_L, that the effect of the follower control is greatly diminished. In fact the follower control oscillated so violently that it eclipsed the behavior of the leader control. Finally, it is worth noting that for the numerical examples presented here, the optimal location of both controls was inside the constraint circles.

Our numerical experience illustrated that the difficulty of the SQP algorithm in solving the hierarchical problem tested here increased with the values of the penalty parameters γ_L and γ_F. This fact is not surprising in light of the fact that similar behavior has been observed for the case of hierarchical control of Burgers' Equation ([13]).

5 Future Research

The freedom to specify multiple targets is extremely important for many practical problems. The high cost of solving multicriteria optimization problems suggests that there may be instances where hierarchical control problem formulations could yield a computational advantage in an affordable way. We plan to investigate the use of this formulation technique to attack more complicated physical phenomena, including those modeled by nonlinear equations. We anticipate the ideas will be fruitful when formulating problems of optimal well placement, contaminant transport, and bioremediation among others.

References

1. M. Berggren. *Control and Simulation of Advection–Diffusion Problems.* Ph.D Thesis, Computational and Applied Mathematics, Rice University, Houston, TX., 1995.
2. P. T. Boggs, A. J. Kearsley, and J. W. Tolle. Optimality conditions for hierarchical control. Technical Report #NISTIR-8321, National Institute of Standards and Technology, 2001.
3. Paul T. Boggs, Anthony J. Kearsley, and Jon W. Tolle. A global convergence analysis of an algorithm for large scale nonlinear programming problems. *SIAM Journal on Optimization*, 9(4):833–862, 1999.
4. Paul T. Boggs, Anthony J. Kearsley, and Jon W. Tolle. A practical algorithm for general large scale nonlinear optimization problems. *SIAM Journal on Optimization*, 9(3):755–778, 1999.
5. C. Carthel, R. Glowinski, and J. L. Lions. On exact and approximate boundary controllabilities for the heat equation. a numerical approach. *Jounal of Optimization Theory and Applications*, 82(3):429–484, 1994.

6. Y. Chen and M. Florian. The nonlinear bilevel programming problem: Formulations, regularity, and optimality conditions. *Optimization*, 32:193–209, 1995.
7. L. C. Cowsar, R. Glowinski, A. J. Kearsley, M. F. Wheeler, and I. Yotov. An optimization approach to multiphase flow. *Journal of Optimization Theory and Applications*, 111(3):473–488, 2001.
8. R. Glowinski. *Numerical Methods for Nonlinear Variational Problems.* Springer-Verlag, New York, 1984.
9. R. Glowinski and J.L. Lions. Exact and approximate controllability for distributed parameter systems i. *Acta Numerica*, pages 269–378, 1994.
10. R. Glowinski and J.L. Lions. Exact and approximate controllability for distributed parameter systems ii. *Acta Numerica*, pages 159–333, 1995.
11. J. Koski H. Eschenauer and A. Osyczka. *Multicriteria Design Optimization.* Springer-Verlag, Berlin, Germany, 1990.
12. J. P. Ignizio. *Goal Programming and Extensions.* Lexington Books, Lexington, Massachusetts, 1976.
13. Anthony J. Kearsley. *The Use of Optimization Techniques in the Solution of Partial Differential Equations from Science and Engineering.* PhD thesis, Rice University, 1996.
14. J. L. Lions. Hierarchical control. *Proceedings of the Indian Academy of Sciences (Mathematical Sciences)*, 104(1):295–304, February 1994.
15. L. Vicente and P. Calamai. Bilevel and multilevel programming: A bibliography review. *Journal of Global Optimization*, 6:1–16, 1994.
16. H. von Stackelberg. *Marktform und Gleichgewicht.* J. Springer, Vienna, 1934.

Part VI

Time-Dependent Problems

A Sequential Quadratic Programming Method for Nonlinear Model Predictive Control

Kenneth R. Muske[1] and James W. Howse[2]

[1] Department of Chemical Engineering, Villanova University, Villanova PA 19085, USA

[2] Modeling, Algorithms and Infomatics Group, Los Alamos National Laboratory, Los Alamos NM 87545, USA

Abstract. A local sequential quadratic programming, or Newton–Lagrange, approach to the solution of the partial differential equation constrained optimization problem arising from the application of model predictive control on nonlinear distributed parameter systems is presented. The nonlinear system model considered is a general form of the initial value advective–diffusion parabolic partial differential equation. This equation form is chosen because it describes a wide range of chemical process systems, however, other initial value partial differential equation forms can be considered in a similar manner. The optimization problem is constructed from the discretized model, constraint, and objective function equations. A finite volume approach is used to discretize the distributed parameter model equations. Inequality constraints on the model states and controls are addressed with an active set method. The nonlinear equations resulting from the first order Karush–Kuhn–Tucker conditions are solved directly using preconditioned Newton–Krylov techniques.

1 Introduction

A local sequential quadratic programming method for the simultaneous solution of the constrained, nonlinear optimization problem obtained from the application of nonlinear model predictive control with partial differential equation-based models is developed. In this work, we will consider the initial value advective–diffusion parabolic partial differential equation of the form

$$\alpha^i(\boldsymbol{x},\boldsymbol{u})\frac{\partial x^i}{\partial t}+\gamma^i(\boldsymbol{x},\boldsymbol{u})\left\{\boldsymbol{v}\cdot\nabla x^i\right\}+\beta^i(\boldsymbol{x},\boldsymbol{u})\nabla^2 x^i+\phi^i(\boldsymbol{x},\boldsymbol{u})=0,\ i=1,\dots,n \tag{1}$$

where $\boldsymbol{x}(\mathcal{S},t)\in\Re^n$ is the vector of model states, $\boldsymbol{u}(\mathcal{S},t)\in\Re^m$ is the vector of controls, and $\mathcal{S}$ represents the set of spatial dimensions. The convective velocity, $\boldsymbol{v}$, and the terms α, β, γ, and ϕ are all possibly nonlinear functions of the state and control. We select this partial differential equation form because it describes mass and energy transport in chemical processes with convection. However, other initial value partial differential equation forms may be considered in the same manner. We will also consider general nonlinear inequality constraints on the states and controls

$$\mathcal{G}^i\big(\boldsymbol{x}(\mathcal{S},t),\boldsymbol{u}(\mathcal{S},t)\big) \leq 0, \qquad i = 1,\dots,p \tag{2}$$

where $\boldsymbol{\mathcal{G}} : (\Re^n \times \Re^m) \to \Re^p$ and the inequality constraints can be spatially and/or temporally dependent.

A model predictive control algorithm determines the future control sequence that makes the future model-predicted state of the system behave in some desired manner. The future state behavior is generally defined by specifying a performance objective, *e.g.* minimal deviation from a prescribed state trajectory, and the future control sequence is generally determined from the solution of an optimization problem. The nonlinear optimization problem arising from the model predictive control formulation in this work is

$$\min_{\boldsymbol{u}(t)} \int_0^{t_f} \mathcal{F}\big(\boldsymbol{x}(\mathcal{S},\tau),\boldsymbol{u}(\mathcal{S},\tau)\big)\,d\tau \;+\; \mathcal{E}\big(\boldsymbol{x}(\mathcal{S},t_f)\big) \tag{3}$$

Subject to:

$$\alpha^i(\boldsymbol{x},\boldsymbol{u})\frac{\partial x^i}{\partial t}+\gamma^i(\boldsymbol{x},\boldsymbol{u})\left\{\boldsymbol{v}\cdot\nabla x^i\right\}+\beta^i(\boldsymbol{x},\boldsymbol{u})\nabla^2 x^i+\phi^i(\boldsymbol{x},\boldsymbol{u}) = 0,\; i = 1,\dots,n$$

$$\mathcal{G}^i\big(\boldsymbol{x}(\mathcal{S},t),\boldsymbol{u}(\mathcal{S},t)\big) \leq 0,\;\; i = 1,\dots,p$$

where Equation 3 is the open-loop objective function in which $\mathcal{F} : (\Re^n \times \Re^m) \to \Re$ is the performance penalty and $\mathcal{E} : (\Re^n) \to \Re$ is the terminal state penalty. The objective is optimized subject to the series of partial differential model equations in Equation 1 and the series of general nonlinear inequality constraints on the states and controls in Equation 2. The result is an open-loop optimal future control sequence for which only the initial part of the sequence is implemented. After one control period, the optimization problem is resolved and the procedure repeated. This technique, referred to as receding horizon control, requires the on-line solution of the optimization problem at each control period. Therefore, an efficient solution technique for partial differential equation constrained optimization problems is required to implement this approach when using distributed parameter process models.

Two general approaches exist to solve this problem. The first is a nested analysis and design (NAND), or black-box, approach in which the decision variables consist of the future control sequence. The advantage of this approach is the relative ease of implementation. The disadvantages are the requirement to resolve the model equations each time the optimization algorithm modifies a decision variable in order to determine the state and the indirect relationship between the decision variables and the state constraints. For systems in which the model equations are expensive to solve and/or the states are highly constrained, this approach can be quite inefficient. In these cases, a simultaneous analysis and design (SAND), or all-at-once approach, in which both the controls and states comprise the decision variable space is a more efficient solution procedure.

We choose an all-at-once approach in this work to address model predictive control applications with distributed parameter process models. These models are typically expensive to solve with state constraints that are typically spatially dependent. For these reasons, black-box approaches can be very inefficient. The motivation for the local sequential quadratic programming, or Newton–Lagrange, method is the application of efficient Newton–Krylov techniques developed for the solution of systems of nonlinear partial differential equations. These techniques, which we have applied to the model solution for black-box nonlinear model predictive control applications [1], are used to solve the system of nonlinear equations obtained from the first order Karush–Kuhn–Tucker conditions. An active set method allows for the direct application in the presence of inequality constraints.

The optimization problem is constructed from a discrete approximation to the model equations, constraint equations, and objective function. The system of partial differential equations is discretized using a finite volume approach in the spatial coordinates and backwards differencing in time. The result is a series of nonlinear algebraic equations in which the unknowns are the discretized control and the state variables at each spatial and temporal node. These same unknowns are used to form a discrete approximation to the general nonlinear inequality constraint equations of the state and control variables and the objective function. An active set method is used to identify the discretized inequality constraint equations that are introduced into the problem as additional equality constraints. The resulting nonlinear system of algebraic equations arising from the first order Karush–Kuhn–Tucker conditions are then solved directly using a preconditioned Newton–Krylov solution technique.

2 Model Discretization

The series of nonlinear partial differential equations that comprise the process model in Equation 1 are first scaled and then discretized using a finite volume representation in space and backwards differencing in time. As discussed in [2], finite volume discretization divides the computational domain into a number of non-overlapping control volumes that completely cover the original problem domain. The control volumes are constructed such that only one grid point lies inside each control volume. The partial differential equations are then integrated over each control volume. These integrals are evaluated by approximating the variation of the state $\boldsymbol{x}$ between each grid point using piecewise profiles. Piecewise linear profiles are typically applied. The terms α, β, γ, and ϕ in Equation 1 are expressed as a function of the state and control vectors at the corresponding grid point for each control volume and typically assumed to be constant over the control volume so that they can be removed from the volume integrals. This procedure results in a set of discrete nonlinear algebraic equations containing the values of $\boldsymbol{x}$ and $\boldsymbol{u}$ at each grid point. Intuitively, these discrete equations define a conservation principle for $\boldsymbol{x}$ over the finite volume of each cell in the same manner as the original partial differential equations for an infinitesimal volume. Backward differencing provides numerical stability of the solution with large time steps.

We illustrate this procedure using the scalar advective partial differential equation describing the component concentration in a plug flow reactor

$$\frac{\partial x}{\partial t} + v\frac{\partial x}{\partial y} + \phi = 0 \tag{4}$$

in which x is the component concentration, y is the axial dimension of the reactor, $v = F/A_c$ is the fluid velocity where the control F is the feed volumetric flow rate and A_c is the reactor cross-sectional area, and $\phi = -Rx^2$ is the source term due to a second order reaction. Making the typical assumptions outlined previously and integrating each term of Equation 4 over the control volume containing the jth spatial node at the kth time step results in

$$\int_{\mathcal{V}} \frac{\partial x}{\partial t}\, d\mathcal{V} = \frac{x_{j,k} - x_{j,k-1}}{\Delta t}\,\Delta V, \qquad \int_{\mathcal{V}} \phi \, d\mathcal{V} = -Rx_{j,k}^2\,\Delta V \tag{5}$$

$$\int_{\mathcal{V}} v\frac{\partial x}{\partial y}\, d\mathcal{V} = \int_{\mathcal{V}} \nabla \cdot (v\,x)\, d\mathcal{V} = \oint_{\mathcal{S}} (v\,x)\cdot \hat{n}\, d\mathcal{S} = v(x_{j-1,k} - x_{j,k})A_c \tag{6}$$

where $x_{j,k}$ is the component concentration of the jth spatial node and kth time step and Gauss' theorem is applied to the volume integral in Equation 6. We use an upwinding scheme to discretize the convective term since this problem is convection dominated. The upwinding scheme is applied by approximating the component concentration of the upstream surface of the control volume as the node concentration of the upstream control volume relative to the direction of the flow and the component concentration of the downstream surface as the node concentration of the control volume. Assuming constant cross-sectional area of the reactor and equal dimensions for each control volume, a series of implicit algebraic equations of the form

$$\left(\frac{1}{\Delta t} - \frac{F_k}{A_c \Delta y}\right) x_{j,k} - Rx_{i,j}^2 - \frac{1}{\Delta t}\, x_{j,k-1} + \frac{F_k}{A_c \Delta y}\, x_{j-1,k} = 0 \tag{7}$$

is obtained for each state variable at every spatial and temporal node where F_k is the control. In this particular example, the result is identical to that obtained using the finite differences. In general, however, finite difference and finite volume discretization will not necessarily produce the same results.

3 Optimization Problem Discrete Representation

The local sequential quadratic programming solution approach for the optimization of constrained distributed parameter systems developed in this work is intended for model predictive control applications. A receding horizon model predictive controller resolves the constrained optimization problem at every control interval using updated information about the current state based on process measurements. The result is a repeated correction to the optimal control sequence that is actually implemented on the process. Therefore, the optimality requirement for the control sequence is somewhat less stringent in model predictive control than is required for an open-loop optimal implementation without feedback. However, the computation time required for the optimization problem must be such that it can be resolved at every control interval. We use these observations as a justification for the discretized approximate problem developed in this work and refer the reader to [3] for an excellent review of model predictive control and survey discussion on the issue of sub-optimality in receding horizon control.

The selection and placement of the spatial and temporal nodes is based on the desired accuracy of the discretized objective function, inequality constraint functions, and model solution. This determination for the model solution can be made locally using grid convergence or an error estimate. A global measure can be based on the conservation error of a conserved quantity in the problem such as mass or energy. The node placement can also be adjusted to satisfy some error criteria if significant changes are possible as discussed in [4]. For most applications, satisfying the accuracy requirement for the model solution will provide sufficient accuracy in the discrete constraint and objective function approximations. Further refinement of the grid beyond the accuracy required for the model equations, however, may be necessary for the discrete approximation to the constraint and objective functions in some applications.

3.1 Model Equations

Applying piecewise linear approximations in each finite volume element to discretize Equation 1 in space and backward differencing to discretize Equation 1 in time results in a series of nonlinear algebraic equations of the form

$$c_{i,j,k}\left(x^{\{1,\dots,n\}}_{\{j-1,j,j+1\},k}, x^{i}_{j,k-1}, u^{\{1,\dots,m\}}_{\{r|r\leq k\}}\right) = 0 \tag{8}$$

where $c_{i,j,k}(\cdot)$ is the equation resulting from the ith state variable at the jth spatial node and kth time step. There is an inequality in the control time index because future controls have no effect on the discretized state variables. The spatial dependence of the controls is accounted for within the discretized model equations. Note that the example discretization in Equation 7 is of this general form.

We construct a vector $\boldsymbol{z}$ that contains all of the state variables at each spatial node and time step and a vector $\boldsymbol{u}$ that contains all of the controls at each time step. The vectors $\boldsymbol{z}$ and $\boldsymbol{u}$ are the decision variables for the all-at-once optimization solution approach in this work. We will consider one spatial dimension in this section. Multiple spatial dimensions can be handled in a similar manner with the addition of an extra index for each additional spatial dimension. The values of the state variables and controls at each spatial and temporal node are represented as

$$z_l = x^i_{j,k}, \qquad w_s = u^q_r \tag{9}$$

in which $x^i_{j,k}$ is the ith state variable value at the jth spatial node and kth time step, $\boldsymbol{z}$ is the vector of discretized model state values, u^q_r is the qth control value at the rth time step, and $\boldsymbol{w}$ is the vector of discretized control values. The index l, used for the discretized state variables, and the index s, used for the discretized control variables, can be determined from the variable, spatial node, and time step indices in a number of different ways. Because the states at each node will typically interact with each other through the source term due to reaction processes, interact with the states in the adjacent spatial nodes due to convection and diffusion transport processes, and interact with the corresponding state at the previous time step due to the backward differencing, a particularly convenient form of the indices for chemical convection/diffusion/reaction processes is the following

$$l = nJ(k-1) + n(j-1) + i, \qquad l \in [1, N] \tag{10}$$

$$s = m(r-1) + q, \qquad s \in [1, M] \tag{11}$$

where J is the total number of spatial nodes and K is the total number of temporal nodes used in the discretization, $N = nJK$ is the total number of discrete model state values, and $M = mK$ is the total number of discrete control values. This choice results in the following ordering of the $\boldsymbol{z}$ vector

$$\boldsymbol{z} = \left[x^{\{1,\dots,n\}}_{1,1}, x^{\{1,\dots,n\}}_{2,1}, \dots, x^{\{1,\dots,n\}}_{J,1}, x^{\{1,\dots,n\}}_{1,2}, \dots, x^{\{1,\dots,n\}}_{J,2}, \dots, x^{\{1,\dots,n\}}_{J,K}\right]^T$$

which provides a reasonably compact sparse diagonal structure for the Jacobian of the Lagrangian function that is required for Newton–Krylov. The series of nonlinear algebraic equations in Equation 8 can then be represented by the following nonlinear equality constraints on the discretized state and control variable values where $\boldsymbol{h} : (\Re^N \times \Re^M) \to \Re^N$.

$$\boldsymbol{h}(\boldsymbol{z}, \boldsymbol{w}) = 0, \qquad h^l(\boldsymbol{z}, \boldsymbol{w}) = c_{i,j,k}(\boldsymbol{x}, \boldsymbol{u}) \tag{12}$$

3.2 Inequality Constraints

A discrete approximation to the inequality constraint equations is expressed as $\boldsymbol{g}(\boldsymbol{z}, \boldsymbol{w}) \le 0$ where $\boldsymbol{g} : (\Re^N \times \Re^M) \to \Re^p$ is defined as follows:

$$\boldsymbol{g}(\boldsymbol{z}, \boldsymbol{w}) = \begin{bmatrix} G^1(\boldsymbol{z}, \boldsymbol{w}) \\ \vdots \\ G^p(\boldsymbol{z}, \boldsymbol{w}) \end{bmatrix}, \qquad G^i(\boldsymbol{z}, \boldsymbol{w}) \approx \mathcal{G}^i\left(\boldsymbol{x}(\mathcal{S}, t), \boldsymbol{u}(\mathcal{S}, t)\right) \tag{13}$$

An active set formulation is used to determine the binding discretized inequality constraints in Equation 13

$$a(z, w) = \begin{bmatrix} G^{\mathcal{A}(1)}(z, w) \\ \vdots \\ G^{\mathcal{A}(P)}(z, w) \end{bmatrix} = 0 \tag{14}$$

in which P is the total number of binding inequality constraints and the set $\mathcal{A}$ indicates the index of the binding or active constraints. Equation 14 represents the additional equality constraints on the discretized model states and controls due to the binding inequality constraints.

3.3 Objective Function

A discrete approximation to the open-loop objective function in Equation 3 is expressed as

$$f(z, w) \approx \int_0^{t_f} \mathcal{F}\big(x(\mathcal{S}, \tau), u(\mathcal{S}, \tau)\big)\, d\tau \; + \; \mathcal{E}\big(x(\mathcal{S}, t_f)\big) \tag{15}$$

where $f : (\Re^N \times \Re^M) \rightarrow \Re$.

4 Lagrangian Formulation

The Lagrangian is formed from the objection function in Equation 15 subject to the nonlinear equality constraints from the model equations in Equation 12 and from the active constraints in Equation 14

$$\mathcal{L} = f(z, w) + \lambda^T h(z, w) + \mu^T a(z, w) \tag{16}$$

where $\lambda \in \Re^N$ is the vector of Lagrange multipliers corresponding to the model equality constraints and $\mu \in \Re^P$ is the vector of Lagrange multipliers corresponding to the active inequality constraints. The first order necessary conditions for a local minimum are determined from the Lagrangian as follows

$$r(z, w, \lambda, \mu) = \begin{bmatrix} \nabla_w \mathcal{L} \\ \nabla_z \mathcal{L} \\ \nabla_\lambda \mathcal{L} \\ \nabla_\mu \mathcal{L} \end{bmatrix} = \begin{bmatrix} \mathcal{L}_w \\ \mathcal{L}_z \\ h \\ a \end{bmatrix} = 0 \tag{17}$$

$$\mu_j \geq 0, \qquad j = 1, \ldots, P \tag{18}$$

provided that the set of equality and active inequality constraint gradients with respect to z and w, $\{\nabla_z h, \nabla_w h, \nabla_z a, \nabla_w a\}$, are linearly independent.

If this condition is satisfied, the vector $[\boldsymbol{z}, \boldsymbol{w}]^T$ is regular [5] or the linear independence constraint qualification holds [6]. The condition on the Lagrange multipliers for the active inequality constraints in Equation 18 is required to ensure that every constraint in the active set is a binding constraint.

The gradient of the Lagrangian with respect to the discrete controls is computed as follows.

$$\mathcal{L}_{\boldsymbol{w}} = \begin{bmatrix} \dfrac{\partial f}{\partial w_1} + \sum_{i=1}^{N} \lambda_i \dfrac{\partial h_i}{\partial w_1} + \sum_{i=1}^{P} \mu_i \dfrac{\partial a_i}{\partial w_1} \\ \vdots \\ \dfrac{\partial f}{\partial w_M} + \sum_{i=1}^{N} \lambda_i \dfrac{\partial h_i}{\partial w_M} + \sum_{i=1}^{P} \mu_i \dfrac{\partial a_i}{\partial w_M} \end{bmatrix} \tag{19}$$

Because future controls have no effect on the current state variables, several of the partial derivatives are zero and need not be considered.

$$\frac{\partial h_i}{\partial w_j} = 0 \;\forall\; j > m\left(\mathsf{int}\left(\frac{i-1}{nJ}\right) + 1\right) \tag{20}$$

An additional reduction in the number of partials, which can be determined *a priori*, may be possible due to the spatial dependence of the controls.

The gradient of the Lagrangian with respect to the discrete states is determined as follows.

$$\mathcal{L}_{\boldsymbol{z}} = \begin{bmatrix} \dfrac{\partial f}{\partial z_1} + \sum_{i=1}^{N} \lambda_i \dfrac{\partial h_i}{\partial z_1} + \sum_{i=1}^{P} \mu_i \dfrac{\partial a_i}{\partial z_1} \\ \vdots \\ \dfrac{\partial f}{\partial z_N} + \sum_{i=1}^{N} \lambda_i \dfrac{\partial h_i}{\partial z_N} + \sum_{i=1}^{P} \mu_i \dfrac{\partial a_i}{\partial z_N} \end{bmatrix} \tag{21}$$

Due to the form of the equality constraints generated from the discretized model equations shown in Equation 8, the number of partials required to compute the gradient of the equality constraints is relatively small.

$$\frac{\partial h_i}{\partial z_j} = 0 \;\forall\; j \neq \{[i-n, \ldots, i+n], i - Jn\} \tag{22}$$

A similar reduction in the number of partials that must be computed for the active constraint gradients may be possible depending on the form of the inequality constraints. In particular, saturation and rate of change constraints on the controls are not a function of the discretized state variables.

Remark 1. The nonzero partial derivatives that must be computed for Equations 19 and 21 can be determined *a priori* and will not change during the implementation of the algorithm.

5 Sequential Quadratic Programming Solution Method

A local sequential quadratic programming approach is developed by the application of Newton–Krylov techniques to solve the system of nonlinear equations arising from the first order optimality conditions in Equation 17. Newton–Krylov provides a robust solution technique that converges rapidly without requiring tuning parameters for systems having disparate eigenvalues in the linear approximation. Disparity in the eigenvalues arises from the different time scales for reaction, convection, and diffusion in the model equations. An active set method is employed for the inequality constraints to allow the direct application of Newton–Krylov techniques on the first order Karush–Kuhn–Tucker system.

We do not consider the second order optimality conditions in this work. Therefore, the optimization approach will only converge to a minimum of the objective if it is started close to a stationary point that satisfies the second order conditions for a minimum. For receding horizon control problems, where the initial starting point is taken as the optimal solution from the previous controller execution and the objective is generally specified as a quadratic penalty on the state deviation from a reference and a quadratic penalty on the controls, it is reasonable to assume that the algorithm will converge to a stationary point that is a minimum. However, convergence toward a minimum of the performance objective is not guaranteed.

5.1 Inexact Newton Method

The residuals of the Lagrangian gradients in Equation 17 are minimized using Newton–Raphson iteration with the linearized system

$$\boldsymbol{J}(\boldsymbol{z}_k, \boldsymbol{w}_k, \boldsymbol{\lambda}_k, \boldsymbol{\mu}_k) \begin{bmatrix} \delta \boldsymbol{w} \\ \delta \boldsymbol{z} \\ \delta \boldsymbol{\lambda} \\ \delta \boldsymbol{\mu} \end{bmatrix} = -\boldsymbol{r}(\boldsymbol{z}_k, \boldsymbol{w}_k, \boldsymbol{\lambda}_k, \boldsymbol{\mu}_k) \tag{23}$$

in which $\boldsymbol{J}$ is the Jacobian of the residual equations in Equation 17 and $\delta\boldsymbol{w}, \delta\boldsymbol{z}, \delta\boldsymbol{\lambda}$, and $\delta\boldsymbol{\mu}$ are the corrections to the controls, model states, and Lagrange multipliers. An inexact Newton algorithm is applied that uses an approximate solution to the linear system at each iteration. It can be shown that superlinear convergence is achievable with these methods provided that the second order sufficiency conditions hold [7]. In this work, preconditioned Krylov subspace projection methods [8], [9] are used to generate an approximate solution to the linear system in Equation 23. The control, model state, and multiplier values are updated at each iteration by taking some fraction of the full Newton step where $\alpha_k \in (0, 1]$.

$$\left[\boldsymbol{w}^T \ \boldsymbol{z}^T \ \boldsymbol{\lambda}^T \ \boldsymbol{\mu}^T\right]_{k+1}^T = \left[\boldsymbol{w}^T \ \boldsymbol{z}^T \ \boldsymbol{\lambda}^T \ \boldsymbol{\mu}^T\right]_k^T + \alpha_k \left[\delta\boldsymbol{w}^T \ \delta\boldsymbol{z}^T \ \delta\boldsymbol{\lambda}^T \ \delta\boldsymbol{\mu}^T\right]^T \tag{24}$$

There are various methods for selecting α_k that enlarge the sphere of convergence for inexact Newton iterations [10]. In this work, the inequality

$$\|(\boldsymbol{r}_k + \alpha_k \boldsymbol{J}_k \boldsymbol{\delta}_k)\|_2 \leq \eta_k \|\boldsymbol{r}_k\|_2, \quad \eta_k \in [0, \eta_{\max}], \quad \eta_{\max} < 1 \tag{25}$$

is enforced where $\|\cdot\|_2$ is the Euclidean norm and η_k is the forcing term. Several methods to determine the forcing term sequence and the effect on the required linear system solution accuracy are discussed in [11].

The Newton iterations are terminated when the criteria

$$\|\boldsymbol{r}(\boldsymbol{w}_k, \boldsymbol{z}_k, \boldsymbol{\lambda}_k, \boldsymbol{\mu}_k)\|_2 < \max\left(\tau_r \ \|\boldsymbol{r}(\boldsymbol{w}_0, \boldsymbol{z}_0, \boldsymbol{\lambda}_0, \boldsymbol{\mu}_0)\|_2 \, , \ \tau_a\right) \tag{26}$$

is satisfied in which $\tau_r \in (0,1)$ is the relative error tolerance and $\tau_a \in (0,1)$ is the absolute error tolerance. The application of this criteria means that the Newton iterations are terminated when either the current residual becomes less than τ_r of its initial value or when it becomes less than τ_a depending on the magnitude of the initial residual.

5.2 Jacobian Determination

The Jacobian of the residual equations is determined as follows.

$$\boldsymbol{J} = \begin{bmatrix} \nabla_{\boldsymbol{w}}^T \mathcal{L}_{\boldsymbol{w}} & \nabla_{\boldsymbol{z}}^T \mathcal{L}_{\boldsymbol{w}} & \nabla_{\boldsymbol{w}} \boldsymbol{h}^T & \nabla_{\boldsymbol{w}} \boldsymbol{a}^T \\ \nabla_{\boldsymbol{w}}^T \mathcal{L}_{\boldsymbol{z}} & \nabla_{\boldsymbol{z}}^T \mathcal{L}_{\boldsymbol{z}} & \nabla_{\boldsymbol{z}} \boldsymbol{h}^T & \nabla_{\boldsymbol{z}} \boldsymbol{a}^T \\ \nabla_{\boldsymbol{w}}^T \boldsymbol{h} & \nabla_{\boldsymbol{z}}^T \boldsymbol{h} & \boldsymbol{0} & \boldsymbol{0} \\ \nabla_{\boldsymbol{w}}^T \boldsymbol{a} & \nabla_{\boldsymbol{z}}^T \boldsymbol{a} & \boldsymbol{0} & \boldsymbol{0} \end{bmatrix} \tag{27}$$

Remark 2. The Jacobian of the residual equations is symmetric. If the Jacobian evaluated at a feasible vector $[\boldsymbol{w}, \boldsymbol{z}]^T$ is full rank, the constraint gradients are linearly independent, $[\boldsymbol{w}, \boldsymbol{z}]^T$ is regular, and the linear independence constraint qualification holds.

The Jacobian may be computed either analytically or numerically and, due to the structure of the node placement, most of the second partials are zero. The functional form and identity of the non-zero second partials can be determined *a priori*. The majority of these second partials are taken on algebraic equations of a form similar to Equation 7. In most model predictive control applications, the objective function is a quadratic function of the states and controls and the inequality constraints consist of maximum and minimum state and control limits. If the state and control dependence of the terms α, β, and γ in Equation 1 are neglected, the resulting approximate second partials are considerably simplified and can be automatically generated based on the original model equation form and the domain discretization. The resulting approximate Jacobian can then be computed relatively easily.

A numerical approximation to the Jacobian can also be considered. It is possible to determine *a priori* which elements of the Jacobian will be effected by a perturbation in any of the unknowns. Therefore, only those elements need to be considered when determining the Jacobian numerically.

5.3 Krylov Subspace Projection Methods

Krylov subspace projection methods are iterative conjugate-gradient like methods for solving the linear system $\boldsymbol{J\delta} = \boldsymbol{r}$ using polynomial approximation. As discussed in [12], there are three broad classes of such methods. We consider the class of methods that minimize the criterion

$$\min_{\boldsymbol{\delta}_l \in \mathcal{K}_l} \|\boldsymbol{r} - \boldsymbol{J\delta}_l\|_\eta \tag{28}$$

over the Krylov subspace

$$\mathcal{K}_l = \text{span}(\boldsymbol{\rho}_0, \boldsymbol{J\rho}_0, \dots, \boldsymbol{J}^{l-1}\boldsymbol{\rho}_0), \tag{29}$$

using an initial linear residual from the initial estimate $\boldsymbol{\rho}_0 = \boldsymbol{r} - \boldsymbol{J\delta}_0$. The estimate and linear residual at the lth iteration can be expressed as a linear combination of basis vectors for the Krylov subspace in which $\{\boldsymbol{p}_j\} \in \mathcal{K}_l$.

$$\boldsymbol{\delta}_l = \boldsymbol{\delta}_0 + \sum_{j=0}^{l-1} \gamma_j \boldsymbol{p}_j, \quad \boldsymbol{\rho}_l = \boldsymbol{\rho}_0 - \boldsymbol{J} \sum_{j=0}^{l-1} \gamma_j \boldsymbol{p}_j \tag{30}$$

The generalized minimum residual (GMRES) algorithm [13] or the transpose-free quasi-minimal residual algorithm (TFQMR) [14] are used to determine $\{\boldsymbol{p}_j\}$ and $\{\gamma_j\}$ in this work. These algorithms generate iterates that minimize the linear residual 2-norm at each iteration,

$$\min \|\boldsymbol{\rho}_l\|_2 = \min \|\boldsymbol{r} - \boldsymbol{J\delta}_l\|_2$$

residuals that are $\boldsymbol{J}$-orthogonal to the previous basis,

$$\boldsymbol{\rho}_j^T \boldsymbol{J} \boldsymbol{p}_n = 0 \;\; \forall \;\; j > n$$

and in the case of GMRES, basis vectors that are $\boldsymbol{I}$–orthonormal.

$$\boldsymbol{p}_n^T \boldsymbol{p}_m = 0 \;\; \forall \;\; n \neq m, \qquad \boldsymbol{p}_n^T \boldsymbol{p}_n = 1 \;\; \forall \;\; n$$

It can be shown that a superlinear convergence rate is achievable with the GMRES algorithm [15], however, this algorithm requires a numerical solution to the least squares problem in Equation 28 and storage of the basis vectors for the Krylov subspace in Equation 29. The result is a computationally intensive iterative procedure that may need restarting with large-dimensional problems due to memory limitations on the storage of the Krylov subspace basis. The TFQMR algorithm computes a much less expensive approximate solution to Equation 28 using a recurrence relation that does not require storage of the Krylov subspace basis. The result is an iterative procedure that typically converges at a slower rate, but may be competitive or superior for large problems because of the reduced computational effort per iteration.

5.4 Preconditioned Krylov Methods

The convergence rate for Krylov subspace projection methods depends on the condition number of the matrix $\boldsymbol{J}$ [12]. In many practical cases, $\boldsymbol{J}$ is so poorly conditioned that these methods will fail to converge. In these cases, preconditioning is required to improve the condition number of $\boldsymbol{J}$. Preconditioning is accomplished by either right or left multiplying the matrix $\boldsymbol{J}$ by a matrix $\boldsymbol{P}$, which resembles $\boldsymbol{J}^{-1}$ in some sense, giving rise to one of the two following preconditioned linear systems.

$$\boldsymbol{P}\boldsymbol{J}\boldsymbol{\delta} = \boldsymbol{P}\boldsymbol{r} \tag{31}$$

$$\boldsymbol{J}\boldsymbol{P}\boldsymbol{u} = \boldsymbol{r}, \quad \boldsymbol{\delta} = \boldsymbol{P}\boldsymbol{u} \tag{32}$$

Equation (31) is the new linear system under left preconditioning and Equation (32) is the new linear system under right preconditioning.

A right preconditioner, incomplete LU factorization [9], is applied prior to the Krylov subspace projection method in this work. Incomplete LU factorization consists of performing Gaussian elimination on the matrix $\boldsymbol{J}$ with a term dropping strategy to eliminate elements from the intermediate matrix LU_m at each elimination step m. The element $LU_m(i,j)$ is dropped at iteration m if

$$|LU_m(i,j)| < \eta \, \frac{\|LU_m(i,\cdot)\|_\infty}{|LU_m(i,i)|} \tag{33}$$

where $LU_m(i,\cdot)$ is the ith row of the matrix LU_m, $LU_m(i,i)$ is the diagonal element of the ith row, $\eta \in (0,\infty)$ is the drop tolerance parameter, and $\|\cdot\|_\infty$ is the infinity norm. The diagonal entries $LU_m(i,i)$ are compensated by subtracting a weighted sum of the dropped terms from the diagonal entry

$$LU_m(i,i) = LU_m^*(i,i) - \varepsilon \sum LU_m(i,j^*) \tag{34}$$

where $LU_m^*(i,i)$ is the diagonal entry computed by elimination, $LU_m^*(i,j^*)$ are the dropped terms, and $\varepsilon \in [0,1]$ is the compensation parameter. This dropping strategy guarantees that the matrix obtained at the final step of the incomplete factorization is sparse, which would generally not be the case for the matrix obtained by Gaussian elimination. The matrix obtained after the last step of the incomplete elimination is used as the right preconditioner.

In order to apply incomplete LU factorization as a preconditioner, the matrices must be assembled. Although the Jacobian is sparse and well-defined, as discussed in Section 5.2, and a dropping strategy is applied, this requirement can preclude the use of incomplete LU factorization as a preconditioner for large dimensional systems. If preconditioning is not applied, however, the Krylov solvers may not converge or may take an unacceptably long time to converge. In this case, iterative methods or approximate inverse techniques that only require matrix-vector products [9] are alternative approaches.

5.5 Block Preconditioning

The linear system in Equation 23 can be expressed as

$$\begin{bmatrix} \boldsymbol{J}_1 & \boldsymbol{J}_2^T \\ \boldsymbol{J}_2 & 0 \end{bmatrix} \begin{bmatrix} \boldsymbol{\delta}_1 \\ \boldsymbol{\delta}_2 \end{bmatrix} = \begin{bmatrix} \boldsymbol{r}_1 \\ \boldsymbol{r}_2 \end{bmatrix} \tag{35}$$

in which

$$\boldsymbol{J}_1 = \begin{bmatrix} \nabla_{\boldsymbol{w}}^T \mathcal{L}_{\boldsymbol{w}} & \nabla_{\boldsymbol{z}}^T \mathcal{L}_{\boldsymbol{w}} \\ \nabla_{\boldsymbol{w}}^T \mathcal{L}_{\boldsymbol{z}} & \nabla_{\boldsymbol{z}}^T \mathcal{L}_{\boldsymbol{z}} \end{bmatrix}, \quad \boldsymbol{\delta}_1 = \begin{bmatrix} \delta \boldsymbol{w} \\ \delta \boldsymbol{z} \end{bmatrix}, \quad \boldsymbol{J}_2 = \begin{bmatrix} \nabla_{\boldsymbol{w}}^T \boldsymbol{h} & \nabla_{\boldsymbol{w}}^T \boldsymbol{a} \\ \nabla_{\boldsymbol{z}}^T \boldsymbol{h} & \nabla_{\boldsymbol{z}}^T \boldsymbol{a} \end{bmatrix}, \quad \boldsymbol{\delta}_2 = \begin{bmatrix} \delta \boldsymbol{\lambda} \\ \delta \boldsymbol{\mu} \end{bmatrix}$$

where $\boldsymbol{J}_1$ is a relatively sparse $(M+N) \times (M+N)$ matrix and $\boldsymbol{J}_2$ is a sparse $(P+N) \times (M+N)$ matrix. Block Gaussian elimination can be applied as a preconditioner to Equation 35 in order to obtain an approximate block diagonal linear system. When an approximate inverse of the matrix $\nabla_{\boldsymbol{w}}^T \mathcal{L}_{\boldsymbol{w}}$ can be determined with little computational cost, it can form the basis of this preconditioner. Block preconditioners derived from particular block-LU factorizations of the Jacobian can also be applied [16].

Block preconditioning is advantageous if the effort required to eliminate the relatively more dense matrix $\boldsymbol{J}_1$ is rewarded by a reduction in the effort required to right precondition the block diagonal matrix using incomplete LU decomposition and/or a reduction in the number of Krylov subspace iterations. The value of block preconditioning is problem dependent. It may not be advantageous when the number of control variables M is small because incomplete LU preconditioning will typically also be required due to the disparate eigenvalues arising from the state equations. Block preconditioning may also destroy the symmetry of the Jacobian, however, symmetry is not required for the Krylov subspace projection methods in Section 5.3.

6 Active Set Method

The inequality constraints in Section 3.2 that are included in the Lagrangian are determined by an active set method. The system in Equation 17 is iteratively solved with updated active sets until

$$\boldsymbol{g}(\boldsymbol{z}, \boldsymbol{w}) \leq 0 \quad \text{and} \quad \mu_j \geq 0 \quad \forall \quad j \in \mathcal{A} \tag{36}$$

in which $\boldsymbol{\mu}$ are the Lagrange multipliers for the active inequality constraints and $\mathcal{A}$ is the current active set. If the conditions in Equation 36 are not satisfied, an inequality constraint left out of the active set was not satisfied, an inequality constraint included in the active set was not a binding constraint, or both situations occurred. In either case, the active set must be adjusted to include the violated constraints and/or remove the non-binding constraints.

In this work, the active set procedure in [17] is adopted. Since an accurate estimate of the Lagrange multipliers is necessary in order to determine whether an active constraint should be removed, elimination of non-binding constraints is performed only after Newton–Krylov convergence when the corresponding Lagrange multiplier is more negative than a specified fraction of the residual 2-norm. Early termination of the Newton iterations for the addition of a violated constraint to the active set is performed only after the 2-norm of the violation is greater than a specified fraction of the residual 2-norm. This check is performed after each Newton iteration.

In order for a solution to Equation 35 to exist, $\boldsymbol{J}_2\boldsymbol{\delta}_1 = \boldsymbol{r}_2$ must have a solution and $\boldsymbol{J}_2$ must be full rank. Therefore, the number of active inequality constraints P cannot exceed the number of controls M and the gradients of the active constraints must be nonsingular. If the active set method determines that $P > M$, the iterations will terminate with no feasible solution. If the active constraint gradients are singular, the linearly dependent constraints are eliminated from the problem. An initial estimate for the current active set is the active set from the previous control calculation. Heuristics based on process knowledge can also be incorporated into the active set update scheme to avoid active constraint sets that cannot physically exist. We note that if the initial active set is correct, only one iteration of Newton–Krylov is required.

The advantage of an active set method is the direct application of Newton–Krylov to the solution of the first order Karush–Kuhn–Tucker system in Equation 17. Disadvantages include the possibility of a large number of iterations when several inequality constraints are present, the difficulty in determining whether a constraint is binding when the corresponding multiplier is close to zero, and *zigzagging*, in which an inequality constraint is repeatedly removed from and then added back to the active set [18].

7 Conclusions

The simultaneous approach in this work requires the iterative solution of $2N + M + P$ nonlinear equations to determine the optimal control profile where N is the number of discretized state values, M is the number of discretized control values, and P is the number of active inequality constraints. The system of nonlinear equations that must be solved at each iteration of a black-box approach consists of only N equations. The most numerically efficient approach will depend on the problem. The simultaneous approach is typically advantageous when the number of controls and inequality constraints are large. However, the local sequential quadratic programming method proposed in this work limits the number of inequality constraints that can efficiently be handled in order to allow for the direct application of Newton–Krylov. If the total number of inequality constraints is small and a good initial active set is available, this technique can be competitive with other simultaneous methods that do not rely on an active set and with interior point methods.

Acknowledgments

This work was performed under the auspices of the Department of Energy under contract W-7405-ENG-36.

References

1. J. W. Howse, G. A. Hansen, D. J. Cagliostro, and K. R. Muske. Solving a thermal regenerator model using implicit Newton-Krylov methods. *Numerical Heat Transfer: Part A*, 38(1):23–44, 2000.
2. C. Hirsch. *Numerical Computation of Internal and External Flows, Volume 1: Fundamentals of Numerical Discretization.* John Wiley & Sons, Ltd., Chichester, United Kingdom, 1988.
3. D. Q. Mayne, J. B. Rawlings, C. V. Rao, and P. O. Scokaert. Constrained model predictive control: Stability and optimality. *Automatica*, 36:789–814, 2000.
4. S. Vasantharajan and L. T. Biegler. Simultaneous strategies for optimization of differential-algebraic systems with enforcement of error criteria. *Comp. Chem. Eng.*, 14(10):1083–1100, 1990.
5. D. P. Bertsekas. *Nonlinear Programming.* Athena Scientific, Belmont, MA, 1995.
6. J. Nocedal and S. J. Wright. *Numerical Optimization.* Springer Series in Operations Research. Springer–Verlag, New York, 1999.
7. R. S. Dembo, S. C. Eisenstat, and T. Steihaug. Inexact newton methods. *SIAM J. Num. Anal.*, 19(2):400–408, 1982.
8. C. T. Kelley. *Iterative Methods for Linear and Nonlinear Equations.* Number 16 in Frontiers in Applied Mathematics. SIAM, Philadelphia, 1995.
9. Y. Saad. *Iterative Methods for Sparse Linear Systems.* PWS Publishing Company, Boston, 1996.
10. P. N. Brown and Y. Saad. Convergence theory of nonlinear Newton-Krylov algorithms. *SIAM J. Optimization*, 4(2):297–330, 1994.
11. S. C. Eisenstat and H. F. Walker. Choosing the forcing terms in an inexact Newton method. *SIAM J. Sci. Comput.*, 14(2):470–482, 1993.
12. N. M. Nachtigal, S. C. Reddy, and L. N. Trefethen. How fast are nonsymmetric matrix iterations? *SIAM J. Matrix Anal. Appl.*, 13(3):778–795, 1992.
13. Y. Saad and M.H. Schultz. GMRES: A generalized minimal residual algorithm for solving nonsymmetric linear systems. *SIAM J. Sci. Stat. Comput.*, 7(3):856–869, 1986.
14. R. W. Freund. A transpose-free quasi-minimal residual algorithm for non-hermitian linear systems. *SIAM J. Sci. Comput.*, 14(2):470–482, 1993.
15. H.A. van der Vorst and C. Vuik. The superlinear convergence behavior of GMRES. *J. Comp. Appl. Mathematics*, 48(3):327–341, 1993.
16. G. Biros and O. Ghattas. Parallel Lagrange–Newton–Krylov–Schur methods for PDE–constrained optimization. Part I: The Krylov–Schur solver. Technical report, Laboratory for Mechanics, Algorithms, and Computing, Department of Civil and Environmental Engineering, Carnegie Mellon University, 2000.
17. D. G. Luenberger. *Linear and Nonlinear Programming.* Addison–Wesley, Reading, MA, 2nd edition, 1989.
18. P. E. Gill, W. Murray, and M. H. Wright. *Practical Optimization.* Academic Press, New York, 1981.

Reduced Order Modelling Approaches to PDE-Constrained Optimization Based on Proper Orthogonal Decomposition

Marco Fahl[1] and Ekkehard W. Sachs[1,2]

[1] Universität Trier, Fachbereich IV - Mathematik, D-54286 Trier, Germany
[2] Virginia Polytechnic Institute and State University, ICAM, Department of Mathematics, Blacksburg, VA, 24060, USA

Abstract. Reduced order modelling techniques can be used in order to circumvent computational difficulties due to large-scale state equations related to PDE-constrained optimization problems. However, if reduced order modelling based on the Proper Orthogonal Decomposition (POD) is performed, it is necessary to include an update mechanism into the optimization procedure in order to guarantee reliable reduced order state solutions during the course of the optimization. Furthermore, specific modelling issues should be taken into account such that sufficiently accurate gradient information is obtained during the optimization process. In this context, we discuss some relevant topics arising from the POD based reduced order modelling approach.

1 Introduction

Appropriate mathematical modelling of control and optimization processes in industry often leads to optimization problems where the governing equations are time-dependent partial differential equations. Consequently, standard discretization schemes for the spatial discretization result in large-scale, and in general, nonlinear systems of ordinary differential equations as state equations. Often, these cannot be used during an optimization process due to their computational complexity. Thus, there is a demand to use reduced order models that serve as low-dimensional approximation models for the large-scale discretized state equations.

While classical spatial discretization approaches (e.g. based on finite elements or finite differences) use standardized discretization schemes that are suitable for large classes of problems, reduced order modelling approaches, e.g. based on the Proper Orthogonal Decomposition (POD) method, typically use a more problem specific discretization. In this case, detailed information of the underlying problem is retained in order to achieve a dimension reduction.

Usually, a POD based reduced order model is constructed for a specific parameter setting, such that it is a suitable model for the corresponding system dynamics. If a single POD based reduced order model is used during the optimization routine, it is possible to encounter unreliable reduced order solutions. This effect can be avoided if the reduced order model is updated in an appropriate way. Consequently, some sort of adaptive technique is required, in which the construction of POD based reduced order models is coupled with the progress in the optimization process. If a trust-region method is used, the range of validity of a reduced order model is automatically restricted. Furthermore, the update decision for the reduced order model is based on information which is obtained during the solution procedure, cf. Arian et al. [1], Fahl [2] . By means of this approach, we are led to a trust-region algorithm based on general, nonlinear model functions with inexact gradient information.

Since accurate gradient information plays a crucial role during the optimization process we discuss a new modelling approach based on POD for the adjoint equations of the problem under consideration. The aspect of improved gradient information in the POD context is also addressed in [3].

The paper is organized as follows: In section 2 we present the proper orthogonal decomposition to obtain a reduced order model for the partial differential equation. The next section contains a description of TRPOD, a trust-region based algorithm which makes use of the POD model in the optimization phase. This way we obtain a systematic approach when to use fully the reduced model and when to update the reduced model. The algorithm can be shown to be globally convergent. In certain cases, the derivative information based on the POD model of the state equation is not accurate enough. In section 4, we develop a separate POD model for the adjoint in order to improve on the gradient information. The last section contains a numerical example, a boundary control problem with the heat equation. The trust region method ceases to work properly after a few iterations due to the lack of accuracy in the gradient information. This is overcome by a separate reduced order model for the adjoint equation.

Notation: Throughout this paper we denote by $L^2 = L^2(\Omega)$ the space of square-integrable real functions on domain Ω and we denote the associated inner product and norm by $(\cdot,\cdot)_{L^2}$ and $\|\cdot\|_{L^2}$. The Euclidean norm of a real vector v will be denoted by $\|v\|$. The solution of the PDE under consideration is denoted by $y(x,t)$, $x \in \Omega$, $t \in (0,T)$. The number of snapshots is denoted by P, the dimension of the POD basis is M, and the dimension of the FEM approximation is N.

2 Reduced Order Modelling based on Proper Orthogonal Decomposition

The reduced order modelling approach based on the Proper Orthogonal Decomposition (also known as Karhunen-Loève decomposition or method of empirical eigenfunctions) consists of computing a small set of orthonormal functions, the POD basis functions, that can be efficiently used to describe the system dynamics under consideration. The introduction to POD in the context of turbulence analysis is credited to Lumley [4], cf. also the monograph [5]. Since then, it has been intensively used for computational studies in many engineering areas where non-stationary partial differential equations occur, see e.g. [6], [7], [8], [9], [10].

We assume that a simulation of the PDE under consideration has been performed such that an input ensemble $\mathcal{Y} = \{y_1, \dots, y_P\}$ of 'snapshots' $y_i(x) = y(x, t_i)$, $x \in \Omega, t_i \in (0,T)$ is given, where Ω denotes the computational domain and $(0,T)$ the time interval of interest. Then, the POD technique uses characteristic features of this input ensemble in the following way: Find orthonormal functions $\{\psi_j\}_{j=1}^M$ that solve

$$\sum_{i=1}^{P} \|y_i - \sum_{j=1}^{M} (y_i, \psi_j)_{L^2}\, \psi_j\|_{L^2}^2 \longrightarrow \min. \tag{1}$$

Definition 1 (POD Basis). Let $\mathcal{Y} = \{y_1, \dots, y_P\}$ be given. We call a solution $\{\psi_1, \dots, \psi_M\}$ to (1) a POD basis of order M and set $\mathcal{Y}^{POD} = \{\psi_1, \dots, \psi_M\}$. Furthermore, we call *span* $\mathcal{Y}^{POD}$ the corresponding POD subspace of *span* $\mathcal{Y}$.

Based on (1) the POD basis elements can be used to approximate the input dynamics in an optimal way. The following result based on [11] holds.

Lemma 2. *Let $\mathcal{Y} = \{y_i\}_{i=1}^P$ denote a given input ensemble based on a simulation run. Let A denote a matrix constructed from this input data where each column of A represents a snapshot at a given instant of time. Let $\mathcal{Y}^{POD} = \{\psi_j\}_{j=1}^M$ denote a POD basis of order M for span $\mathcal{Y}$.*

Then, the POD representation error, $\varepsilon(M)$, is given by

$$\varepsilon(M) = \sum_{i=1}^{P} \|y_i - \sum_{j=1}^{M} (y_i, \psi_j)\, \psi_j\|^2 = \sum_{j=M+1}^{P} \sigma_j^2, \tag{2}$$

where σ_j $(j = 1, \dots, P)$ denote the singular values of A.

A commonly used criterion for choosing M, the POD subspace dimension, based on heuristic considerations is the so-called energy criterion. For a predefined 'percentage of energy' $\bar{\epsilon} > 0$, M is chosen such that

$$\frac{\sum_{j=1}^{M} \sigma_j^2}{\sum_{j=1}^{P} \sigma_j^2} \geq \bar{\epsilon} \tag{3}$$

holds.

After a POD subspace of order M has been computed, the POD basis functions are used in a Galerkin projection of the governing equations onto this subspace. Consequently, if a small number of POD basis elements suffices to approximate the input snapshots (as representatives of the system dynamics), the resulting ODE system is of small dimension and can be used as a reduced order model for the system dynamics (instead of, e.g., a high order finite element model).

The POD based reduced order model is especially suited for approximate state solutions which correspond to the parameter setting of the input ensemble or similar parameters. The same reduced order model might deliver no reliable information on the system behaviour for a completely different parameter setting. Since we are interested in solving an optimization problem with a specified cost functional, we can monitor the POD based reduced order model's quality by comparing the performance of a reduced order solution to the performance of the original large-scale solution with regard to the optimization goal, which leads to a trust-region approach for POD based reduced order modelling for optimization purposes.

3 A trust-region framework for POD based Optimization

In the following, we describe the idea of combining POD based reduced order modelling with a trust-region methodology. For this purpose we consider the unconstrained optimization problem

$$\min_{u \in \mathbb{R}^n} \hat{f}(y^N(u), u) \tag{4}$$

where $u \in \mathbb{R}^n$ denotes the vector of decision variables and $y^N(u)$ denotes a finite element approximation of the weak solution of the state equation for given u.

If the evaluation of

$$f(u) = \hat{f}(y^N(u), u) \tag{5}$$

is computationally expensive, it is attractive to replace $y^N(u)$ by a corresponding reduced order solution, $y^M(u)$, since the reduced order solution can be computed fast. According to the classical trust-region philosophy we (locally) define

$$m_k(u_k + s) = \hat{f}(y^M(u_k + s), u_k + s) \tag{6}$$

as a model function for

$$f(u_k + s) = \hat{f}(y^N(u_k + s), u_k + s) \tag{7}$$

on the trust-region $\|s\| \le \delta_k$ for given $u_k \in \mathbb{R}^n$. In the definition of the model function (6), $y^M(u_k + s)$ denotes the POD based reduced order solution at $u_k + s$ computed with the reduced order model at u_k. We recall that the superscript M indicates the number of POD basis functions that have been used to build this reduced order model. Furthermore, we remark that

$$m_k(u_k) \neq f(u_k) \qquad \text{and} \qquad \nabla m_k(u_k) \neq \nabla f(u_k),$$

i.e. both the model function value as well as the model gradient at the trust-region center point are approximations to the corresponding values of the actual objective f.

Based on the definition (6) of the model function m_k we obtain the following trust-region algorithm for the solution of PDE-constrained optimization problems based on POD.

Basic TRPOD Algorithm

Initialization: Choose $0 < \eta_1 < \eta_2 < 1$, $0 < \gamma_1 \leq \gamma_2 < 1 \leq \gamma_3$, an initial trust-region radius $\delta_0 > 0$ and an initial iterate $u_0 \in \mathbb{R}^n$. Compute a snapshot set $\mathcal{Y}_0$ corresponding to u_0 and compute $f(u_0)$. Set $k = 0$.

1. Compute a POD basis $\mathcal{Y}_k^{POD}$ and build a POD based reduced order model.

2. Compute a descent step, s_k, for the model function $m_k(u_k+s)$ within the trust-region, $\|s\| \leq \delta_k$.

3. Compute a snapshot set $\mathcal{Y}_{k+}$ corresponding to $u_k + s_k$ and compute $f(u_k+s_k)$. Set

$$\rho_k = \frac{f(u_k) - f(u_k + s_k)}{m_k(u_k) - m_k(u_k + s_k)}.$$

4. Update the trust-region radius:

 – If $\rho_k \geq \eta_2$: Set $u_{k+1} = u_k + s_k$, $\mathcal{Y}_{k+1} = \mathcal{Y}_{k+}$ and choose $\delta_{k+1} \in [\delta_k, \gamma_3\delta_k]$. Set $k = k+1$ and GOTO 1.
 – If $\eta_1 \leq \rho_k < \eta_2$: Set $u_{k+1} = u_k + s_k$, $\mathcal{Y}_{k+1} = \mathcal{Y}_{k+}$ and choose $\delta_{k+1} \in [\gamma_2\delta_k, \delta_k)$. Set $k = k+1$ and GOTO 1.
 – If $\rho_k < \eta_1$: Set $u_{k+1} = u_k$ and choose $\delta_{k+1} \in [\gamma_1\|s_k\|, \gamma_2\|s_k\|]$. Set $k = k+1$ and GOTO 2.

In Fahl [2] the following convergence result can be found where results of Carter [12] and Toint [13] are used for its proof.

Theorem 3 (Convergence). *Let $\{u_k\}$ denote the sequence of iterates. Let $f : \mathbb{R}^n \to \mathbb{R}$ be continuously differentiable on an open convex set U with $U_0 = \{u \in \mathbb{R}^n \,|\, f(u) \leq f(u_0)\} \subset U$ and let f be bounded below with a Lipschitz continuous gradient ∇f. Furthermore, we assume that each model m_k*

is differentiable on an open convex set containing $\{u \in \mathbb{R}^n \mid \|u - u_k\| \leq \delta_k\}$ *for given* $\delta_k > 0$ *and that the model gradient satisfies the error condition*

$$\|\nabla f(u_k) - \nabla m_k(u_k)\| \leq \zeta \|\nabla m_k(u_k)\| \tag{8}$$

for some $\zeta \in (0, 1 - \eta_2)$, η_2 *according to the TRPOD algorithm.*

If, in addition, the curvature of each model function m_k *is bounded, then*

$$\lim_{k \to \infty} \|\nabla f(u_k)\| = 0$$

follows.

4 Reduced Order Modelling for the Adjoint

According to the convergence results presented in Theorem 3, we emphasize the importance of sufficiently accurate model gradients in order to achieve a good performance of the TRPOD algorithm. In the context of POD based reduced order modelling the error condition

$$\|\nabla f(u_k) - \nabla m_k(u_k)\| \leq \zeta \|\nabla m_k(u_k)\| \tag{9}$$

is hard to satisfy, especially when $\|\nabla m_k(u_k)\|$ approaches zero. Here, it is worth noting that $\|\nabla m_k(u_k)\| \longrightarrow 0$ implies the asymptotic consistency condition $\|\nabla f(u_k) - \nabla m_k(u_k)\| \longrightarrow 0$

When using POD methods, one should expect errors in the gradient. In the TRPOD approach presented so far, the POD basis functions are derived using snapshot information of the state equation. These basis functions are used for the discretization of the state equation. On the other hand, derivatives of the model function m_k have to be computed during the solution of a trust-region subproblem. If we compute the model gradient ∇m_k using, e.g., an adjoint equation approach, then we use the adjoint equation of the reduced order state equation for this computation. Hence, this implies that we discretize the original adjoint equation using basis functions that are especially suited for the discretization of the state equation. Therefore, it is not clear that $\nabla m_k(u_k)$ based on the POD model using snapshot information for the state $y(u_k)$ is a good approximation of $\nabla f(u_k)$ unless we make sure that adjoint information of the 'full' problem is taken into account.

Based on this consideration we present a separate POD based reduced order model for the adjoint equation in the TRPOD framework. We are aware of the fact that this can lead to the computation of inconsistent gradients [15], but numerical results seem to justify this approach.

In order to distinguish the reduced order model for the state y from the reduced order model for the adjoint π we use the terms y-POD ROM and π-POD ROM. Furthermore, we denote by $\nabla^{(y)} m_k(u)$ the gradient of m_k at u where the reduced order adjoint that corresponds to the y-POD ROM has been used for the computation of an approximate adjoint solution. Conversely, we denote by $\nabla^{(\pi)} m_k(u)$ the gradient of m_k at u where a separate π-POD ROM has been used for the computation of an approximate adjoint solution. We obtain the following trust-region algorithm that makes use of two separate POD based reduced order models, one for the computation of an approximation to the state solution and one for the computation of an approximation to the adjoint solution. Although this strategy improves the numerical performance, it still does not satisfy condition (8) needed for the convergence theory.

Modified TRPOD Algorithm

Initialization: Choose $0 < \eta_1 < \eta_2 < 1$, $0 < \gamma_1 \le \gamma_2 < 1 \le \gamma_3$, an initial trust-region radius $\delta_0 > 0$ and an initial iterate $u_0 \in \mathbb{R}^n$. Compute a snapshot set $\mathcal{Y}_0$ for the state corresponding to u_0 and compute $f(u_0)$. Compute a snapshot set Π_0 for the adjoint corresponding to u_0 and compute $\nabla f(u_0)$. Set $k = 0$.

1. Compute the POD bases $\mathcal{Y}_k^{POD}, \Pi_k^{POD}$ and the corresponding y-POD ROM and π-POD ROM.

2. Compute a descent step, s_k, for the model function $m_k(u_k{+}s)$ within the trust-region, using the y-POD ROM to obtain state information and the π-POD ROM to obtain adjoint information.

3. Compute a snapshot set $\mathcal{Y}_{k+}$ for the state corresponding to $u_k + s_k$ and compute $f(u_k{+}s_k)$. Set

$$\rho_k = \frac{f(u_k) - f(u_k + s_k)}{m_k(u_k) - m_k(u_k + s_k)}.$$

4. Update the trust-region radius:

– If $\rho_k \ge \eta_2$: Set $u_{k+1} = u_k + s_k$, $\mathcal{Y}_{k+1} = \mathcal{Y}_{k+}$ and choose $\delta_{k+1} \in [\delta_k, \gamma_3\delta_k]$. Compute a snapshot set Π_{k+1} for the adjoint corresponding to u_{k+1}. Set $k = k + 1$ and GOTO 1.

– If $\eta_1 \le \rho_k < \eta_2$: Set $u_{k+1} = u_k + s_k$, $\mathcal{Y}_{k+1} = \mathcal{Y}_{k+}$ and choose $\delta_{k+1} \in [\gamma_2\delta_k, \delta_k)$. Compute a snapshot set Π_{k+1} for the adjoint corresponding to u_{k+1}. Set $k = k + 1$ and GOTO 1.

– If $\rho_k < \eta_1$: Set $u_{k+1} = u_k$ and choose $\delta_{k+1} \in [\gamma_1\|s_k\|, \gamma_2\|s_k\|]$. Set $k = k + 1$ and GOTO 2.

5 Numerical Example: Control of Heat Equation

For the illustration of the presented concepts, we consider an optimal control problem involving the linear heat equation on $\Omega = (0,1)$:

$$\text{Minimize} \quad f(u) = \tfrac{1}{2}\int_0^T (y(u;0,t) - q(t))^2 \, dt + \tfrac{\gamma}{2}\int_0^T u(t)^2 \, dt \tag{10}$$

where $y(x,t) = y(u;x,t)$ is the solution of

$$y_t(x,t) - \kappa \, y_{xx}(x,t) = 0, \qquad 0 < x < 1, \, 0 < t < T \tag{11}$$

$$y(x,0) = y_0(x), \quad 0 < x < 1 \tag{12}$$

with boundary conditions

$$\kappa \, y_x(1,t) = \alpha(y(1,t) - u(t)), \quad 0 < t < T \tag{13}$$

$$\kappa \, y_x(0,t) = 0, \qquad 0 < t < T. \tag{14}$$

In (10) – (14), $y(x,t)$ denotes the temperature in $x \in \Omega$ at time $t \in (0,T)$. Here, the parameters κ and α denote the heat conductivity and heat transfer coefficient, respectively, and $\gamma > 0$ is a regularization parameter. The cost functional is of tracking-type, where $q(t)$, $t \in (0,T)$ denotes a desired temperature profile at $x = 0$. The control, $u(t)$ for $t \in (0,T)$, is applied at $x = 1$. The problem parameters where chosen as $\kappa = 0.3$, $\alpha = 10^3$, $\gamma = 0.005$, $T = 1$, $q(t) \equiv 1$, $u_0 \equiv 0.001$.

The gradient of f is given by,

$$(\nabla f(u))(t) = \kappa \, \alpha \, \pi(1,t) + \gamma \, u(t) \tag{15}$$

where $\pi(x,t)$ is the solution of the adjoint equation

$$-\pi_t(x,t) - \kappa \, \pi_{xx}(x,t) = 0, \quad 0 < x < 1, \, 0 < t < T$$

$$\pi(x,T) = 0, \quad 0 < x < 1$$

with boundary conditions

$$\kappa \, \pi_x(1,t) = -\alpha \pi(1,t), \qquad 0 < t < T$$

$$\kappa \, \pi_x(0,t) = q(t) - y(0,t), \quad 0 < t < T.$$

The trust-region algorithm parameters were set to: $\eta_1 = 0.25, \eta_2 = 0.75$, $\gamma_1 = \gamma_2 = 0.5, \gamma_3 = 2$ and $\delta_0 = 10$. For the computation of the snapshot sets for the state equation and the adjoint equation we used a FEM discretization with piecewise linear basis functions for the spatial discretization ($\delta x = 0.02$) and a backward Euler method for the time discretization ($\delta t = 0.01$).

Table 1. Results obtained with the basic TRPOD algorithm

k	$f(u_k)$	$m_k(u_k)$	δ_k	$\|s_k\|$	ρ_k	ζ_k	$\|\nabla^{(y)} m_k(u_k)\|$	M_y
0	0.504493	0.504496	10.00	9.72	1.02	0.0749	0.0586	2
1	0.141687	0.142525	20.00	10.52	1.07	0.0326	0.0188	3
2	0.075494	0.077617	40.00	6.69	1.12	0.3228	0.0032	3
3	0.069595	0.068544	80.00	1.63	1.89	0.3708	0.0029	3
4	0.067584	0.067063	160.00	0.60	-1.73	1.0498	0.0012	3
5	0.067584	0.067063	0.30	0.28	-0.87	1.0498	0.0012	3
6	0.067584	0.067063	0.14	0.14	-0.25	1.0498	0.0012	3
7	0.067584	0.067063	0.07	0.07	-0.07	1.0498	0.0012	3
8	0.067584	0.067063	0.03	0.03	0.00	1.0498	0.0012	3
9	0.067584	0.067063	0.02	0.02	0.03	1.0498	0.0012	3
10	0.067584	0.067063	0.01	0.01	0.05	1.0498	0.0012	3
11	0.067584	0.067063	0.00	0.00	0.05	1.0498	0.0012	3
12	0.067584	0.067063	0.00	0.00	0.06	1.0498	0.0012	3

Table 2. Results obtained with the modified TRPOD algorithm

k	$f(u_k)$	$m_k(u_k)$	δ_k	$\|s_k\|$	ρ_k	ζ_k	$\|\nabla^{(\pi)} m_k(u_k)\|$	M_y	M_π
0	0.504493	0.504496	10.00	9.73	1.02	0.0389	0.0585	2	2
1	0.140841	0.141630	20.00	11.50	1.11	0.1118	0.0188	3	1
2	0.095115	0.098160	40.00	6.51	0.95	0.0460	0.0114	3	2
3	0.072678	0.071419	80.00	5.20	2.92	0.1010	0.0032	3	2
4	0.067598	0.069022	160.00	2.86	-0.07	0.1013	0.0020	3	2
5	0.067598	0.069022	1.43	1.24	0.66	0.1013	0.0020	3	2
6	0.066819	0.067148	0.62	0.61	1.02	0.2751	0.0015	4	2
7	0.066411	0.066720	1.24	1.21	1.45	0.1042	0.0011	4	2
8	0.065938	0.066290	2.49	1.01	0.94	0.3741	0.0011	4	2
9	0.065725	0.066009	4.98	0.42	1.75	0.1738	0.0010	4	2
10	0.065582	0.065888	9.96	1.15	1.83	0.3553	0.0008	4	2
11	0.065437	0.065737	19.91	0.23	15.82	0.1103	0.0007	4	2
12	0.065404	0.065712	39.83	0.00	0.48	0.3983	0.0006	4	2

In Table 1 and Table 2 we compare the performance of the basic TRPOD algorithm with the modified TRPOD algorithm which differ in the computation of reduced order adjoint information.

There we present at each iteration, k, the objective function value, $f(u_k)$, the corresponding model function value, $m_k(u_k)$, the current trust-region radius, δ_k, the norm of the computed step, $\|s_k\|$, the ratio of achieved reduction to predicted reduction, ρ_k, and the POD subspace dimension for the state approximation, M_y. In Table 2 also the POD subspace dimension for the adjoint approximation, M_π, with a separate reduced order model is given. In all cases, the POD subspace dimension was chosen according to an energy criterion with $\bar{\varepsilon} = 0.999$, cf. (3). For both algorithms we also give the norm of the gradient, i.e. $\|\nabla^{(y)} m_k(u_k)\|$ and $\|\nabla^{(\pi)} m_k(u_k)\|$, respectively. Furthermore, the value of

$$\zeta_k = \|\nabla f(u_k) - \nabla^{(\cdot)} m_k(u_k)\| / \|\nabla^{(\cdot)} m_k(u_k)\|$$

according to (8) is shown. For this purpose, we have to compute the gradient of f using the full adjoint in order to compute the quantity ζ_k, but we do not use $\nabla f(u_k)$ otherwise.

The numerical results illustrate the previous discussion. The basic TRPOD algorithm leads to good results concerning the decrease in the objective function value. It is obvious that the gradient error deteriorates as indicated by the value of ζ_k, when the reduced order gradient approaches 0, such that no new acceptable step can found after iteration $k = 3$.

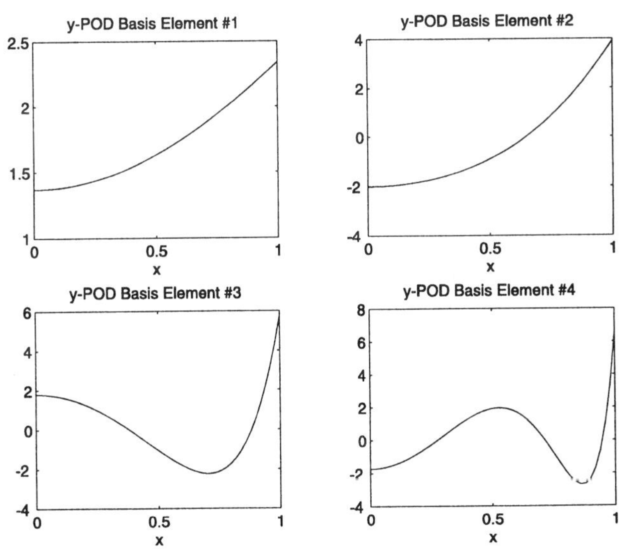

Fig. 1. POD Basis Elements for $\mathcal{Y}_0$

The modified TRPOD algorithm leads to better results. Here, the value of ζ_k indicates that we obtain a good approximation $\|\nabla^{(\pi)} m_k(u_k)\|$ to the true gradient $\nabla f(u_k)$, even when the reduced order gradient approaches 0.

In Figs. 1 and 2 the first 4 POD basis elements for the state equation are compared to the POD basis elements for the adjoint equation at iteration k. It is obvious that each set of POD basis elements correponds to different dynamics. Consequently, the corresponding reduced order models for the adjoint lead to different results which are depicted in Fig. 4 in comparison to the full adjoint solution based on a FEM discretization (Fig. 3). This illustrates that there are significant differences between the reduced order adjoint solution based on the y-POD or the π-POD ROM even with an increasing number of POD basis elements.

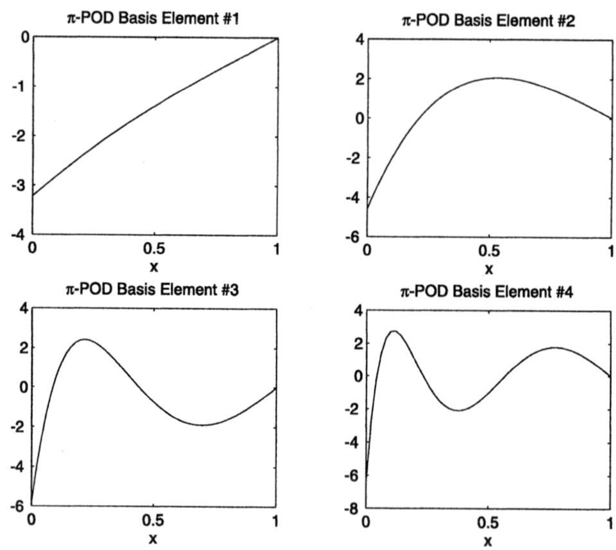

Fig. 2. POD Basis Elements for Π_0

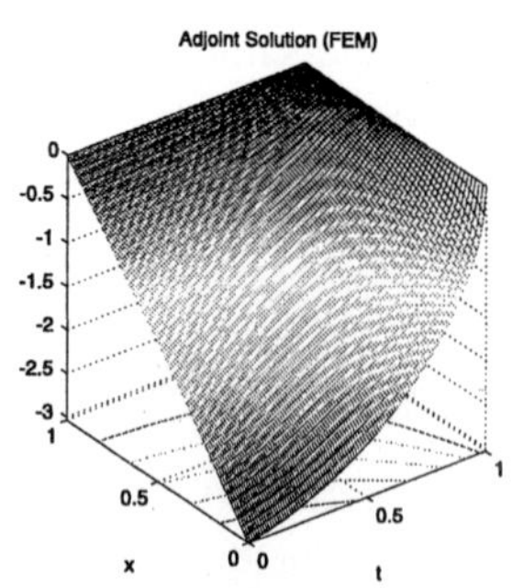

Fig. 3. FEM Adjoint Solution in Iteration $k = 0$ (see Plate 9 on page 348)

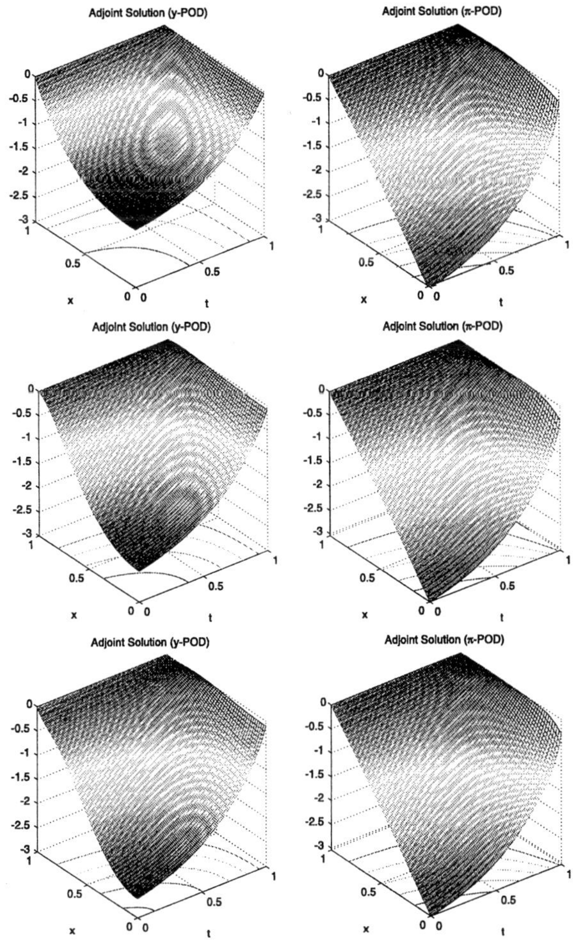

$M = 2$

$M = 3$

$M = 4$

Fig. 4. POD Based Adjoint Solutions in Iteration $k = 0$ (see Plate 10 on page 349)

Since the POD technique has been especially successful for the optimal control of the Navier-Stokes equation, we expect similar results for these type of problems. Numerical tests for this extension are in progress.

6 Conclusions

In this paper we illustrate the application of reduced order models based on proper orthogonal decomposition to boundary control problems with partial differential equations. It leads to a low dimensional model which is updated based on strategies known from trust region methods. We show how to overcome the difficulty that the required accuracy in the gradient information breaks down. Since the gradient is computed with an adjoint equation, we propose to introduce a separate reduced order model for the adjoint equation. Numerical results confirm the success of this approach.

References

1. Arian, E., Fahl, M., Sachs, E.W.: Trust-region proper orthogonal decomposition for flow control. ICASE Report No. 2000-25, ICASE, NASA Langley Research Center, Hampton, 2000
2. Fahl, M.: Trust-region methods for flow control based on reduced order modelling. PhD thesis, Universität Trier, Trier, 2000
3. Dikowy, F., Volkwein, S.: Nonlinear boundary control for the heat equation utilizing proper orthogonal decomposition. Report, Universität Graz, 15 pages, 2001
4. Lumley,J.L.: Coherent structures in turbulence. In: R.E. Meyer (ed.), Transition and Turbulence, Academic Press, New York, 1981, pp. 215–242
5. Holmes, P., Lumley, J.L., Berkooz, G.: Turbulence, coherent structures, dynamical systems and symmetry. Cambridge University Press, Cambridge, 1996
6. Ly, H.V., Tran, H.T.: Proper orthogonal decomposition for flow calculations and optimal control in a horizontal CVD reactor. Techreport CRSC-TR-98-13, North Carolina State University, Raleigh, 1998
7. Atwell, J.A., King, B.B.: Proper orthogonal decomposition for reduced basis controllers for parabolic equations. ICAM Report No. 99-01-01, Virginia Polytechnic Institute and State University, Blacksburg, 1999
8. Tang, K.Y., Graham, W.R., Peraire, J.:, Active Flow Control using a reduced order Model and optimum control. AIAA 96-1946, 1996
9. Kunisch, K., Volkwein, S.: Galerkin proper orthogonal decomposition methods for parabolic problems. Bericht Nr. 171, Spezialforschungsbereich F 003 - Kontrolle und Optimierung, Karl-Franzens-Universität Graz, Graz, 1999
10. Banks, H.T., Joyner, M.L., Wincheski, B., Winfree, W.P.:, Evaluation of material integrity using reduced order computational methodology. Techreport CRSC-TR-99-30, North Caroliona State University, Raleigh, 1999
11. Eckart, C., Young, G.: The approximation of one matrix by another of lower rank. Psychometrica **1** (1936) 211–218
12. Carter, R.G.: On the global convergence of trust region algorithms using inexact gradient information. SIAM J. Numer. Anal. **28** (1991) 251–265
13. Toint, P.L.: Global convergence of a class of trust-region methods for nonconvex minimization in Hilbert space. IMA J. Numer. Anal. **8** (1988) 231–252
14. Moré, J.J.: Recent developments in algorithms and software for trust region methods. In: A. Bachem and M. Grötschel and B. Korte (eds.), Mathematical programming - The state of the art, Springer, Berlin, 1983, pp. 258-287
15. Borggaard, J., Burns, J.: A PDE sensitivity equation method for optimal aerodynamic design. J. Comput. Phys. **136** (1997) 366–384
16. Alexandrov, N.M., Dennis, J.E., Lewis, R.M., Torczon, V.: A trust-region framework for managing the use of approximation models in optimization. Struct. Optim. **15** (1997) 16–23
17. Chang, K.J., Haftka, R.T., Giles, G.L., Kao, P.-J.: Sensitivity-based scaling for approximating structural response. J. of Aircraft **30** (1993) 283–288

Adaptive Simulation, the Adjoint State Method, and Optimization

Mark S. Gockenbach[1] and William W. Symes[2]

[1] Michigan Technological University, Houghton, MI 49931
[2] Rice University, Houston, TX 77251

Abstract. Adaptive grids in inverse and control problems can lead to computed objective functions that are nonsmooth, even though the underlying problem is well-behaved. This leads to the question of how to compute the linearization of the scheme—how should a nonsmooth function be differentiated? The C++ class `afdtd` uses automatic differentiation techniques to implement an abstract marching scheme in an object-oriented fashion, making it possible to use the resulting simulator to solve inverse or control problems. The class takes a complete specification of a *single step* of the scheme, and assembles from it a complete simulator, along with the linearized and adjoint simulations. The result is a (nonlinear) operator in the sense of the Hilbert Class Library, a C++ package for optimization. Moreover, `afdtd` supports locally "frozen" grids, allowing the implementation of an operator that is piecewise smooth in spite of the use of adaptivity.

1 Introduction

Many optimization problems are based on the solution of systems of differential equations. A general class of examples is provided by least-squares problems of the form

$$\min_{c_\epsilon} \|F_\epsilon(c_\epsilon) - d^{obs}\|^2.$$

Often d^{obs} is either a data or target vector and c_ϵ is a (discretized) coefficient or other parameter in a system of differential equations. The operator F_ϵ is the (approximate) solution operator: it takes a value of c_ϵ and simulates the result of the experiment. The subscript ϵ signifies that F_ϵ and c_ϵ are discretized versions of some underlying (continuous) F and c, with the discretization depending on a tolerance ϵ. The idea of the least-squares problem is to choose the value of c_ϵ that produces a simulated value of d as close as possible to the observed or desired value d^{obs}.

For reasons of efficiency, the differential equations are often solved by *adaptive* schemes. In an adaptive method, the grid used to compute $F_\epsilon(c_\epsilon)$ from c_ϵ can change each time the value of c_ϵ changes. These changes in the grid can lead to nonsmoothness of the computed operator F_ϵ. Nonetheless, F can be well-behaved, and it may make sense to solve the least-squares problem for F_ϵ.

The goal of this project is to study optimization problems depending on adaptive marching (or time stepping) simulations, and to determine what modifications in optimization practice are necessary to effectively compute solutions. In this preliminary report, we present an example of an adaptive marching scheme that leads to a nonsmooth objective function. We then present an abstract framework for adaptive schemes and describe a C++ class, called `afdtd`, which automatically assembles such a simulation, along with the linearization and its adjoint, from a complete description of a single step. This class forms a nonlinear operator in the sense of the Hilbert Class Library (HCL), a C++ optimization package. We conclude by returning to the example and discussing the idea of a "frozen grid" derivative.

2 An example

The following system of ordinary differential equations (see [2]) describes the orbit of a satellite of the earth-moon system:

$$\begin{aligned} \ddot{x} &= 2\dot{y} + x - \tfrac{(1-\mu)(x+\mu)}{r_1(x,y)^3} - \tfrac{\mu(x-1+\mu)}{r_2(x,y)^3} \\ \ddot{y} &= -2\dot{x} + y - \tfrac{(1-\mu)y}{r_1(x,y)^3} - \tfrac{\mu y}{r_2(x,y)^3}, \end{aligned} \tag{1}$$

where

$$\begin{aligned} r_1(x,y) &= \sqrt{(x+\mu)^2 + y^2}, \\ r_2(x,y) &= \sqrt{(x-1+\mu)^2 + y^2}. \end{aligned}$$

In these equations, $(x(t), y(t))$ are the coordinate of the satellite at time t. The origin of the (moving) coordinate system is the center of mass of the earth-moon system, and the x-axis is the line through the centers of the earth and the moon. The center of the moon is at the point $(1-\mu, 0)$ and the center of the earth is at $(-\mu, 0)$, where $\mu = 1/82.45$ is the ratio of mass of the moon to the mass of the earth. The unit of length is the distance from the center of the earth to the center of the moon.

If the satellite satisfies the following initial conditions, then its orbit, shown in Figure 1, is periodic with period $T \doteq 6.19216933$:

$$\begin{aligned} x(0) = 1.2,\ \dot{x}(0) = 0, \\ y(0) = 0,\ \dot{y}(0) \doteq -1.04935751. \end{aligned} \tag{2}$$

The direction of the satellite changes rapidly when its orbit brings it close to the earth, requiring a small time step for accurate integration. On the other hand, during much of the orbit a much larger time step is acceptable. For this reason, it is advantageous to employ an adaptive method. The Runge-Kutta-Fehlberg 4-5 (RKF45) scheme (see [2]) was used to produce Figure 1.

We now pose a simple control problem based on (1). We regard the quantity μ as a control parameter and ask the question: For what value of μ is the

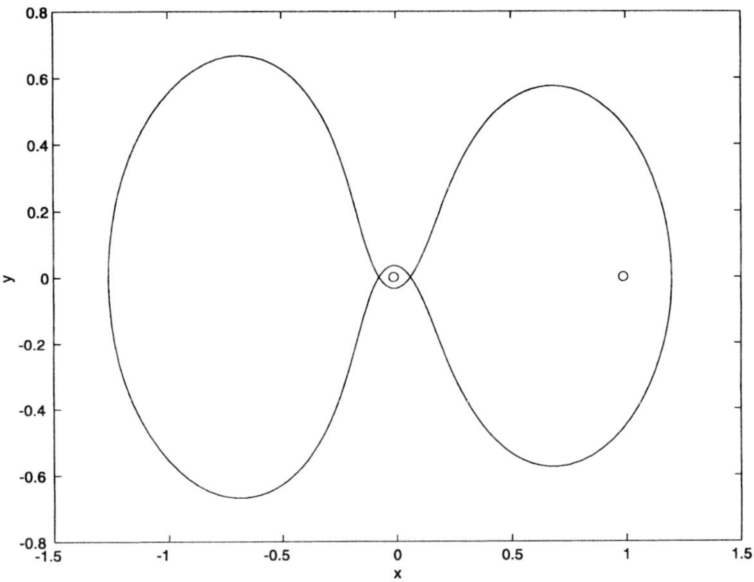

Fig. 1. The orbit of the satellite modeled by (1) and (2).

solution to (1), (2) periodic with period T? (Since μ represents the ratio of the masses of the moon and the earth, this control problem is best regarded as a purely synthetic exercise!)

Define $F : (0, \infty) \to \mathrm{R}^4$ by $F(\mu) = \mathbf{u}(T)$, where $\mathbf{u}$ is the solution to the first-order system derived from (1) in the usual fashion ($u_1 = x, u_2 = \dot{x}, u_3 = y, u_4 = \dot{y}$). We define a discretized operator $F_\epsilon : (0, \infty) \to \mathrm{R}^4$ so that $F_\epsilon(\mu)$ is the estimate of $\mathbf{u}(T)$ produced by RKF45 using a relative error tolerance of ϵ in the step control algorithm (all other parameters to the algorithm are fixed). We then define the least-squares objective function $J : (0, \infty) \to \mathrm{R}$ by

$$J(\mu) = \frac{1}{2} \left\| F(\mu) - \mathbf{u}(0) \right\|^2,$$

where the norm is the Euclidean norm. The discretized version J_ϵ is defined similarly, replacing F by F_ϵ.

In Figure 2, we graph F_ϵ, for $\epsilon = 10^{-3}$, for μ near $\mu^* = 1/82.45 \doteq 1.2129 \cdot 10^{-2}$. The graph shows that the computed solution is nonsmooth on the scale of the order of 1% of μ^* (that is, this is the size of the interval shown in the graph). We also graph, in Figure 3, the forward differences of J_ϵ, using a step of 10^{-6}. (Of course, the finite difference estimate of the derivative can be stabilized by increasing the step size. The point here is simply to emphasize that the function is nonsmooth on an appreciable scale, and to provide a comparison for Figure 5.)

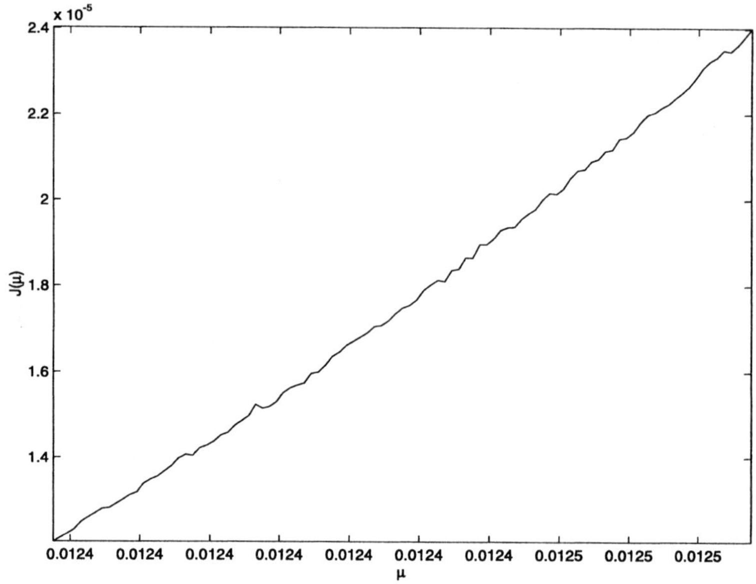

Fig. 2. Part of the graph of J_ϵ.

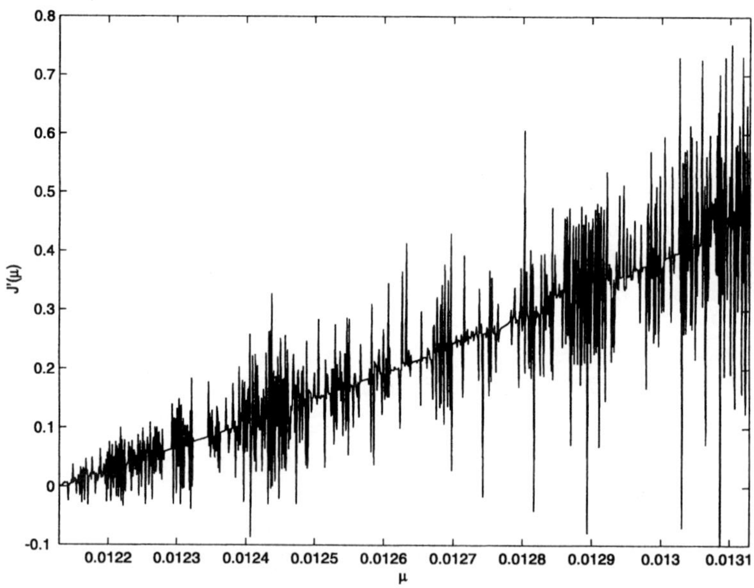

Fig. 3. Forward difference approximation to the derivative of J_ϵ.

3 An abstract framework for adaptive marching algorithms

We now consider explicit, adaptive time-stepping methods. To make the notation slightly less dense, we will not explicitly indicate the dependence of the operators and functions on ϵ.

We must allow for the spatial grid's changing when the time step changes, which considerably complicates matters. It is almost inevitable that the simulated field will be sampled on a uniform grid; therefore, even if the marching algorithm is a single step scheme, we must carry along several time levels of the simulated field.

We therefore define the *extended state space* to be the product of k possibly different factors—each representing the space of discretized functions on a given grid. The extended state space will be denoted by

$$\tilde{U}_h = \mathcal{U}_{h_1} \times \mathcal{U}_{h_2} \times \cdots \times \mathcal{U}_{h_k},$$

where

$$h = (h_1, h_2, \dots, h_k).$$

The space $\mathcal{U}_{h_i}$ corresponds to the spatial grid, and therefore depends on the time step h_i, since the spatial grid may need to be refined or coarsened when the time step changes. We assume that $\mathcal{U}_{h_i}$ depends only on the time step (and perhaps on the set of admissible controls, although we do not explicitly indicate this).[1]

We will frequently write $h = (\tilde{h}, \Delta t)$, where $\tilde{h} = (h_1, \dots, h_{k-1})$. This distinguishes the recent time steps (which have already been taken and are fixed) from the current time step, which may or may not be found acceptable.

We denote by C the space of controls, and by C_0 the subset of admissible controls. We require a family of *stencil operators*

$$H_{t,\tilde{h},\Delta t} : C_0 \times \tilde{U}_{(\tilde{h},\Delta t)} \to \tilde{U}_{(\tilde{h}',\Delta t)}, \ \tilde{h}' = (h_2, h_3, \dots, h_{k-1}, \Delta t).$$

A stencil operator represents the step taken on the grid corresponding to $\mathcal{U}_{\Delta t}$; the operator is defined on the extended state space in the event that the method is a multistep method or to indicate the updating of the earlier time levels for a single step method.

Since the spatial grid can change, we need a family

$$G_{h,\Delta t} : \tilde{U}_h \to \tilde{U}_{h'}, \ h' = (h_1, h_2, \dots, h_{k-1}, \Delta t)$$

[1] This assumption is adequate for some problems, particularly for hyperbolic equations in which the grids must be adapted to maintain stability. In other cases, a more general formulation, in which the spatial grid depends directly on the control c, may be necessary.

of *re-gridding* operators. The effect of $G_{h,\Delta t}$ is to interpolate the latest time slice to put it on a finer grid (if $\Delta t < h_k$) or coarser grid (if $\Delta t > h_k$). If $\Delta t = h_k$, then $G_{h,\Delta t}$ is assumed to be the identity operator.

Finally, we denote by $\mathcal{D}$ the data space, and we assume there is a family

$$S_{t,h} : \tilde{U}_h \to \mathcal{D}$$

of *sampling operators*. The output of the simulator will be a data set produced from the simulated field by sampling, as described in Equation (3) below.

We now assume that the sequence of time steps actually taken in the integration is $\Delta t_0, \Delta t_1, \ldots, \Delta t_s$; below we explain our assumptions on how these steps are chosen. We then define

$$F : C_0 \to \mathcal{D}$$

by

$$F(c) = \sum_{i=0}^{s} S_{t_i,h^{(i)}} U^{(t_i)}, \tag{3}$$

where

$$U^{(t_i)} \in \tilde{U}_{h^{(i)}},\ i = 0, 1, \ldots, s$$

is given by

$$U^{(t_{i+1})} = H_{t_i,\tilde{h}^{(i)},\Delta t_i}\left(c, G_{h^{(i)},\Delta t_i} U^{(t_i)}\right),\ i = 0, 1, \ldots, s-1.$$

The initial field $U^{(t_0)}$ is taken formally to be the zero vector; nonzero Cauchy data must be incorporated into the stencil operator $H_{t_0,0,\Delta t_0}$. This gives the flexibility to give the Cauchy data as a constant (in the definition of $H_{t_0,0,\Delta t_0}$) or as part or all of the control variable c.

Of course, this characterization of the calculation of the field U is given after the fact; in practice, the adaptive marching algorithm must determine the sequence $\Delta t_1, \Delta t_1, \ldots, \Delta t_{s-1}$ as the integration proceeds. The ith step is taken in one of three ways (the reader will recall that $h_k^{(i)}$ is the length of the step just taken, while Δt_i is the length of the step about to be taken):

1. $h_k^{(i)}$ was acceptable, and $\Delta t_i = h_k^{(i)}$ is not too large. Then
$$U^{(t_{i+1})} = H_{t_i,\tilde{h}^{(i)},\Delta t_i}\left(c, U^{(t_i)}\right).$$
2. $h_k^{(i)}$ was acceptable, and $\Delta t_i = h_k^{(i)}$ is too large. Then an acceptable $\Delta t_i < h_k^{(i)}$ is computed, and
$$U^{(t_{i+1})} = H_{t_i,\tilde{h}^{(i)},\Delta t_i}\left(c, G_{h^{(i)},\Delta t_i} U^{(t_i)}\right).$$

3. $h_k^{(i)}$ was too small. Then an acceptable Δt_i is computed (probably satisfying $\Delta t_i > h_k^{(i)}$), and

$$U^{(t_{i+1})} = H_{t_i,\tilde{h}^{(i)},\Delta t_i}\left(c, G_{h^{(i)},\Delta t_i} U^{(t_i)}\right).$$

In all three cases, we take $t_{i+1} = t_i + \Delta t_i$ and $h^{(i+1)} = (h_2^{(i)}, h_3^{(i)}, \ldots, h_k^{(i)}, \Delta t_i)$.

3.1 Linearization

To apply gradient-based optimization to a least-squares problem involving the operator F, we need to differentiate the function

$$J(c) = \frac{1}{2}\|F(c) - d^{obs}\|^2,$$

where the norm is induced by the inner product on the data space $\mathcal{D}$. We have

$$DJ(c)\delta c = (DF(c)\delta c, F(c) - d^{obs})_{\mathcal{D}} = \left(\delta c, DF(c)^*\left(F(c) - d^{obs}\right)\right)_{\mathcal{C}},$$

where $DF(c)^*$ is the adjoint of the derivative $DF(c)$ of F. It then follows that

$$\nabla J(c) = DF(c)^*\left(F(c) - d^{obs}\right).$$

Therefore, it is necessary to compute the linearization $DF(c)$ and its adjoint.

The linearization of the operator F is straightforward. We have

$$DF(c)\delta c = \sum_{i=0}^{s} S_{t,h^{(i)}} \delta U^{(t_i)},$$

where

$$\delta U^{(t_0)} = 0$$

and, for $i = 0, 1, \ldots, s-1$,

$$\delta U^{(t_{i+1})} = D_U H_{t_i,\tilde{h}^{(i)},\Delta t_i}\left(c, G_{h^{(i)},\Delta t_i} U^{(t_i)}\right) G_{h^{(i)},\Delta t_i} \delta U^{(t_i)} + D_c H_{t_i,\tilde{h}^{(i)},\Delta t_i}\left(c, G_{h^{(i)},\Delta t_i} U^{(t_i)}\right)\delta c.$$

Let us write

$$\tilde{U} = \tilde{U}_{h^{(0)}} \times \tilde{U}_{h^{(1)}} \times \cdots \times \tilde{U}_{h^{(s)}}.$$

We can represent $DF(c)$ as the composition of three operators, $DF(c) = SM^{-1}B$. We now define these three operators. First, we define $S : \tilde{U} \to \mathcal{D}$ by

$$S\delta U = \sum_{i=0}^{s} S_{t_i,h^{(i)}} \delta U^{(t_i)}.$$

Next, we write

$$A^{(i)} = D_U H_{t_i,\tilde{h}^{(i)},\Delta t_i}\left(c, G_{h^{(i)},\Delta t_i} U^{(t_i)}\right) G_{h^{(i)},\Delta t_i}$$

and

$$K^{(i)} = D_c H_{t_i,\tilde{h}^{(i)},\Delta t_i}\left(c, G_{h^{(i)},\Delta t_i} U^{(t_i)}\right).$$

Let $B : C \to \tilde{U}$ be defined by

$$B\delta c = \left(0, K^{(0)}\delta c, K^{(i)}\delta c, \ldots, K^{(s-1)}\delta c\right),$$

and write $\delta F = B\delta c$. We then have

$$\begin{aligned} \delta U^{(t_0)} &= 0, \\ \delta U^{(t_{i+1})} - A^{(i)}\delta U^{(t_i)} &= \delta F^{(i+1)}, \; i = 0, 1, \ldots, s-1. \end{aligned}$$

We can write these equations as $M\delta U = \delta F$, where $M : \tilde{U} \to \tilde{U}$ is the block operator

$$M = \begin{bmatrix} I & & & & \\ -A^{(0)} & I & & & \\ & -A^{(1)} & I & & \\ & & \ddots & \ddots & \\ & & & -A^{(s-1)} & I \end{bmatrix}.$$

3.2 Adjoint

The above results show that $DF(c)^* = B^* M^{-*} S^*$, where M^{-*} denotes the adjoint of the solution operator M^{-1}. We now exhibit the three individual adjoints.

It is straightforward to show that

$$(S^*\delta D)^{(i)} = S^*_{t_i,h^{(i)}}\delta D$$

and

$$B^*\delta F = \sum_{i=1}^{s} \left(K^{(i-1)}\right)^* \delta F^{(i)}.$$

Now, $M^{-*} = (M^{-1})^* = (M^*)^{-1}$, and

$$M^* = \begin{bmatrix} I & -\left(A^{(0)}\right)^* & & & \\ & I & -\left(A^{(1)}\right)^* & & \\ & & I & -\left(A^{(2)}\right)^* & \\ & & & \ddots & \ddots \\ & & & I & -\left(A^{(s-1)}\right)^* \\ & & & & I \end{bmatrix}.$$

Therefore, if $\delta W = M^{-*}\delta V$ or, equivalently, $M^*\delta W = \delta V$, we see that

$$\delta W^{(s)} = \delta V^{(s)},$$
$$\delta W^{(i-1)} = \delta V^{(i-1)} + \left(A^{(i-1)}\right)^* \delta W^{(i)}, \; i = s, s-1, \dots, 1.$$

Putting all of this together, we obtain the following algorithm for computing $\delta c = DF(c)^*\delta D = B^* M^{-*}\delta D$:

- Step s:

$$\delta W^{(s)} \leftarrow S^*_{t_s,h^{(s)}}\delta D,$$
$$\delta c \leftarrow \left(K^{(s-1)}\right)^* \delta W^{(s)}.$$

- Step $i = s-1, s-2, \dots, 1$:

$$\delta W^{(i)} \leftarrow S^*_{t_i,h^{(i)}}\delta D + \left(A^{(i)}\right)^* \delta W^{(i+1)},$$
$$\delta c \leftarrow \delta c + \left(K^{(i-1)}\right)^* \delta W^{(i)}.$$

This algorithm is fairly straightforward when presented at this high level, that is, in terms of the stencil, re-gridding, and sampling operators. As we will see below, the C++ class `afdtd` implements it at this level of abstraction, taking advantage of the high level constructs present in the HCL optimization package.

However, there is a logistical problem inherent in this algorithm, essentially the same problem familiar to implementors of reverse mode automatic differentiation: the field U is computed by forward time stepping, while the adjoint state field δW is computed by backward time stepping. Since U is needed during the computation of δW, we face the problem of how to make the various time levels $U^{(t_i)}$ available to the scheme for δW when needed.

There are two straightforward approaches to this problem, both of which present efficiency difficulties. First, one can simply compute and store the entire field U prior to marching backward in time. However, if the individual time levels $U^{(t_i)}$ are large, this can require a prohibitive amount of storage. On the other hand, one can simply begin marching backward in time; whenever a time level $U^{(t_i)}$ is needed, it can be computed by forward marching from time t_0. Thus $s-1$ steps are needed to deliver $U^{(t_{s-1})}$, $s-2$ steps to deliver $U^{(t_{s-2})}$, and so forth, for a total number of time steps of $O(s^2/2)$. This is a prohibitive amount of (re-)computation, when it is considered that the original simulation only requires s time steps.

It should be noted that, with the technique of computing and storing the entire time history of U, the storage costs increase by a factor of $O(s)$ over the storage cost of the original simulation. If the time levels of U are re-computed when needed, then the computation cost increases by a factor of $O(s)$ over the cost of the original simulation.

The need for extra storage and re-computation can be balanced, as explained by Griewank [4], so as to increase both the storage and computation costs by only $O(\log s)$. This is accomplished by the technique of *checkpointing*, saving strategically chosen time levels $U^{(t_i)}$ and re-computing from the most recent checkpoint rather than from t_0. By applying this idea in a recursive fashion, logarithmic growth in both storage and computation is achieved. Such a checkpointing scheme is implemented in the class `afdtd`.

3.3 The class `afdtd`

We now describe the implementation and use of the C++ class `afdtd`; at the time of this writing, a preliminary version has been implemented, though it may require refinement in design and/or implementation. We begin with a brief description of the HCL optimization package.

3.4 The Hilbert Class Library

The Hilbert Class Library (HCL) [3, 5] is a library of C++ classes representing the basic mathematical objects (vectors, functions, operators, etc.) used in formulating and solving optimization problems. It also includes a number of state-of-the-art algorithms for large-scale optimization (currently emphasizing unconstrained optimization). HCL allows optimization algorithms to be implemented in an abstract fashion. Practically speaking, this means that the optimization code does not deal directly with the data structures or interfaces used by application code. Therefore, optimization algorithms implemented using HCL classes can be used with application code of arbitrary complexity.

In addition to classes representing fundamental mathematical objects and classes implementing algorithms, HCL includes "tool" classes for building up common constructs from simpler objects. For example, a least-squares object function of the form

$$J(c) = \|F(c) - d^{obs}\|^2 \tag{4}$$

can be automatically created from a class implementing the operator F (such a class would be an example of an `HCL_Op` object) and a vector class containing the data (or target) vector d^{obs}. Therefore, to solve

$$\min_c J(c) \tag{5}$$

using an HCL optimization algorithm, one would have to:

- choose or implement vector classes for the model (or control) vector c and the data vector d. (HCL contain several vector classes implementing common data structures: simple in-core arrays, out-of-core (disk) storage, regular rectangular grids. If a user needs a specialized data structure, he or she could implement a new vector class, which would essentially mean implementing the data storage and the basic vector arithmetic operations. The ability of HCL to accommodate user-defined vector classes is one of its strengths.)

- implement an operator class representing F. This implies writing the same code to compute $F(c)$, $DF(c)$, and $DF(c)^*$ that would be required by any optimization package (this code could be written in Fortran, for example), and then "wrapping" it in a C++ class derived from the HCL base class `HCL_Op`.

The class `HCL_Op` is the abstract base class for nonlinear operators in the HCL framework; it mandates that a nonlinear operator F must have the following properties:

- it can identify its domain and range (as vector spaces);
- it can compute $F(c)$ and $DF(c)$. The image $F(c)$ is a vector (that is, an instance of `HCL_Vector`) and the derivative $DF(c)$ is a linear operator (an instance of `HCL_LinearOp`).

More details about HCL and its implementation can be found in the above references. For the purposes of this discussion, the important point is that, to use an HCL optimization algorithm to solve a least-squares problem of the form (4), (5), the basic requirement is the creation of an HCL operator class for F.

3.5 Implementation of `afdtd`

The purpose of the class `afdtd` is to implement the abstract framework for adaptive marching schemes described in Section 3. It therefore acts as a bridge between the problem-specific operator families

$$H_{t,\tilde{h},\Delta t},\ G_{h,\Delta t},\ S_{t,h}$$

and HCL.

The class `afdtd` is a concrete (i.e. fully implemented) class derived from `HCL_Op`. In order to create an instance of `afdtd` implementing the marching algorithm for a certain problem, a user must create three classes, one each for the stencil, re-gridding, and sampling operator families. For each of these operators, there is an abstract (i.e. unimplemented) base class, which is the C++ technique for describing the functionality of an object (and effectively providing a template for a user to follow in implementing a specific instance). Therefore, associated with `afdtd` are three classes:

- `aStencilOp`: Base class for $H_{t,\tilde{h},\Delta t}$;
- `aSampleOp`: Base class for $S_{t,h}$;
- `aGridOp`: Base class for $G_{h,\Delta t}$.

The product space $\tilde{U}_h$ (the extended state space) is represented by any instance of `HCL_ProductSpace` (the HCL base class for product vector spaces); the temporal grid information h is represented in a `LocalMarchingGrid` class. This class implements a 4-tuple $(t, \tilde{h}, \Delta t, \Delta t')$ $(h = (\tilde{h}, \Delta t))$ and also serves as a node in a doubly linked list (so that the sequences $t_0, t_1, \ldots, t_s$, $h^{(0)}, h^{(1)}, \ldots, h^{(s)}$, and $\Delta t_0, \Delta t_1, \ldots, \Delta t_{s-1}$ can be easily represented). Thus, to represent $\tilde{U}_h$ requires two objects, a product space and a "local marching grid", which are not formally linked.

3.6 `aStencilOp`

Most of the work associated with applying **`afdtd`** to a specific marching scheme lies in creating a stencil operator. This entails coding the basic step of the scheme, as well as the partial derivatives of the step with respect to the control and extended state variables. The adjoints of the linearizations must be provided.

All of these capabilities must be methods (member functions) of the stencil class. Here is a summary of these methods:

- **`Advance`**: Takes c and $G_{h^{(i)},\Delta t_i} U^{(t_i)}$, and computes $U^{(t_{i+1})}$, as well as returning a time step expected to be acceptable for the next step. Has an integral return value, which signals whether the step just taken was too large, too small, or acceptable.
- **`PartialDerivImageAddControl`**: Computes the partial derivative of

$$H_{t,\tilde{h},\Delta t}$$

  with respect to c, acting on a perturbation to c, and adds the result to the output.
- **`PartialDerivImageState`**: Computes the partial derivative of $H_{t,\tilde{h},\Delta t}$ with respect to U, acting on a perturbation to U, and assigns the result to the output.
- **`PartialDerivAdjImageAddControl`**: Computes the adjoint of the partial derivative of $H_{t,\tilde{h},\Delta t}$ with respect to c, acting on a perturbation, and adds the result to the output.
- **`PartialDerivAdjImageState`**: Computes the adjoint of the partial derivative of $H_{t,\tilde{h},\Delta t}$ with respect to U, acting on a perturbation, and adds the result to the output.
- **`SetMarchingGrid`**: Allows the calling program (**`afdtd`**) to set $(t, \tilde{h}, \Delta t)$.

Note that the various operations must be implemented according to how they are used in the algorithm (for example, we need an "add-to" version of the derivative with respect to c, but an "assign-to" version of the derivative with respect to U).

3.7 aSampleOp

The class `aSampleOp` is derived from `HCL_LinearOp`; the methods are the usual ones for a linear operator in HCL (computing the image of the linear operator acting on an input vector, and similarly for the adjoint), along with a `SetMarchingGrid`, which allows the calling program to specify (t, h).

3.8 aGridOp

The class `aSampleOp` is derived from `HCL_LinearOp`, and has the same methods as `aSampleOp`.

3.9 Methods of afdtd, Evalafdtd

Like all operator classes in HCL, the main method in `afdtd` is `Evaluate`, which produces an "evaluation" object representing $(F(c), DF(c))$. The class `afdtd` is derived from `HCL_Op`. The computations are all implemented in the evaluation class, `Evalafdtd`, and the two classes have mechanisms for "freezing" and "unfreezing" the computational grid, as we now explain.

During the computation of $F(c)$, a computational grid is determined adaptively. This grid, according to the assumptions we made earlier (specifically, that the space representing the spatial grid, $\mathcal{U}_{h_i}$, is determined by the time step h_i alone), is completely determined by the sequences $t_0, t_1, \ldots, t_s$, $h^{(0)}, h^{(1)}, \ldots, h^{(s)}$, and $\Delta t_0, \Delta t_1, \ldots, \Delta t_{s-1}$. These sequences are recorded in a linked list of `LocalMarchingGrid` objects. (Obviously, there is some redundant information among the three sequences.)

The `afdtd` object contains a flag, indicating whether a previous computational grid is to be re-used in the next computation. Specifically, `afdtd` has the following methods:

- `SetGrid`: Sets the grid (for "frozen grid" mode); takes as input the root node of a linked list of `LocalMarchingGrid` objects.
- `FreezeGrid`: Tells the class to freeze the grid using the next grid computed.
- `UnFreezeGrid`: Returns the class to adaptive computations.

The `Evalafdtd` class then has, in addition to the usual image methods, the following private methods:

- `FrozenGridImage` : Computes the image of the operator. This version of the Image method assumes that the grid is frozen.
- `AdaptiveImage` computes the image of the operator. This version of the Image method determines the grid adaptively.
- `AdaptiveDerivImage` : Computes the image of the derivative using an adaptive grid.
- `FrozenDerivImage` : Computes the image of the derivative using a frozen grid.

The usual image methods call the appropriate private method, depending on an internal flag.

The most complicated method in `Evalafdtd` is `DerivaAdjImage`, which computes the action of $DF(c)^*$ acting on an input perturbation. This method implements Griewank's optimal checkpointing technique to make the computation of the adjoint affordable. The `DerivAdjImage` method must first step through the forward simulation in order to record the "snapshot" of the state field at the checkpoints. Because the growth in the amount of computation is only logarithmic in the number of time steps, for all practical purposes the cost of the adjoint calculation for F (and hence of the computation of the gradient of J) is a small multiple of the cost of a simulation.

The typical use envisioned for `afdtd` is as follows: Stencil, sampling, and re-gridding operators are created for common schemes. The stencil operator depends in turn on the details of the particular problem. For a specific problem, the user must communicate to the stencil operator of the desired scheme the details of the problem. For example, the problem in Section 2 was coded using a stencil operator that implements the RKF45 algorithm. The specific vector field embodied in equations (1), along with the necessary linearizations and adjoints, was then provided to the RKF45 stencil operator in the form of Fortran subroutines. To address another system of ODEs, only these (relatively simple) subroutines would change.

4 Using `afdtd` in optimization

One advantage of the class `afdtd` is that it does a lot of the work involved in coding a time stepping simulation and linking it to an optimization algorithm. It also ensures that the adjoint state calculation is implemented efficiently, and relieves the user of the need to code the adjoint, a notoriously tedious task. (The user must still provide some adjoint code, but this code would implement a much simpler adjoint, pertaining to one step of the marching algorithm or even just to the vector field under consideration.)

Another possible advantage of `afdtd`, which remains to be investigated, is its ability to compute a meaningful derivative through the "frozen grid" mechanism. To demonstrate the possibilities, we repeat the computations from Section 2, except using the frozen grid feature. We divide the interval $[\mu^*, \mu^* + 10^{-3}]$ into 10 subintervals, and compute J_ϵ using a single grid on each subinterval. In Figure 4, we display the same part of the graph of J_ϵ as is given in Figure 2. The resulting graph is smooth, except at the midpoint of the interval shown, where the grid changes. The finite differences displayed in Figure 5, which should be compared to Figure 3, shows that the resulting objective function is smooth except at the nine points where the grid changes.

An obvious advantage of this implementation is the ability to compute an accurate derivative. Since the grids can be frozen, it is also possible to

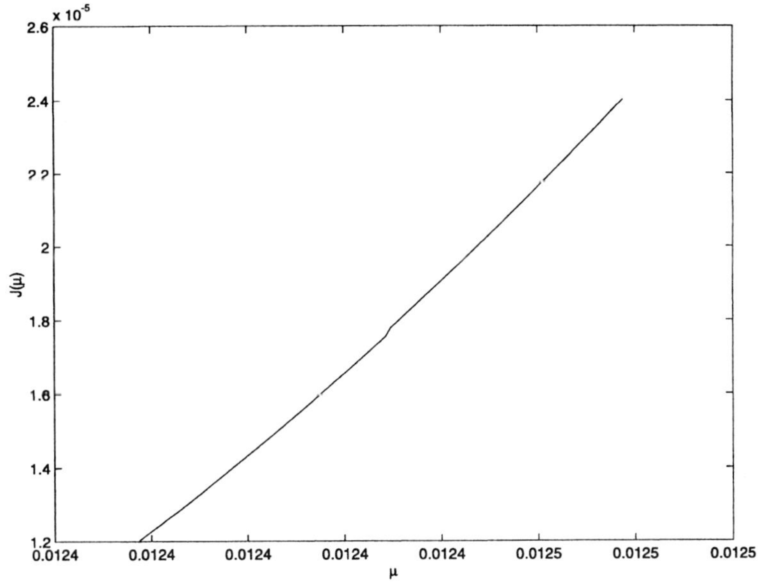

Fig. 4. Part of the graph of J_ϵ (frozen grids).

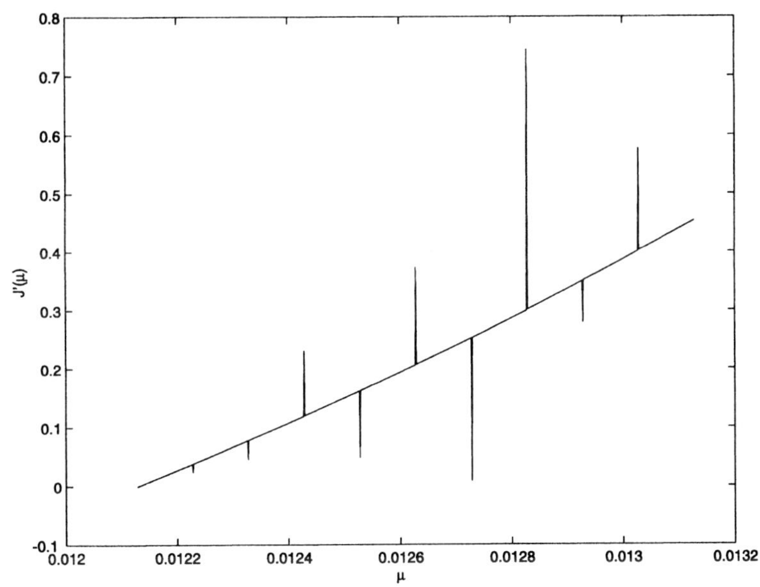

Fig. 5. Forward difference approximation to the derivative of J_ϵ (frozen grids).

check the derivative for accuracy using finite differences, a not insignificant point, since ensuring the correctness of derivatives provided to optimization code is essential and not always trivial. Finally, it may be possible to design optimization algorithms that take advantage of the frozen grid idea to improve convergence, particularly local convergence. This last idea remains to be investigated, however.

5 Other approaches to differentiation of adaptive schemes

Some authors have advocated independent or joint (adaptive) discretization of the systems of partial differential equations for the state and for the derivative and/or adjoint derivative of the state with respect to the control ("optimize then discretize", [1, 6]). This approach has two drawbacks, serious enough in our opinion to be ultimately fatal. First, simple tests for code verification are essentially precluded. For example, independently discretized derivative and adjoint derivative codes generally do not define linear maps adjoint to each other, and the deviation from adjointness can easily be on the same order as discretization error, and therefore difficult to separately bound. Likewise, the difference between calculated derivative and finite difference approximations does not go to zero with finite difference step, and the estimated rate of convergence does not appear to tend to a limit. We believe that such checks on code validity are extremely important, especially for complex applications joining many numerical modules. However the "optimize then discretize" approach makes simple checks, such as those just described, unavailable. Second, the discrepancies introduced in this way between the function and its derivatives and adjoint derivatives are sufficient to derail some optimization algorithms (as reported, for example, in [9]).

Other authors have suggested differentiating the state equation, using an AD package, on the grid determined by the state computation itself—see for example [8]. This approach is closely related to ours, in that the computed derivative is actually the derivative of the computed state, or would be if the state were considered as defined locally in the control with fixed (control-independent) grid. The same authors also carry out mesh adaptation sporadically for efficiency reasons, as do a number of multistep codes for ODEs [7]. Such sporadic re-meshing is necessary to extend the definition of the state with fixed grid locally in the control.

Computing derivatives in the context of adaptive simulation is, in our opinion, an open research issue. In particular, there is currently no satisfactory theory on which to base such calculations.

Acknowledgments

This work was supported in part by NSF grants DMS-9627355, DMS-9973423, and DMS-9973308, by the Los Alamos National Laboratory Computer Science Institute (LACSI) through LANL contract number 03891-99-23, and by The Rice Inversion Project.

References

1. J. Borggaard and J. Burns. A PDE sensitivity equation method for optimal aerodynamic design. *J. of Comp. Phys.*, pages 366–384, 1997.
2. George E. Forsythe, Michael A. Malcolm, and Cleve B. Moler. *Computer Methods for Mathematical Computations.* Prentice Hall, Englewood Cliffs, 1977.
3. M. S. Gockenbach, M. J. Petro, and W. W. Symes. C++ classes for linking optimization with complex simulations. *ACM Transactions on Mathematical Software*, 25(2):191–212, 1999.
4. A. Griewank. Achieving logarithmic growth of temporal and spatial complexity in reverse automatic differentiation. *Optimization Methods and Software*, 1:35–54, 1992.
5. Mark S. Gockenbach and William W. Symes. The Hilbert Class Library: a library of abstract C++ classes for optimization and inversion. *Computers and Mathematics with Applications*, 32:1–13, 1996.
6. S. Li, L. Petzold, and J. Hyman. Solution adapted mesh refinement and sensitivity analsysi for parabolic partial differential equation systems. Technical report, University of California - Santa Barbara, www.engineering.ucsb.edu/ cse, 2001.
7. Lawrence F. Shampine and Marilyn Kay Gordon. *Computer Solution of Ordinary Differential Equations.* W. H. Freeman, San Francisco, 1975.
8. R. Serban, S. Li, and L. Petzold. Adaptive algorithms for optimal control of time-dependent partial differential-algebraic equation systems. Technical report, University of California - Santa Barbara, www.engineering.ucsb.edu/~cse, 2001.
9. W.W. Symes and J. Meza. Conjugate residual methods for almost symmetric linear systems. *J. Optimization Theory and Applications*, 72(3):415–440, 1992.

Part VII

Frameworks for PDE-Constrained Optimization

The SIERRA Framework for Developing Advanced Parallel Mechanics Applications

James R. Stewart and H. Carter Edwards

Sandia National Laboratories
P.O. Box 5800, MS 0826
Albuquerque, NM 87109

Abstract. SIERRA is an object-oriented computational framework currently under intense development at Sandia National Laboratories. The SIERRA framework is a set of software services and parallel data structures upon which many mechanics applications can be written. The main goal is to bring together distributed mesh management, field management (i.e., the variables), and mechanics algorithm support services to facilitate rapid code development and code reuse. The motivation behind SIERRA is that designing and implementing modern applications in a multiphysics, parallel environment is very difficult, and involves advanced computer science as well as engineering mechanics. By providing the common, mostly computer science related services in a framework, they can be developed once and then used by many different mechanics codes. Although SIERRA is written in C++, the mechanics codes using SIERRA can be written in any language such as Fortran or C.

1 Introduction

Modern engineering mechanics software development has evolved from a relatively simple process to a very sophisticated and complex process. The simple process was usually controlled by one or two skilled numerical analysts who designed and implemented every aspect of the simulation code. This "old way" of developing engineering software is rarely possible now. The class of problems addressed by simulation has dramatically grown in scope, complexity and size. The computer platforms are also bigger and more diverse in terms of hardware architectures and operating systems. Finally, modern programming languages such as Java, C++, and Fortran 90 have matured, opening the door to sophisticated object-oriented designs and data structures. These issues and others have led to a modern software development process which involves teams of specialists and often involves multiple simulation codes.

A relatively new concept which is rooted in the idea of specialization is the software framework. In engineering mechanics applications, consolidating the common, mostly computer science components into one framework eliminates redundancy of development effort, simplifies debugging and testing, and promotes code re-use. The other key role of a framework is to provide a common underlying structure, leading to architectural similarity among the applications, The result is a cultural change in the way these codes are written. In effect, the tradeoff is between complete ownership of the software (the "old way") versus the ability to leverage capability others have developed so that one can focus on a particular mechanics-specific or algorithmic area.

SIERRA is an engineering mechanics software framework developed at Sandia National Laboratories. It is designed and implemented in accordance with a wide set of mechanics application code developer requirements, such as: Discretizations on unstructured meshes (finite element and finite volume methods); multiple element types (hexahedra, tetrahedra, shells, wedges, beams, etc.); multiple high-fidelity physics models and the ability to "loosely" couple different models to create a single multiphysics application (e.g., fluid-structure interaction); parallel scalability up to thousands of processors (including dynamic load balancing); parallel IO (input-output) for multiple file formats; multiple platform support, including the ASCI[1] supercomputers at Los Alamos National Laboratory, Lawrence Livermore National Laboratory, and Sandia; user input parsing; linear solver interfaces; and h–adaptivity.

Several other computational mechanics frameworks have also recently been developed. The *Trellis* framework of Beall and Shephard [1] supports a variety of analysis techniques such as finite element and partition of unity methods. Its design is fully object-oriented, and is centered around a geometry-based environment with direct linkages of the mesh to CAD geometrical representations. Trellis also supports various forms of adaptivity, including h– and p–adaptivity. Yu and Kumar [2] have developed an object-oriented framework for implementing the finite element method. They provide a multi-level class derivation hierarchy with pure virtual functions, providing a common interface for typical finite element components and operations. Also provided is an interface for file input and output which supports multiple file formats. Bangerth and Kanschat [3] have created an object-oriented finite element framework called Deal.II, which supports a variety of functionality including adaptivity and error estimation, multiple element types, multiple spatial dimensions, various output file formats, and multi-threading for shared-memory platforms.

A very different kind of computational mechanics framework is currently under development by de Sturler, et al. [4]. This framework focuses specifically on the coupling of codes and load balancing in a parallel environment, utilizing the Charm++ load balancing system developed by Kale and Krishnan [5]. It is the most "lightweight" framework of those discussed here in that it requires minimal re-write of existing codes (this is a design requirement). For this reason the framework moves away from the ideas of team specialization discussed previously, and allows the application code developer to maintain almost complete ownership of his or her existing code. There are

[1] Accelerated Strategic Computing Initiative, a U.S. Department of Energy program for stockpile stewardship.

obviously situations where this is desirable and advantageous. However, the set of services provided by such a framework will be rather limited.

There are two primary factors which differentiate the various frameworks. First is the design of the framework itself, which defines its ease of use and the extendability of the framework services. A good design also implies good maintainability and testability of the code. This is why object-oriented designs are generally used. The second differentiating factor is the different set of requirements each framework is trying to address. Each framework offers functionality to deliver its own set of requirements. SIERRA is a very ambitious and "heavyweight" framework which has a relatively large number of services and features. More importantly, SIERRA is continually evolving. In the future SIERRA may also be expanded to support, e.g., meshless methods, hierarchical *p*–version finite elements as well as *p*– and *hp*–adaptivity, and structured grids.

The SIERRA services are broken down into the general categories of basic and advanced. The basic services include the management of the *mesh*, the *fields* (i.e., the variables), and *mechanics* (i.e., mechanics algorithms and supporting data) in a distributed parallel environment. SIERRA adheres to a single-program-multiple-data (SPMD) paradigm implemented with the message passing interface (MPI). Advanced services include transfer operators for coupled simulations, adaptive mesh refinement and unrefinement, and interfaces to master elements. Using SIERRA, a mechanics code developer can focus primarily on the mechanics algorithms instead of things like advanced data structures, IO, input parsing, load balancing, low-level parallelism, etc. SIERRA employs an object-oriented design and is implemented in C++. However, the application mechanics algorithms can be implemented in other languages such as C or Fortran 77. This may be desirable for performance reasons as well as for easier integration of existing legacy software which is already written in such a language.

SIERRA clearly provides a separation of concerns allowing a specialized team environment for developing engineering mechanics software. Also part of the SIERRA system is a set of software development *tools* which handle software configuration management, automated regression testing, multiplatform builds and execution, and web-based source code and documentation browsing. By working in a common source code repository, SIERRA application developers are encouraged to share software designs, actual source code, and software engineering practices. SIERRA tool development efforts are guided by the software process improvement principles defined in the Capability Maturity Model (CMM)[2] [6].

In Section 2, we present a list of mechanics application codes which have been built using SIERRA. A more detailed description of several of SIERRA's basic and advanced services is given in Sections 3 and 4, respectively. Concluding remarks are given in Section 5.

---

[2] ® Carnegie Mellon University Software Engineering Institute.

2 SIERRA Applications

A number of mechanics application codes have been developed or are currently under development using SIERRA, as illustrated in Fig. 1. It is important to note that each code is its own executable, implying that each code is responsible for its own `main()` function. SIERRA is *not* a stand-alone executable; its services are delivered in the form of libraries which are linked to each specific code. Therefore each code using SIERRA has its own name and capabilities, and interfaces to the framework according to the services it needs. The application codes shown in Fig. 1 are:

1. Calore — finite element thermal diffusion and enclosure radiation [7];
2. Opera — finite element coupled fluid/solid free and moving boundary problems, including species transport;
3. Salinas — finite element linear structural dynamics and eigenanalyses;
4. Premo — finite volume compressible fluid dynamics;
5. Krino — level sets;
6. Tempo — quasistatics preloading followed by explicit dynamics release (i.e., the coupling of Adagio and Presto);
7. Presto — finite element explicit dynamics [8];
8. Adagio — finite element nonlinear quasistatics of solids [9];
9. Andante — finite element implicit dynamics;
10. Calagio — full, loose coupling of thermal diffusion and nonlinear quasistatics (i.e., the coupling of Calore and Adagio);
11. Syrinx — finite element thermal radiation transport;
12. Fuego — finite volume (nearly) incompressible fluid dynamics.

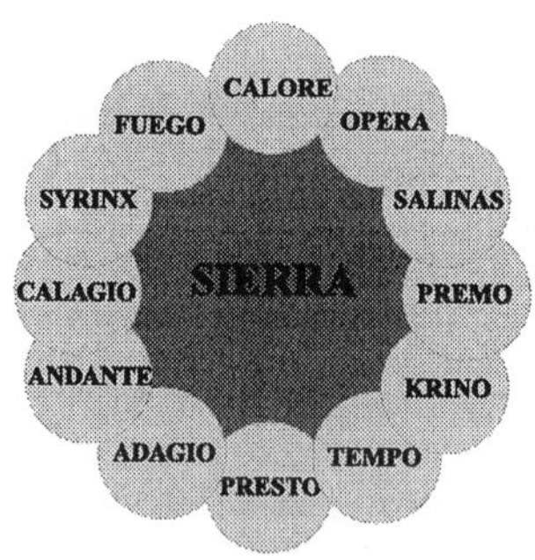

Fig. 1. A large variety of application codes have been developed or are currently under development using the SIERRA framework.

Figure 1 also implies that applications can overlap in that they can share code amongst themselves, outside the framework. This has been made possible as a result of all application developers using the same code repository, a common architecture of each code, and a new culture of maximizing code re-use. Examples of non-framework, shared code include parser command handlers for common input file commands, material property libraries, and master elements (SIERRA defines an *interface* to master elements, but the elements themselves are the responsibility of the mechanics code developers).

3 Basic Framework Services

The set of framework services considered "basic" are those which are used by all (or nearly all) mechanics applications. A few of the basic services are further grouped into the SIERRA "kernel", which forms the core of the SIERRA design. Basic services which are not part of the SIERRA kernel include user input parsing and parallel IO (input-output). It should be noted that pre-processing (mesh generation, parallel mesh decomposition, etc.) and post-processing (visualization, etc.) are not part of the SIERRA services.

The SIERRA kernel services are *mesh* management, *field* management, and *mechanics* management, and their interaction in a distributed parallel environment. For this reason the parallel communication services are also part of the SIERRA kernel. The pieces of the kernel can be thought of as separate entities as illustrated in Fig. 2, but their collaborations form a cohesive interacting unit. We now describe each kernel service in more detail.

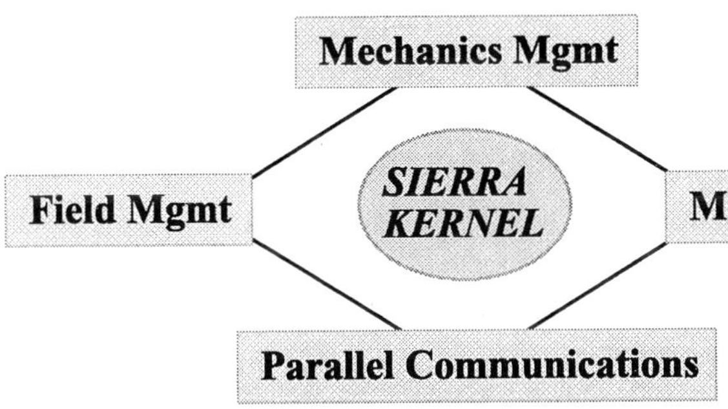

Fig. 2. The SIERRA kernel services, which collaborate to form a a single interacting unit. These services form the core of the SIERRA design.

3.1 Mesh Management

The SIERRA mesh management services utilize the following concepts and definitions:

Definition 1. A *mesh extent* is a well-defined subset of a mesh.

A mesh extent can be assembled in arbitrary ways. For example, mesh extents can be defined according to a particular usage, association with a particular geometric feature such as a boundary, or a particular property such as the part of the mesh that is shared among two or more processors.

Definition 2. A *mesh object* is an individual member, or entity, of a mesh.

Examples of mesh objects are nodes, edges, faces, and elements. The list of mesh object types is easily extendable. SIERRA also defines a *constraint* mesh object. An example is a node-face pair in contact for which a contact constraint must be applied.

Definition 3. A *mesh object relation* defines the pair-wise connectivity between mesh objects, the purpose of that connection, and the orientation of that connection.

Each mesh object contains a dynamically-sized list of its mesh object relations, which allows complete generality and extendability of the mesh data structure. The simplest example of the use of relations is for storing the element-node connectivity. A trilinear hexahedron would have eight node relations in its relations list; a linear tetrahedron would have four node relations. A hexahedron which has been adaptively refined[3] into eight child hexahedra would have eight element relations in its list. Furthermore, a "CHILD" designation would be assigned to each relation to indicate the purpose of the connection. A hexahedron might also have six face relations; the corresponding face objects would have a positive surface normal computed according to its own face-node connectivity. Each of the hexahedron's face relations would then have an "orientation" designator to indicate whether that face's normal points *into* or *out of* the hexadradron.

Through the use of mesh object relations, SIERRA provides a mesh data structure of arbitrary richness. The extent to which this data structure gets populated is controlled by the mechanics application developer, not the framework. Normally when a mesh is read in from an input mesh file, the element objects are created along with their node relations. The mechanics application developer can request that face and/or edge relations also be stored. If face objects are only required for computing surface fluxes for a particular boundary condition, for example, then it is possible to construct *only those faces which lie on that boundary.* In this case only those elements adjacent to this surface would have any face relations, since any "interior faces" would not exist.

The application developer can also construct "reverse" connectivities (i.e., relations), for example the list of elements connected to a node. In an unstructured mesh the number of elements connected to each node would in general be different; each node's list would be sized appropriately. Another

[3] Adaptive mesh refinement is one of the "advanced" SIERRA services.

example of the richness and abstract nature of mesh object relations is for the *constraint object* discussed above. A constraint pairing a node and a face in contact would be represented as a constraint mesh object with one node relation and one face relation.

Definition 4. A *mesh object roster* is a data structure used for managing mesh objects of a particular type.

A single physics type (as defined by a SIERRA *region*) would contain one mesh object roster for each mesh object type (for example, one *element* roster, one *face* roster, etc.). Rosters support dynamic retrieval, creation, modification, and destruction of mesh objects. The dynamic creation and destruction capabilities are required to support mechanics with dynamic mesh extents and dynamic load balancing. Examples of mechanics with dynamic mesh extents could include large deformation contact mechanics or mechanics that erode mesh objects. Mechanics application developers can access all mesh objects (of a given type) by iterating an entire roster. Alternatively, the iteration can be restricted to a particular mesh extent.

Each roster is a *distributed container* [10] of mesh objects. The mesh objects managed by a roster are distributed among the processors used by an application. All mesh objects have a *global identifier* that is unique within the containing roster and across all processors. Each mesh object is assigned to be "owned" by a single processor; those that are duplicated among processors are globally shared. Globally unique identifiers ensure consistency of shared mesh objects, and are also used to support parallel communications.

3.2 Field Management

Definition 5. A *field* is any variable that has a value associated with each mesh object in a mesh extent.

A field may be thought of as an array of values where each value is associated with a particular mesh object. Recall that a mesh extent may define a dynamic subset of a mesh; therefore, this conceptual array of values may not be defined over the entire mesh and its length may change dynamically. SIERRA is responsible for allocating the memory and storing each field according to its specification. This specification, defined by the mechanics application developer for each field, includes:

1. Text name (e.g., coordinates, temperature);
2. (Possibly mechanics-oriented) data type (e.g., integer, real, vector-3D, tensor-symmetric-3D);
3. Mesh object type to which the field is associated (e.g., node, element);
4. (One or more) *mechanics* that define(s) the mesh extent of the field;
5. Data partition;
6. Data topology.

The mechanics-oriented data types define arrays of real values. For example, a "vector-3D" type is an array of three real numbers. The *data partition* defines a field's usage with respect to the application's state. Examples are *model* and *state*. SIERRA provides for a multiple-state architecture through its *state* data partition. For example, a time integration algorithm might require values from previous states (i.e., time steps). The final item in the field specification, the *data topology*, defines how a field is associated with a mesh object. For example, it may be required by a mechanics algorithm to access a field at the Gauss quadrature points within an element. In this case, the data topology for that field would tell SIERRA to dimension an array according to the number of Gauss points.

Definition 6. A *registrar* is a data structure used for managing the aggregation of field specifications for a particular mesh object type.

A single physics type (as defined by a SIERRA *region*) would contain one registrar for each mesh object type (e.g., one *element* registrar, one *face* registrar, etc.). Thus all the nodal variables, for example, may be accessed by iterating the node registrar. It is important to note that an actual field value cannot be accessed until a given field (obtained from the registrar) and mesh object are paired. In addition, through the assignment of a mesh extent to a given field (see Item 4 in the field specification list), the mechanics application developer may wish to define a variable only on a subset of the mesh, such as boundary. The allocation of the proper amount of memory for a given field is handled dynamically by SIERRA.

3.3 Mechanics Management

Definition 7. A SIERRA *mechanics* is a collection of algorithms and supporting data which implements a particular set of physics.[4]

The definition of what constitutes a "mechanics" is purely a design decision on the part of the mechanics application code developer. There is much flexibility here. For example, a mechanics may implement a particular partial differential equation (PDE), and each of its mechanics algorithms may implement one term of the PDE. Mechanics objects can also be *nested*. For example, the coefficients of a particular term in a PDE may be determined by another algorithm which is part of a nested mechanics. Thus the definition above is recursive: A mechanics is a collection of algorithms and *other mechanics*.

The core of the SIERRA framework design builds on the concept of mechanics. Figure 3 shows the building blocks — *region*, *procedure*, and *domain*. Also depicted is the integration of the *mesh*, *fields*, and *mechanics* within each

[4] The *algorithms* and the *data* are actually maintained in separate classes by SIERRA — the *mechanics* and *mechanics instance* classes, respectively.

region. Finally, Fig. 3 shows the SIERRA *transfer* object which allows fields to be transferred between regions.[5]

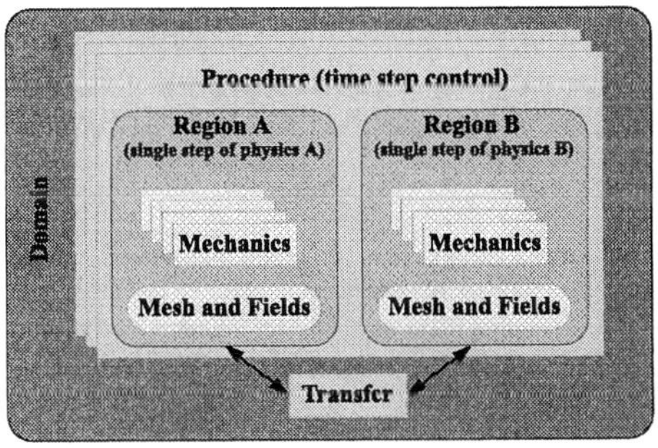

Fig. 3. The building blocks of the SIERRA framework design: There is exactly one *domain* per application. The *region* and *procedure* classes are derived from the *mechanics* class. The region is composed of a *mesh*, *fields*, and other *mechanics*. Multiple regions may be coupled using *transfer* algorithms.

Definition 8. A *region* is the aggregation of the *mechanics*, *fields*, and *mesh* required to perform a single set of (tightly coupled) physics for a single time step.

The *region* class is derived from the *mechanics* class. Thus a region `is-a` mechanics (with nested mechanics objects), plus the mesh object rosters, field registrars, etc. A mechanics application code developer is *not* required to derive any mechanics-specific classes from the SIERRA region class. All that is required is that a mechanics-specific version of the SIERRA region is instantiated.

Definition 9. A *procedure* is the aggregation of one or more regions, and is typically responsible for advancing the mechanics simulation through a sequence of state updates (i.e., time steps).

A procedure is also responsible for invoking the SIERRA transfer algorithm if two or more regions are to be coupled within a time step. As with the region class, the *procedure* class is also derived from the *mechanics* class. Thus a procedure `is-a` mechanics (with nested region objects), plus a list of any transfer objects and any timing and time stepping information. A mechanics application code developer is *not* required to derive any mechanics-specific classes from the SIERRA procedure class. All that is required is that a mechanics-specific version of the SIERRA procedure is instantiated.

[5] The transfer object is an advanced SIERRA framework service and is discussed in Sect. 4.2.

Definition 10. A *domain* is the aggregation of one or more procedures. There is exactly one domain within a mechanics application code.

The domain is not a derived class. It essentially holds things that are "global", such as application-defined material property libraries. The domain is responsible for sequencing the procedures, as well as invoking any transfers algorithms for possible inter-procedural application coupling. In the simplest case, only one procedure is executed and no transfers are required.

Workset Services. A critical basic SIERRA service is the *workset*, which brings together each of the *mechanics*, *field*, and *mesh* services. A workset is a flat array of field values needed for a particular mechanics algorithm. Although the use of a workset by a mechanics code is optional, we nonetheless consider it a "basic" framework service since it is fundamental for algorithm performance. Features of a workset are given below.

1. Field values are gathered from the mesh objects (which in general are *not* contiguous in memory) into the workset by the framework.
2. Multiple mesh object types are accounted for. An element calculation, for example, usually needs nodal field values; these values would all be present in the "element workset".
3. A workset may be delivered in Fortran or C order, so that existing mechanics algorithms may easily be transferred to SIERRA from legacy codes.
4. A workset is sized according to the cache of the hardware platform. An "element workset" would contain data for as many elements as would (conservatively) fit into cache. This is done to prevent page faulting which impedes performance.

A workset can be thought of as a data delivery system — the framework delivers the right data to the right algorithms in an efficient form for computation. As mentioned previously, the use of a workset is optional, since all the field and mesh data is available directly to any mechanics algorithm using the mesh object rosters and registrars.

3.4 Parallel Communications

The SIERRA communication services provide the communications required for scalable single-program-multiple-data (SPMD) multi-mechanics applications in an "as transparent as possible" level of abstraction. Only a small number of mechanics algorithms are required to be aware of the application's parallelism. These parallel-aware algorithms typically belong to a region or other "control" mechanics. The "nearly transparent" level of abstraction insulates the mechanics algorithms from low-level message passing, i.e., calls to the message passing interface (MPI).

Part of the communication services provides the standard global-scalar reduction operations such as global sum, minimum, or maximum of ordinary scalar values. These services support mechanics' global-reduction operations such as the computation of total energy or a time step determination based on some global stability condition (such as a time step computed using the smallest element size).

The more complex communication operations require the communication of information between pairs of specific mesh objects. Since these communication oeprations are typically sparse, the pairs of mesh objects represent a small fraction of the total number of mesh objects in a mesh. This, in turn, only requires a given processor to communicate with a fraction of the total number of processors used by an application. The communication services supporting these operations are separated into three major components:

1. A *communication specification* (i.e., *CommSpec*);
2. A *field map*;
3. Communication operations — *copy*, *sum* (i.e., assemble), *construct*, and *synchronize*.

The *Commspec* is a well-defined mapping between two potentially distributed mesh extents. Furthermore, the mesh extents may be associated with meshes of *different* regions. The *CommSpec* mapping pairs the "sending" mesh object with the "receiving" mesh object, which in general lies on another processor. *CommSpecs* are used extensively in several advanced SIERRA services such as transfers.

The *field map* is a one-to-one mapping between sets of fields. The purpose of the map is to provide a pairing of the "sending" field with the "receiving" field for a given communication operation. As with the *CommSpec*, the fields may be associated with *different* regions.

The *Commspec* and field map must be supplied as arguments to the *copy* and *sum* communication operations. The *construct* operation takes a partially constructed *CommSpec* (the list of "sending" objects) and constructs the "receiving" objects on the destination processors. This service is used extensively to support, for example, dynamic load balancing. The *synchronize* operation is used to restore global consistency to a mesh which may have been locally modified. This is required, for example, in adaptivity where elements are locally refined and unrefined.

4 Advanced Framework Services

SIERRA provides an evolving list of advanced services which depend on many of the basic services described in the previous section. By categorizing a service as "advanced", it is implied that usage of the service by a mechanics application code is optional. Examples of advanced services include an interface to linear solvers through the Finite Element Interface [11], an interface to master elements, adaptive mesh refinement and unrefinement, and element erosion (sometimes referred to as element "death"). In this section we focus briefly on two other advanced services — multiple concurrent domain decompositions and transfer operators.

4.1 Multiple Concurrent Domain Decompositions

In a distributed parallel environment each processor owns a different piece of the problem (with small regions of overlap, or ghosting). This decomposition is usually constructed to balance the computational load while minimizing inter-processor communication, which amounts to parallel "overhead". In SIERRA, such a decomposition is referred to as the *primary decomposition*, The primary decomposition is obtained through a pre-processing step (outside of SIERRA); the decomposed mesh is subsequently read in as input.

In many situations the primary decomposition will either impede scalability or cause an algorithm to fail. For these reasons SIERRA provides a mechanism to dynamically create a *secondary decomposition*, whereby a copy of the mesh (or appropriate mesh subset) is created according to the requirements of the algorithm. Perhaps the algorithm operates on only a subset of the mesh, such as a surface, which may in turn reside on only a subset of the processors (in the primary decomposition). If this agloritthm was executed in the primary decomposition, the remaining processors would sit idle. The secondary decomposition would dynamically balance the computational load by equally distributing the surface among all the processors.

For other algorithms such as the parallel transfer, which requires extensive geometric searching, the algorithm may *fail* in the primary decomposition (since on-processor geometric alignment of the two meshes is required). In this case a single secondary decomposition would be created for *both* regions. This decomposition, sometimes called a *geometric rendezvous* decomposition, would align the two meshes so that searching and interpolation can proceed. A simple illustration of multiple concurrent domain decompositions is given in Fig. 4, which shows the overlay of a primary and a secondary decomposition among four processors for a $2D$ mesh.

The secondary decomposition is actually a copy of the of the appropriate SIERRA *region(s)*, including appropriate subsets of the fields as well as the mesh. Specifically, the SIERRA services are used to

1. define a secondary load balanced decomposition for the subset of mesh objects;
2. construct a communication map (i.e., a *CommSpec*) between the primary decomposition and secondary decomposition;
3. construct a copy of the appropriate region subset in the secondary decomposition; and
4. communicate field values between the primary and secondary regions.

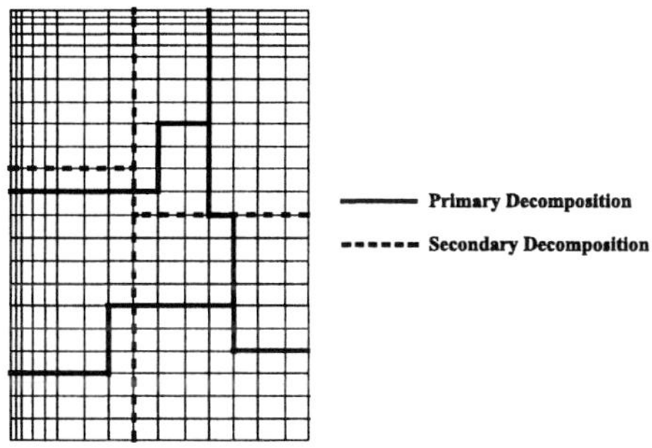

Fig. 4. The overlay of the *primary* decomposition and the *secondary* decomposition among four processors for a 2*D* mesh. In this example, the primary decomposition is topologically based, while the secondary decomposition is geometrically based, making it suitable for geometric searching algorithms.

Although more memory is required, the secondary decomposition is *temporary* — it is created dynamically, it exists for the lifetime of the algorithm, then the memory is reclaimed. Furthermore, as mentioned above, often these are not full copies but only copies of small parts of the region.

4.2 Parallel Transfer Services

Transfer operators handle the transfer of fields between two regions, allowing loosely coupled multi-physics applications to be developed (see Fig. 3). The SIERRA transfer services make extensive use of the secondary decomposition, as discussed in the previous section. In general, we must handle the case where the two meshes involved in the transfer (referred to as the "source" and "destination" meshes) have been *independently* generated and decomposed among the processors, thus causing geometric mis-alignment on any given processor. This is solved by constructing a secondary decomposition, often referred to as a geometric-rendezvous decomposition, to provide full geometric overlap of the two meshes on all processors.

One possible transfer algorithm is illustrated in Fig. 5. In this example we are transferring a nodal field from the source primary mesh to the destination primary mesh (the meshes are assumed to discretize the same volumetric domain). First, a rendezvous decomposition is constructed using some geometric decomposition algorithm such as Recursive Coordinate Bisection (RCB). Load balance of the search algorithm is assured since we include both the elements of the source mesh and the nodes of the destination mesh in the construction of the decomposition (i.e., in the generation of the RCB "cuts"). In order to construct the source and destination secondary regions, primary-to-secondary *CommSpecs* are created. In this algorithm we wish to perform the nodal field interpolation in the (destination) secondary region; therefore we require that both mesh *and* nodal field data be copied into the source secondary region. Note that we only require mesh data to be copied into the destination secondary region, since the field data is yet to be determined.

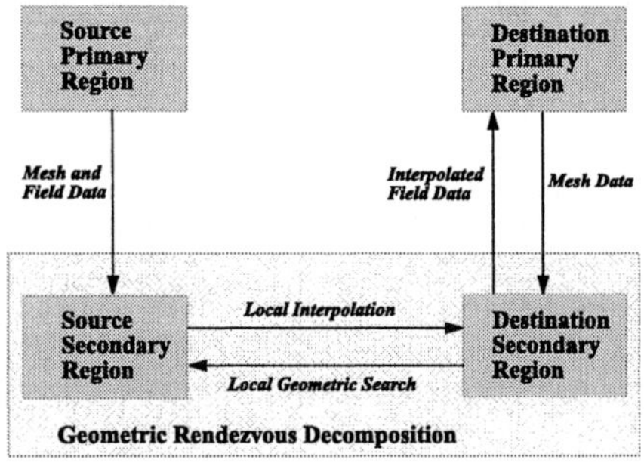

Fig. 5. Parallel transfer algorithm for a field transferred from the source region to the destination region.

In the secondary regions, all search and interpolation operations are entirely local. The search phase maps nodes in the destination region to the elements in the source region which contain the nodes. The nodal shape functions are then used to interpolate the fields onto the destination nodes. Finally, the destination primary-to-secondary *CommSpec* is reversed, allowing the interpolated nodal values to be communicated back to the destination primary region. This algorithm has been implemented in the framework and is currently used by several multi-physics applications.

Note that the high-level abstractions of the *CommSpec* and secondary decompositions allow variations of this algorithm to be easily designed. For example, by composing the *CommSpecs* in particular ways, one may choose to communicate the search data (i.e., the element-node pairs) back to the source *primary* region for interpolation. Then again using composition of mappings, a *CommSpec* can be constructed to send the interpolated values directly from the primary source to the primary destination region. The advantage of such an algorithm, in terms of load balance and communication overhead, is mechanics-application dependent.

5 Concluding Remarks

SIERRA is an object-oriented computational mechanics framework which provides an extensive set of services which are common to mechanics application codes. By providing these mostly computer science related services in a framework, application code developers can focus on their mechanics algorithms. The result is to eliminate redundant development efforts, maximize code re-use, and increase the testability and maintainability of the code. In addition, the application codes built using SIERRA have similar architectures, and they are developed using the same software engineering processes.

At the core of the "basic" services is the SIERRA kernel, which consists of *mesh*, *field*, and *mechanics* management, along with low-level parallel communications in a single-program-multiple-data (SPMD) environment. The basic services are used by virtually all application codes built using SIERRA. The "advanced" services are optional and provide many high-level features. In this paper we described two of these advanced services — multiple concurrent domain decompositions and parallel transfer operators. At least a dozen mechanics applications have been written so far using SIERRA, and more are planned. SIERRA is continually evolving to support a growing number of application code developer requirements.

References

1. Beall, M.W., Shephard, M.S.: An object-oriented framework for reliable numerical simulations. Eng. Comput. **15** (1999) 61–72
2. Yu, L., Kumar, A.V.: An object-oriented modular framework for implementing the finite element method. Comput. Struct. **79** (2001) 919–928
3. Bangerth, W., Kanschat, G.: Concepts for object-oriented finite element software — the deal.II library. IWR Heidelberg: Preprint 99-43 (1999) SFB 359
4. de Sturler, E., Hoeflinger, J., Kale, L., Bhandarkar, M.: A new approach to software integration frameworks for multi-physics simulation codes. Proc. Work. Conf. 8 Arch. Scient. Soft. (October 2000)
5. Kale, L.V., Krishnan, S.: Charm++: Parallel programming with message-driven objects. In: Parallel Prog. using C++ (eds: Wilson, G.V., Lu, P.) MIT Press (1996) 175–213
6. Paulk, M.C., Weber, C.V., Curtis, B., Chrissis, M.B.: The cability maturity model: guidelines for improving the software process. Carnegie Mellon Univ. Soft. Engin. Inst., Addison-Wesley, Reading, MA (1994)
7. Bova, S.W.: An approach to coupled, multiphysics thermal analysis. In: Proc. 1st MIT Conf. Comput. Fluid Struct. Mech., Elsevier, Amsterdam (2001)
8. Koteras, J.R., Gullerud, A.S., Porter, V.L., Sherzinger, W.M., Brown, K.H.: PRESTO: Impact dynamics with scalable contact using the SIERRA framework. In: Proc. 1st MIT Conf. Comput. Fluid Struct. Mech., Elsevier, Amsterdam (2001)
9. Mitchell, J.A., Gullerud, A.S., Scherzinger, W.M., Koteras, J.R., Porter, V.L.: ADAGIO: Non-linear quasi-static structural response using the SIERRA framework. In: Proc. 1st MIT Conf. Comput. Fluid Struct. Mech., Elsevier, Amsterdam (2001)
10. Edwards, H.C.: A parallel infrastructure for scalable adaptive finite element methods and its application to least squares C^∞ collocation. Ph.D. dissertation, Dept. Comp. Appl. Math., Univ. Texas Austin (May 1997)
11. Clay, R.L., Mish, K.D., Otero, I.J., Taylor, L.M., Williams, A.B.: An annotated reference guide to the Finite-Element Interface (FEI) specification Version 1.0. Sandia Report SAND99-8229 (January 1999)

rSQP++ : An Object-Oriented Framework for Successive Quadratic Programming

Roscoe A. Bartlett[1] and Lorenz T. Biegler[1]

Department of Chemical Engineering, Carnegie Mellon University

Abstract. An Object-Oriented (OO) framework called rSQP++ is currently being developed for Successive Quadratic Programming (SQP). It is designed to support many different SQP algorithms and to allow for external configuration of application-specific linear algebra objects such as matrices and linear solvers. In addition, it is possible for a client to modify the SQP algorithms to meet other specialized needs without having to touch any of the source code within the rSQP++ framework or even having to recompile existing SQP algorithms. Much of this is accomplished through a set of carefully constructed interfaces to various linear algebra objects such as matrices and linear solvers. The initial development of rSQP++ was done in a serial environment and therefore issues related to the use of massively parallel iterative solvers used in PDE-constrained optimization have not yet been addressed. In order to more effectively support parallelism, rSQP++ needs the addition and integration of an abstract vector interface to allow more flexibility in vector implementations. Encapsulating vectors away from algorithmic code would allow fully parallel linear algebra, but could also greatly restrict the kinds of operations that need to be performed. The difficulty in developing an abstract vector interface and a proposed design for a remedy are discussed.

1 Introduction

Nonlinear Programming (NLP) is an important tool in many areas of engineering and design. Application areas where large scale NLPs arise include DAE- and PDE- constrained optimization as well as many others. A standard form for constrained NLPs is

$$\min \quad f(x) \tag{1}$$

$$\text{s.t.} \quad c(x) = 0 \tag{2}$$

$$x^L \leq x \leq x^U \tag{3}$$

where: $x, x^L, x^U \in \mathbb{R}^n$, $f(x) \in \mathbb{R}^n \to \mathbb{R}$, $c(x) \in \mathbb{R}^n \to \mathbb{R}^m$

A new object-oriented (OO) framework for building Successive Quadratic Programming (SQP) algorithms, called rSQP++, is currently being implemented in C++. The goals for rSQP++ are quite ambitious. The rSQP++ framework is being designed to incorporate many different SQP algorithms and to allow external configuration of specialized linear algebra objects such as matrices and linear solvers. Data-structure independence has been recognized as an important feature missing in current optimization software [12].

In addition, it is possible for the client to modify the SQP algorithms to meet other specialized needs without having to touch any of the source code within the rSQP++ framework.

SQP methods are attractive because they generally require fewer iterations and function and gradient evaluations to solve a problem compared to other methods. Another attractive property of SQP methods is that they can be adapted to exploit the structure of an NLP. A variation of SQP, known as reduced space SQP (rSQP), works well for NLPs where there are few degrees of freedom and many constraints. Another advantage of rSQP is that the decomposition used for the equality constraints only requires solves with a basis of the constraint Jacobian and possibly its transpose (see Section 3.1).

2 Motivation for developing specialized NLP solvers and next generation optimization software

Much research has been, and continues to be, devoted to finding efficient methods to solve large-scale NLPs. Several methods have been proposed with successful implementations: Generalized Reduced Gradient (GRG) (CONOPT), Successive Linearly Constrained (SLC) (MINOS), Sequential Augmented Lagrangian (SAL) (LANCELOT), and Successive Quadratic Programming (SQP) (SNOPT , SOCS) [12]. More recently, Interior Point (IP) methods have been investigated and show promise (e.g. NITRO, LOQO). Many of these methods and even some implementations have been studied and refined for more than a decade. Many users rely on them through the use of modeling environments like GAMS and AMPL and these solvers are also embedded into many other applications.

With so many different well refined implementations for NLP solvers available, why would anyone want to develop their own? It seems that the overriding motivation for developing specialized optimization software is a desire to better exploit the properties of specialized classes of NLPs. For the most part, all of the aforementioned NLP solvers can only exploit the general sparsity of an NLP and use direct linear solvers that require explicit nonzero entries from the Jacobian and Hessian matrices. These solvers generally can not take advantage of opportunities for problem specific specialized linear algebra or possibly even parallelism. One example of such an application area is DAE-constrained optimization of chemical processes. In this type of optimization problem, the model, consisting of Differential and Algebraic Equations (DAEs), can be discretized over the time domain using orthogonal collocation on finite elements [4]. The Jacobian of the discretized equations has a special structure that can be exploited to significantly reduce the runtime and storage compared to general solvers (see Section 5). Another example of a specialized application area is PDE-constrained optimization using the Finite Element Method (FEM). Large-scale simulation codes have been developed for solving nonlinear PDEs with millions (even billions) of discretized state variables.

This has been made possible due to advances in massively parallel iterative linear solvers and multi-processor computers [11]. Most current optimization software can not take advantage of these developments.

While the desire to better exploit the properties of specialized NLPs is clear, in practice, general NLP solvers are often used instead. The reason for this is simple. These general NLP solvers have become very reliable due to significant development work, testing and refinement on many different optimization problems. The Herculean task of implementing a reliable specialized NLP solver is usually not worth the effort. There are many reasons why developing an NLP solver is so difficult. While implementing the basic computations for generating search directions may be fairly straightforward in some simplified cases, more elaborate means are generally necessary. In addition, sophisticated globalization strategies are needed to achieve convergence from remote starting points. Also, as with any numerical code, special attention needs to be paid to finite precision floating point errors that are amplified by ill conditioning. Another point is that every refined NLP implementation includes many potentially complex heuristics to deal with exception conditions. Even for a conceptually simple algorithm, the tasks of validating and debugging code are substantial.

What is required is a different approach to implementing the next generation of optimization software. Of paramount importance is the need to allow a sophisticated user to specialize the algorithm to exploit "their" special class of NLPs. This might include problem specific data structures, linear solvers, and opportunities for parallelism. This might also include altering the logic of the algorithm. But of course, for this software to be useful, it needs to include those elements of an NLP solver that are, for the most part, independent of a specific class of NLPs and are difficult to implement well. These include such things as problem-independent QP solvers, globalization methods (e.g. line searches, trust regions etc.), quasi-Newton methods, as well as many other refinements (e.g. heuristics). Also, if users are going to be allowed to specialize components in the algorithm, the optimization software needs to include efficient but optional run-time tests to validate the various computations. This includes validating functions and gradients, linear solvers and every other major computation that has clear well-defined post conditions. Trying to determine why an algorithm fails to find a desired solution without such built-in testing and validation is an extremely difficult task even for an expert.

3 Successive Quadratic Programming (SQP)

Successive Quadratic Programming (SQP) has proven to be an attractive framework for the development of specialized NLP algorithms. The main feature of an SQP algorithm is the solution of a Quadratic Programming

(QP) subproblem at each iteration such as:

$$\min \; g^T d + \tfrac{1}{2} d^T W d \tag{4}$$
$$\text{s.t.} \; A^T d + c = 0 \tag{5}$$
$$x^L - x^k \le d \le x^U - x^k \tag{6}$$

where: $d = x^{k+1} - x^k \in \mathbb{R}^n$ $\quad g = \nabla f(x^k) \in \mathbb{R}^n$
$W = \nabla_x^2 L(x^k, \lambda^k) \in \mathbb{R}^{n \times n}$ $\quad A = \nabla c(x^k) \in \mathbb{R}^{n \times m}$

Globalization methods are used to insure (given a few assumptions are satisfied) the SQP algorithm will converge to a local solution from remote starting points. The major classes of globalization methods include line search and trust region methods [6]. Because SQP is similar to applying Newton's method to the KKT optimality conditions, it can be shown to be quadratically convergent near the solution of the NLP [5]. One difficulty, however, is that in order to achieve quadratic convergence the exact Hessian of the Lagrangian W is needed which requires exact second order information for $\nabla^2 f(x)$ and $\nabla^2 c_j(x)$, $j = 1 \ldots m$. For many NLP applications this information is not readily available. Also, for large problems (4)–(6) can be extremely expensive, or nearly impossible, to solve directly. These and other difficulties with SQP have motivated the research of large-scale decomposition methods for SQP. One class of these methods is reduced space SQP (rSQP).

3.1 Reduced Space Successive Quadratic Programming (rSQP)

In a rSQP method, (4)–(6) is decomposed into two smaller subproblems that, in many cases, are easier to solve. To accomplish this, a range-space / null-space decomposition is computed for $A \in \mathbb{R}^{n \times m}$ (assuming it is full rank). This decomposition is defined by a null space matrix Z and range space matrix Y with the following properties:

$$Z \in \mathbb{R}^{n \times (n-m)}, Y \in \mathbb{R}^{n \times m} \text{ s.t. } A^T Z = 0, \begin{bmatrix} Y & Z \end{bmatrix} \text{ nonsingular} \tag{7}$$

By using (7), the search direction d can be broken down into $d = Yp^y + Zp^z$, where $p^y \in \mathbb{R}^m$ and $p^z \in \mathbb{R}^{n-m}$ are the range space and null space steps respectively. By substituting $d = Yp^y + Zp^z$ into (4)–(6), we obtain the range space (8) and null space (9)–(10) subproblems.

Range Space Subproblem	**Null Space (QP) Subproblem**
$p^y = -R^{-1} c^d \quad (8)$	$\min \; (g^r + w)^T p^z + \tfrac{1}{2}(p^z)^T [Z^T W Z] p^z \quad (9)$ $\text{s.t.} \; b^L \le Zp^z \le b^U \quad (10)$
$R \equiv [A^T Y] \in \mathbb{R}^{m \times m}$	
$g^r \equiv Z^T g \in \mathbb{R}^{n-m}$	$w \equiv Z^T W Y p^y \in \mathbb{R}^{n-m}$
$b^L \equiv x^L - x^k - Yp^y \in \mathbb{R}^n$	$b^U \equiv x^U - x^k - Yp^y \in \mathbb{R}^n$

Using this decomposition, the Lagrange multipliers λ for (5) do not need to be computed in order to compute the steps $d = Yp^y + Zp^z$. However, it is useful to compute these multipliers in the context of some globalization methods and at the solution of the NLP since they represent sensitivities. An expression for computing λ can be derived by applying (7) to $Y^T \nabla L(x, \lambda, \nu)$ to yield $\lambda = -R^{-T} Y^T (g + \nu)$.

There are many details that need to be worked out in order to implement an rSQP algorithm and there are opportunities for significant variability. Some significant decisions need to be made: how to compute the range/null space decomposition that defines the matrices Z, Y and R, and how the reduced Hessian $Z^T W Z$ and the cross term w in (9) are calculated (or approximated).

There are several different ways to compute decomposition matrices Z and Y that satisfy (7). Several choices for Z and Y have been investigated that are appropriate for large-scale rSQP [8]. One class of decomposition is based on a variable reduction. In a variable reduction decomposition, the variables are partitioned into dependent (basic) x^D and independent x^I sets ($x^T = [(x^D)^T \; (x^I)^T]$) such that the Jacobian of the constraints A^T is partitioned (assuming some permutations) as shown below, where C is a square nonsingular matrix known as the basis matrix. This partitioning is used to define a variable reduction null space matrix Z. Two choices for the range space matrix Y are termed the coordinate and orthogonal approaches and are shown below.

Variable Reduction	**Coordinate**	**Orthogonal**	
$A^T = [C \; N]$, $Z \equiv \begin{bmatrix} D \\ I \end{bmatrix}$	$Y \equiv \begin{bmatrix} I \\ 0 \end{bmatrix}$, $R = C$	$Y \equiv \begin{bmatrix} I \\ -D^T \end{bmatrix}$, $R = C(I + DD^T)$	(11)
where: $C \in \mathbb{R}^{m \times m}$, $N \in \mathbb{R}^{m \times (n-m)}$, $D \equiv -C^{-1} N \in \mathbb{R}^{m \times (n-m)}$			

The orthogonal decomposition ($Z^T Y = 0$) is more numerically stable and has other desirable properties in the context of rSQP [8]. However, the amount of dense linear algebra required to compute the factorizations needed to solve for linear systems with R is $O((n-m)^2 m)$ floating point operations (flops) which can dominate the cost of the algorithm for larger $(n-m)$. Therefore, for larger $(n-m)$, the coordinate decomposition ($Z^T Y \neq 0$) is preferred because it is cheaper. The downside is that it is also more susceptible to problems associated with a poor selection of dependent variables and ill conditioning in the basis matrix C that can result in greatly degraded performance.

Another important decision is how the compute the reduced Hessian $B \approx Z^T W Z$. When quasi-Newton is used, limited memory as well as several different dense approaches (e.g. update matrix or some form of its factors) are possible and the best choice depends on the entire context of the algorithm.

In some cases, computing the exact reduced Hessian $B = Z^T W Z$ or using it implicitly in some manner is computationally feasible.

In addition to variations that affect the convergence behavior of the rSQP algorithm (i.e. range/null decompositions, approximations used for the reduced Hessian, and many different types of merit functions and globalization methods) there are also many different implementation options. For example, linear systems such as (8) can be solved using direct or iterative solvers, and the reduced QP subproblem in (9)–(10) can be solved using a variety of methods (active set vs. interior point) and software [9].

4 An object-oriented approach to SQP (rSQP++)

Most numerical software (optimization, non-linear equation solvers, etc.) consists of an iterative algorithm that primarily involves common linear algebra operations. Mathematicians use a precise notation for these linear algebra operations when they describe an algorithm. For example, $y = A\ x$ denotes matrix-vector multiplication irrespective of the special properties of the matrix A or the vectors y and x. Such elegant and concise abstractions are usually lost, however, when the algorithm is implemented in most programming environments and implementation details such as sparse data structures obscure the conceptual simplicity of the operations being performed. Modern software engineering modeling and development methods, collectively known as Object-Oriented Technology (OOT), can provide powerful abstraction tools for dealing with these types of issues [1], [7]. In addition to abstracting linear algebra operations, Object-Oriented Programming (OOP) languages like C++ can be used to abstract any special type of quantity and operation. Also, OOT can be used to abstract larger chunks of an algorithm and provide for greater reuse. There are primarily two advantages to using data abstraction: it improves the clarity of the program, and it allows the underlying implementation of the operations to be changed and optimized without affecting the design of the application or even requiring recompilation of much of the code.

There are many types of challenges in trying to build a framework for SQP that allows for maximal sharing of code, and at the same time is understandable and extensible. Specifically, three types of variability are discussed: (a) algorithmic variability, (b) implementation variability and (c) NLP-specific specializations.

(a) First, we need to come up with a way of modeling and implementing iterative algorithms that will allow for steps to be reused between related algorithms and for existing algorithms to be extended. For example, how the same basic line-search globalization code be used in many different, but related, optimization algorithms? This type of higher level algorithmic modeling and implementation is needed to make the steps in our rSQP algorithms more independent so that they are easier to maintain and reuse. A framework called *GeneralIterationPack* has been developed for these types of iterative algorithms and serves as the backbone for rSQP++.

(b) The second type of variability is in allowing for different implementations of various parts of the rSQP algorithm. There are many examples where different implementation options are possible and the best choice will depend on the properties of the NLP being solved. One example is whether to represent $D = -C^{-1}N$ in (11) explicitly or implicitly. Another example is the implementation of the Quasi-Newton reduced Hessian $B \approx Z^T W Z$. The choice for whether to store B directly or its factorization (and in what form) or both depends on the choice of QP solver used to solve (9)–(10). Yet another example is allowing different implementations for the QP solver.

(c) A third source of variability involves the use of the special properties of an application area. For example, very specialized iterative linear solvers can be developed for many different types of PDE-constrained optimization problems. Abstract interfaces to matrices have been developed that serve as the foundation for facilitating the type of implementation and NLP-specific linear algebra variability described above. In addition, these abstract interfaces help manage some of the algorithmic variability such as the choice of different range/null space decompositions.

Figure 1 shows a coarse grained UML [1] object diagram for a rSQP++ algorithm configured for the solution of an NLP. At the core is a set of algorithmic objects. The `rSQPAlgorithm` object acts as the center hub for the algorithm and its main job is to fire off a set of steps in sequential order and perform major loops. One or more `rSQPStep` objects perform the actual computations in the algorithm. The `rSQPStep` objects operate on iteration quantity objects (`IterQuantity`) that are stored in the `rSQPState` object. In addition to simple linear execution of an algorithm, more sophisticated control strategies can be performed. This design allows step classes to be shared in many different related algorithms and also provides for modifications of the algorithm by adding, removing and replacing `rSQPStep` and `IterQuantity` objects. In other words, the behavior of the algorithms is not fixed and can be modified at runtime. In this way, users can modify the rSQP algorithms without touching any of the base source code in rSQP++.

Also shown in Figure 1 are `DecompositionSystem` and `NLP` objects. These objects provide the keys to specializing the linear algebra for a particular NLP. The `DecompositionSystem` interface (Figure 2) abstracts the choice for the range/null space decomposition away from the optimization algorithm. The `DecompositionSystemVarReduct` node subclass is for variable reduction decompositions. A `BasisSystem` object is used to abstract the variable reduction matrices including the basis matrix C. The `DecompositionSystemVar-ReductOrthogonal` and `DecompositionSystemVarReductCoordinate` subclasses implement the orthogonal and coordinate decompositions.

The `NLP` interface (Figure 3) is used to abstract the application. The base `NLP` interface provides basic information such as variable bounds and the initial guess and computes $f(x)$ and $c(x)$. The `NLPFirstOrderInfo` specializa-

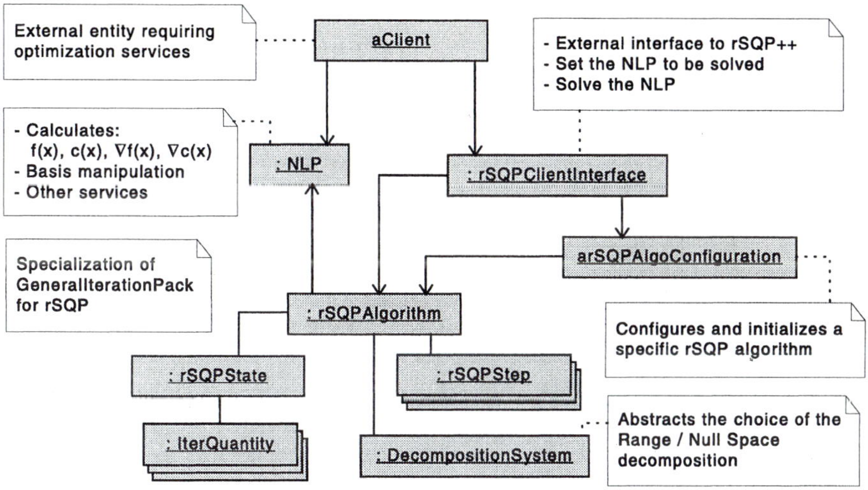

Fig. 1. UML Object Diagram: Coarse grained object diagram for an rSQP++ algorithm configured for the solution of an NLP

tion is for NLPs that can compute $\nabla f(x)$ and $\nabla c(x)$. The matrix $A = \nabla c(x)$ is represented as an abstract matrix object of type `MatrixWithOp` and can therefore be implemented by any appropriate means. Through this matrix interface, the optimization algorithm can perform only simple operations like matrix vector multiplication $v = op(A)u$. It is only in conjunction with a compatible `BasisSystem` object that the algorithm can perform all the needed computations. The `NLPSecondOrderInfo` interface is for NLPs that can compute the Hessian of the Lagrangian $\nabla^2 L$ which is also abstracted as a `MatrixWithOp` object. In this way the core rSQP++ code is independent of the specialized data-structures and solvers for an NLP. By configuring the algorithm with `NLP` and `BasisSystem` objects and `MatrixWithOp` objects for $\nabla c(x)$ and possibly $\nabla^2 L$, specialized data structures and linear algebra for an NLP can be accommodated.

The `NLPFirstOrderInfo` interface assumes that matrix vector multiplications with $A = \nabla c(x)$ and its transpose can be performed. The `BasisSystem` interface assumes that linear systems involving the basis matrix C and its transpose can be solved for arbitrary right hand sides. For many applications, these requirements can not be met. For these applications, the `NLPrSQP-TailoredApproach` interface (Figure 3) is defined and it is used by the algorithm to extract the bare minimum information (i.e. $\nabla f(x)$, $p^y = -C^{-1}c$ and $D = -C^{-1}N$). With this information, the coordinate and orthogonal range/null decompositions can both be used. A more detailed introduction to rSQP++ can be found in [2].

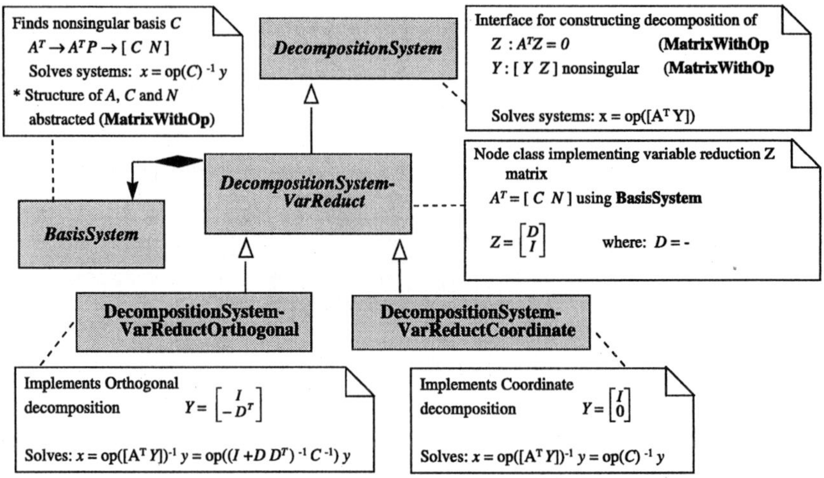

Fig. 2. UML Class Diagram: Range/null decomposition system classes

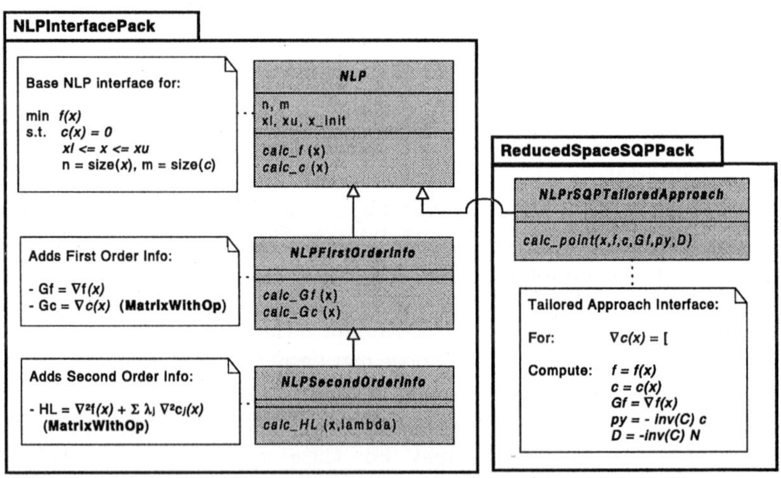

Fig. 3. UML Class Diagram: Interfaces to Nonlinear Programs (NLPs)

5 Tailored linear algebra for specialized NLPs

Figure 4 shows the special structure of a DAE system discretized using orthogonal collocation on finite elements. A specialized implementation of the basis matrix C for this structure called the elemental decomposition [4] was compared to a generic implementation using a general sparse linear solver (MA28). The Tennessee Eastman challenge problem was used as the example with 80 finite elements for a total NLP size of $n = 43,950$, $n - m = 2,640$ [3]. Figure 4 shows a dramatic reduction of the average CPU time for an rSQP iteration. In addition, storage savings using the elemental decomposition allowed the solution of larger problems. The elemental decomposition was implemented by externally configuring the rSQP++ algorithm with specialized `BasisSystem`, `MatrixWithOp` (for $\nabla c(x)$) and `NLP` objects.

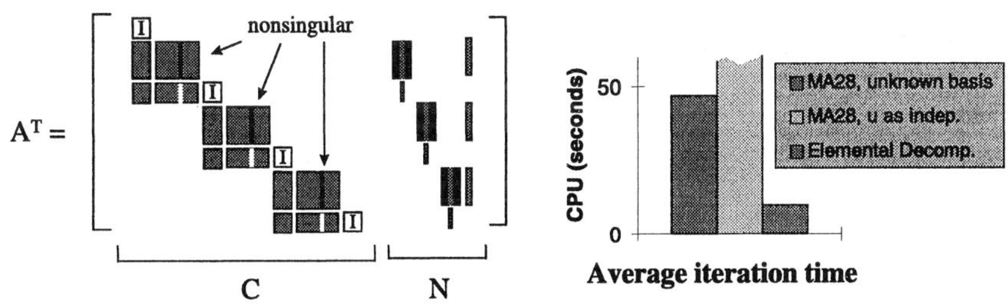

Fig. 4. Tennessee Eastman example: Tailored linear algebra for DAE optimization. Note: associate the legond entries from top to bottom with left to right in the bar graph (see Plate 11 on page 349).

A "Tailored Approach" interface (`NLPrSQPTailoredApproach`) has also been implemented for MPSalsa[1] which uses Aztec[2] as the linear algebra library. This interface was used to optimize a CVD reactor model with $m = 31,995$ discretized state variables and $n - m = 1$ decision variables. The rSQP++/MPSalsa implementation converged[3] with the solution in 1.56 hrs while a non-invasive "black box" approach took 22 hrs [10]. In this implementation, multiple processors where also utilized with fair results.

6 Moving from serial to distributed memory: vector interfaces

There are many different types of objects provided by a numerical linear algebra library such as vectors, matrices, and linear solvers. We contend that

---

[1] `http://www.cs.sandia.gov/CRF/MPSalsa`

[2] Aztec, `http://www.cs.sandia.gov/CRF/aztec1.html`

[3] Single Intel PIII 500 MHz processor, RedHat Linux 6.1

vectors are what primarily "glue" optimization software to the specialized implementations of higher level linear algebra components (e.g. matrices, linear solvers) as well as to other types of objects. For example, in an NLP solver, vector objects represent the estimate of the unknown variables $x^k \in \mathrm{I\!R}^n$, the residual of the nonlinear constraints $c(x^k) \in \mathrm{I\!R}^m$, the gradient of the objective or other functions $\nabla f(x^k) \in \mathrm{I\!R}^n$, as well as many other intermediate vectors (i.e. search directions, Lagrange multipliers etc.). Many of these vectors are passed around by the optimization code to various other objects to be used to transform other objects. For instance, vectors are passed to an NLP interface to compute function (scalar $f(x^k)$ and vector $c(x^k)$) and gradient (vector $\nabla f(x^k)$ and matrix $\nabla c(x^k)$) values. Right hand side vectors are passed to linear solvers to compute solution vectors. In addition to having the optimization software just shuffle vector objects around to other objects so they can be operated on or transformed, the algorithm also needs to perform various reduction (e.g. norms, dot products etc.) and transformation (e.g. vector addition, scaling etc.) operations with these vectors. If all of the linear algebra operations are implemented in a shared memory environment, then simple Fortran style arrays for dense and sparse vectors could be used, and the optimization software can trivially and cheaply access any arbitrary vector element. For the most part, utilizing parallelism in a shared memory environment is relatively easy if the optimization software calls standard packages like the BLAS and LAPACK. Multiple processors can be utilized by simply linking the object code to multi-threaded libraries. This is possible partly because the optimization software can communicate with these various implementations using simple serial Fortran style arrays.

The difficulty arises, however, when we move to a distributed memory environment. Now we can no longer deal with simple Fortran style serial arrays and instead have to consider distributed data structures and communication mechanisms. This can be seen in several different packages for distributed memory numerical linear algebra (ScaLAPACK[4], Aztec, PETSc[5], SuperLU[6] etc.). In such a distributed environment, direct and cheap access to every vector element is no longer a luxury an optimization code can enjoy. Trying to maintain an optimizer for every different distributed linear algebra library and flavor of parallelism (i.e. library based MPI[7], PVM[8] or compiler generated PGHPF[9], MPC++[10]) and also deal with serial vectors as well as specialized implementations (e.g. out-of-core) is a difficult task. What is needed is a set of abstract vector interfaces that can encapsulate these de-

[4] Scalable LAPACK, `http://www.netlib.org/scalapack/index.html`

[5] Portable Extendable Toolkit for Scientific Computation, `http://www.mcs.anl.gov/petsc`

[6] Supernodal LU, `http://www.nersc.gov/~xiaoye/SuperLU`

[7] Message Passing Interface, `http://www.mpi-forum.org/docs/docs.html`

[8] Parallel Virtual Machine, `http://www.epm.ornl.gov/pvm/pvm_home.html`

[9] Portland Group High Performance Fortran, `http://www.pgroup.com`

[10] Massively Parallel C++, `http://pdswww.rwcp.or.jp/mpc++`

tails away from the optimization algorithm, so that the implementations of the library and algorithm can be varied independently.

Because rSQP++ was developed initially to use serial direct solvers, issues related to the use of massively parallel iterative solvers used in PDE-constrained optimization were not addressed. In order to effectively support parallelism, rSQP++ needs the addition and integration of an abstract vector interface. Encapsulating vector implementations away from the core rSQP++ algorithmic code would allow for fully parallel linear algebra but could also greatly restrict the kinds of vector operations that could be performed. The primary goal of this discussion is to present a new strategy for implementing abstract vector interfaces that will allow an optimization developer to define and implement arbitrary vector reduction and transformation operations in a way that does not require vectors to reveal how they are implemented (parallel or serial, out-of-core etc.).

Perhaps the primary distinction between vectors and other linear algebra objects is the large number of operations that need to be performed with them. In addition to the 15 operations that are part of the Basic Linear Algebra Subroutines (BLAS), many other types of operations need to be performed as well. For example, in addition to many of the standard BLAS-like operations, some of the other vector reduction and transformation operations that the interior-point QP solver OOQP[11] must perform are shown below.

$$y_i \leftarrow \begin{cases} y^{\min} - y_i & \text{if } y_i < y^{\min} \\ y^{\max} - y_i & \text{if } y_i > y^{\max} \\ 0 & \text{if } y^{\min} \le y_i \le y^{\max} \end{cases} \qquad \text{for } i = 1...n \tag{12}$$

$$\alpha \leftarrow \{\max \alpha \mid x + \alpha d \ge \beta\} \tag{13}$$

Some examples of other non-standard vector reduction operations that may be performed in rSQP++ include the following.

$$\gamma \leftarrow d^{\max} + \frac{1}{\rho} ln \left\{ \frac{1}{n} \sum_{i=1}^{n} exp\left\{\rho\left((x_i - b_i) - d^{\max}\right)\right\} \right\} \tag{14}$$

$$\gamma \leftarrow max \left\{ \frac{|d_i|}{1 + |x_i|}, i = 1...n \right\} \tag{15}$$

$$\{y_k, k\} \leftarrow max\{y_i, i = 1...n\} \tag{16}$$

Many of these operations are not really vector operations in the mathematical sense.

It seems that the most common approach taken by developers of linear algebra libraries has been to provide a large set of primitive vector operations, and then the user can perform more complex operations by stringing together a set of primitives. For most vector operations this is feasible as long as the vector interface includes the needed primitives. For example, the

[11] http://www-unix.mcs.anl.gov/~wright/ooqp/

vector reduction operation (15) could be performed with two temporary vectors $u, v \in \mathbb{R}^n$ and five primitive vector operations as $|d_i| \rightarrow u_i$, $|x_i| \rightarrow v_i$, $v_i + 1 \rightarrow v_i$, $u_i/v_i \rightarrow u_i$, $max\{u_i\} \rightarrow \gamma$. Many of the other example vector operations shown above can be performed using primitives. It is difficult to see how (12) and (13) could be implemented with general primitive vector operations, however. At the very least it would take a number of temporary vectors and several operations. The number of primitive operations that need to be included in a vector interface in order to implement most of the needed vector operations is very large. For example, the HCL[12] vector interface contains more than 50 operations and still can not accommodate some of the above operations. In addition, using these non-overlapping operations can result in a performance bottleneck in a parallel application. For example, in ISIS++[13], the operation `Dist_Vector::QMR3norm2dot(...)` which performs $\{\alpha,\gamma,\xi,\rho,\epsilon\} \leftarrow \{(x^T x)^{1/2},(v^T v)^{1/2},(w^T w)^{1/2},w^T v,v^T t\}$ was added to the base vector class to overcome a bottleneck in the QMR solver[14].

Before a new approach to implementing vector interfaces is presented, note that the majority of vector implementations store and manipulate vector elements in continuous chunks (sub-vectors). For instance, most serial vectors are stored and manipulated as a whole 1-D array. For a parallel vector, elements are distributed between a set of processors and operations are performed in parallel on the local sub-vectors. In an out-of-core vector, the elements are stored in a set of files and the elements are read into RAM as sub-vectors in order to participate in vector operations. For efficient implementation in all of these cases, these sub-vectors must be of reasonably large size (i.e. hundreds to thousands of elements) so that the low level numeric operations performed with these elements dominate the runtime cost.

6.1 Vector reduction/transformation operators

A new approach is to allow clients of abstract vector objects to define an operator object, give it to a vector object through an abstract vector interface and then have the vector implementation apply the operator through an abstract operator interface. The idea is to allow a client to create a vector reduction/transformation operator that is equivalent to the following element-wise operators.

$$op(i, v_i^1 \ldots v_i^p, z_i^1 \ldots z_i^q) \rightarrow z_i^1 \ldots z_i^q \tag{17}$$
$$op(i, v_i^1 \ldots v_i^p, z_i^1 \ldots z_i^q) \rightarrow \beta \tag{18}$$
$$op(\beta^1, \beta^2) \rightarrow \beta^2 \tag{19}$$

[12] Hilbert Class Library, `http://www.trip.caam.rice.edu/txt/hcldoc/html/index.html`
[13] ISIS++, `http://z.ca.sandia.gov/isis`
[14] Example provided by Ben Allan, `baallan@california.sandia.gov`

where $v^1 \dots v^p \in \mathbb{R}^n$ are a set of p non-mutable input vectors ($p = 0$ allowed), $z^1 \dots z^q \in \mathbb{R}^n$ are a set of q mutable input/output vectors ($q = 0$ allowed) and β is a reduction target object which may be a simple scalar, a more complex non-scalar (i.e. (16)), or NULL. In the most general case the client can define an operator that will simultaneously perform multiple reduction and transformation operations involving a set of vectors. Simpler operations can be formed by setting $p = 0$, $q = 0$ or $\beta = \text{NULL}$. For example, reduction operations over one vector argument such as vector norms ($||v||$) are defined with $p = 1$, $q = 0$ and $\beta = \{\text{scalar}\}$. With this design, all of the standard BLAS operations, all of the example vector operations in (12)–(16), and many more vector operators can all be expressed. The key to optimal performance is that the vector implementation applies a user defined operator not element-wise as shown above, but instead on an entire set of sub-vectors (for elements $i = a \dots b$) at once

$$op(a, b, v^1_{a:b} \dots v^p_{a:b}, z^1_{a:b} \dots z^q_{a:b}, \beta) \rightarrow z^1_{a:b} \dots z^q_{a:b}, \beta \tag{20}$$

In this way, as long as the size of the sub-vectors is sufficiently large, the cost of performing a function call to invoke the operator will be insignificant compared to the cost of performing the computations within the operator. In a parallel distributed vector, $op(\dots)$ is applied to the local sub-vectors on each processor (no communication). The only communication between processors is to reduce the intermediate reduction objects $op(\beta^1, \beta^2) \rightarrow \beta^2$ (unless $\beta = \text{NULL}$, and no communication is required). A prototype implementation called `RTOP` has been developed for these operator interfaces in C. A paper and a set of html documentation describing this design in more detail can be found at the web site[15]. Support of more sophisticated client/server runtime configurations and other issues are discussed as well.

This design provides a more feasible means by which unusual vector operations can be efficiently implemented. The developers of vector implementations only need to implement a single `apply(op,...)` method that accepts user-defined reduction/transformation operator objects. It is then the optimization algorithm developer's responsibility to implement strange or specialized vector operators, but they can do so in a way that is independent of any particular vector implementation.

7 Summary and future work

Many application areas give rise to large-scale NLPs with specialized properties that can not be exploited by most of the current generation of optimization software. Noted examples include DAE- and PDE- constrained optimization. The next generation of optimization software needs to allow these specializations but also include those components of successful implementations that provide reliability and efficiency. The rSQP++ framework is

[15] `http://dynopt.cheme.cmu.edu/roscoe/RTOp`

being developed in an attempt to address these needs in the context of SQP. With rSQP++, the user can externally configure the algorithm with specialized linear algebra objects and can perform other modifications without having to recompile any of the base rSQP++ source code.

In order for rSQP++ to fully exploit parallelism (i.e. PDE-constrained optimization), an abstract vector interface must be developed and incorporated. A new design for abstract vector interfaces based on vector reduction/transformation operators would allow developers of optimization algorithms to efficiently implement specialized vector operations, while also allowing great flexibility in the implementation of vectors (i.e. serial, parallel, out-of-core, etc.).

References

1. Booch, G., J. Rumbaugh, and I. Jacobson. "The Unified Modeling Language User Guide." *Addison-Wesley*, New York (1999)
2. Bartlett, R. "An Introduction to rSQP++ : An Object-Oriented Framework for Reduced Space Successive Quadratic Programming" Technical Report, Department of Chemical Engineering, Carniege Mellon University (2000)
3. Bartlett, R. "New Object-Oriented Approaches to SQP for Large-Scale Process Optimization" Parallel and Large-Scale Computing: algorithms and Applications, AICHE Annual Meeting, Los Angeles (2000)
4. Biegler, L., A. Cervantes and A. Waechter. "Advances in Simultaineous Strategies for Dynamic Optimization." CAPD Technical Report B-01-01, Department of Chemical Engineering, Carniege Mellon University (2001)
5. Nocedal, J. and M. Overton. "Projected Hessian Updating Algorithms for Nonlinear Constrained Optimization". *SIAM J. Numer. Anal.*, Philadelphia, **22**, 821 (1985)
6. Nocedal, J. and Stephen Wright. "Numerical Optimization". *Springer*, New York, (1999)
7. Rumbaugh, J. et. al. "Object-Oriented Modeling and Design." *Prentice Hall*, Englewood Cliffs, New Jersey (1991)
8. Schmid, C. and L.T. Biegler. "Acceleration of Reduced Hessian Methods for Large-Scale Nonlinear Programming." *Comp. Chem. Eng.* **17**, 451 (1993)
9. Schmid, C. and L.T. Biegler. "Quadratic Programming Methods for Reduced Hessian SQP." *Comp. Chem. Eng.* **18**, 817 (1994)
10. van Bloemen Waanders, B., A. Salinger, R. Pawlowski, L. Biegler, R. Bartlett "Simultaneous Analysis and Design Optimization of Massively Parallel Simulation Codes using an Object Oriented Framework." Tenth SIAM Conference on Parallel Processing for Scientific Computing (SIAG/SC) (PP01) March (2001)
11. Womble, D.E, S.S. Dosanjh, B. Hendrickson, M.A. Heroux, S.J. Plimpton, J.L. Tomkins and D.S. Greenberg. "Massively Parallel Computing: A Sandia Perspective." *Parallel Computing*, **25**, 1853-1876 (1999)
12. Wright, S. "Optimization Software Packages." *ANL/MCS-P8xx-0899*, Mathematics and Computer Science Division, Argonne National Laboratory (1999)

Sundance Rapid Prototyping Tool for Parallel PDE Optimization

Kevin R. Long

Sandia National Laboratories, Livermore, CA 94551

Abstract. High-performance algorithms for PDE-constrained optimization often require application of operators and solution of systems of equations that are different from those used in a single solution of the PDE; consequently, exploration of a research idea entails startup costs for modification to the PDE solver. A software tool to enable rapid development of parallel codes for large-scale, complex PDEs on realistic problems would be a useful aid to research in this area. As part of Sandia's research efforts in PDE-constrained optimization, we are developing Sundance, an environment in which a parallel PDE solver is accessed via a high-level problem description, using abstract concepts such as functions, operators, and regions. With this high-level problem description, it is possible to specify a variational formulation of a PDE and its discretization method in a small amount of user-level code. It is then straightforward to obtain operators such as Jacobians and Hessians for use in optimization algorithms.

1 Introduction

Although there has been much research in algorithms and software for solving PDE-constrained optimization problems, the field is hampered by the the fact that both PDE software and optimization software tend to be difficult to use; working with both more than compounds the problem. Most PDE software has been developed with the idea of solving a specific example of a problem from a class of problems supported by the code, e.g., fluid flow. There is often no way to get access to the internal operators so that they can be used in non-traditional ways, nor is there a means of computing derivatives of the PDEs with respect to some of the parameters. On the other hand, efficient optimization codes are usually gradient-based and thus require derivatives of the objective function and constraints with respect to the parameters over which the optimization is taking place. In particular, optimization codes typically want a routine to evaluate the function value and Jacobian of the objective function and the Jacobians of the (vector-valued) constraints. Thus, a PDE tool with facilities that allow a high level of expression of the PDE system along with complete control of the creation and use of the resulting operators would greatly aid the research and development of algorithms for PDE based optimization.

Sundance is being developed to provide these capabilities. In particular, Sundance accepts a system of coupled PDEs and boundary conditions written in symbolic form that is close to the form in which a scientist or engineer would normally write them. Each function or variation appearing in this symbolic description is annotated with a specification of the finite-element basis with which that object will be discretized. This information, along with a mesh, is then used by Sundance to assemble the implied discretized operators. At this point, the user could simply ask Sundance to solve the system, or request certain evaluations to be made.

For example, if the linear system representing the PDE is $Au = f$, then Sundance could be called to compute $Au - f$, i.e. the residual on this system for a specified value of u. Sundance allows these calculations to be made and controlled by the optimization algorithm. An important feature of the Sundance system is that a Sundance calculation can be controlled remotely by an optimization code running on another machine.

Given the features and ease of use of Sundance, we believe that Sundance can be used to evaluate quickly many algorithmic ideas. For example, it is relatively straightforward to create reduced-physics or coarse-grained approximations to the Hessian matrices that arise in sequential quadratic programming algorithms. Sundance can also be used in other rapid prototyping applications where the user can experiment easily with various models; for example, we envision that Sundance might be used by solver and preconditioner researchers to experiment with alternate formulations or discretizations, or by applications developers to prototype mathematical formulations of a physical problem.

We must emphasize that for performance reasons, the high-level objects used for problem specification are not used for numerical calculations. Rather, they are used to marshal a set of internal objects that can be used for efficient calculations.

In the remainder of the paper, we describe more of the details of the Sundance system. In Section 2 we describe the symbolic problem specification language and in Section 3 we show examples of how it can be used to set up and modify problems with a small amount of code. In Section 4 we show how to use Sundance to solve a simple PDE-constrained optimization problem. Finally, in Section 5 we summarize the current state of Sundance and near-term and long-term plans.

2 An Abstract Interface to PDE Problems

High-level tools such as Matlab are often used for prototyping; however, most such existing commercial tools are not suitable for high-performance computing. One approach that provides both high-level ease of use and high performance is an abstract interface for vector spaces, vectors, and operators. If an optimization algorithm can be coded in terms of such abstract objects, then that algorithm can be used with any underlying high-performance representation that implements the abstract operator interface. There are now several projects underway to provide an abstract interface for linear and nonlinear operators, notably HCL [3] and TSF [4].

Abstract operator interfaces such as HCL provide a means for us to *use* operators in a high-level way. With Sundance we attempt to extend the realm of abstraction to encompass the *creation* of discrete operators from a high-level specification. With Sundance to create operators and representation-independent algorithms to use them, a complete optimization problem can be coded in a very high-level way.

As an example, consider the specification of a heat conduction problem for temperature $T(\mathbf{x})$ on a domain Ω. The governing equation on the interior of Ω is Poisson's equation

$$\nabla^2 T = f(\mathbf{x}), \tag{1}$$

with appropriate boundary conditions. Assume that on some boundary subdomain $\partial\Omega_1$ we will have Dirichlet boundary conditions $T = T_0$, and on the remainder, $\partial\Omega_2$, we have Neumann boundary conditions $\frac{\partial T}{\partial n} = G$. Mixed (Robin) boundary can be handled easily in Sundance but for explanatory simplicity we will restrict the present discussion to Dirichlet and Neumann types.

There are a number of ways to formulate the above system as a weak problem; for example Galerkin's method or first-order least squares. Different formulations will be represented with different symbolic expressions for the weak form; because Sundance can represent essentially arbitrary symbolic expressions, it can accommodate a variety of finite-element formulations.

Proceeding with a Galerkin formulation with variation δT, we have

$$\int_\Omega -\nabla\delta T \cdot \nabla T + \int_\Omega \delta T f + \int_{\partial\Omega_2} \delta T G = 0, \tag{2}$$

augmented by the Dirichlet boundary condition $T = T_0$ on a boundary subdomain $\partial\Omega_1$. Equation 2 is an infinite-dimensional statement of the problem; to form a discrete problem we also need to specify a discrete mesh, a set of elemental basis functions, and a quadrature method.

We now explain how the concepts above are encapsulated in Sundance code. Figure 1 shows a code fragment in which a Galerkin formulation of the heat conduction equation is specified. The first two lines,

```
Expr varT = new Variation(Lagrange(2));
Expr T = new Unknown(Lagrange(2));
```

define symbolic objects for the unknown function T and variation δT. These are represented with the class `Expr`, which is a handle [2] for expression subtypes; classes `Unknown` and `Variation` are two of those subtypes. The `Lagrange` arguments are a specification of the finite-element basis that will be used to discretize those functions. The handle does automatic memory management with reference-counting; the user allocates objects with `new`, but need never worry about deallocation with `delete`. This system gives Sundance code something of a Java look and feel.

Next, we can create a symbolic object to represent the gradient operator.

```
Expr T = new Unknown(Lagrange(2));
Expr varT = new Variation(Lagrange(2));

Expr dx = new Derivative(0);
Expr dy = new Derivative(1);
Expr grad = List(dx, dy);

Expr x = new CoordExpr(0);
Expr y = new CoordExpr(1);
Expr f = y + sin(3.0*x);
Expr laplacian = -(grad*varT)*(grad*T);
Expr source = varT*f;

Expr eqn = Integral(AllMaximal(), laplacian, GaussLegendre(2))
+ Integral(AllMaximal(), source, GaussLegendre(4));

EssentialBC bc = EssentialBC(someEdge, v*(u-u0));
```

Fig. 1. Fragments of Sundance code for Galerkin solution of the heat diffusion equation

```
Expr dx = new Derivative(0);
Expr dy = new Derivative(1);
Expr grad = List(dx, dy);
```

Here, `dx` and `dy` represent the partial derivatives in the x and y directions (indicated with the integers 0 and 1, respectively), and in the third line we combine them into a vector {`dx, dy`}.

In the next lines, we create coordinate functions `x` and `y`, representing the functions x and y. We then use these to create a source term $f(x, y) = y + \sin(3x)$, where the arithmetic operators on `Expr` have been overloaded appropriately. Had our source term been more complicated, we could have created a `UserDefExpr` object that would link to a user's C++, C, or Fortran code to evaluate an arbitrary function of position.

Next, we use expression operations to build the weak Laplacian and source terms. Note that the gradient operator is applied with the overloaded multiplication operator.

Finally, we use the previously defined objects to assemble an integral form of the problem and boundary conditions. Note that each term can have its own quadrature family, allowing fine-grained control over the integration. The Dirichlet boundary conditions are represented with the `EssentialBC` object; boundary conditions specified in this way are imposed by overriding the internal equations for the selected boundary nodes. Other methods of imposing boundary conditions, for instance the weak boundary conditions arising in a least-squares formulation, can be handled by putting the boundary conditions into an `Integral` object rather than an `EssentialBC` object.

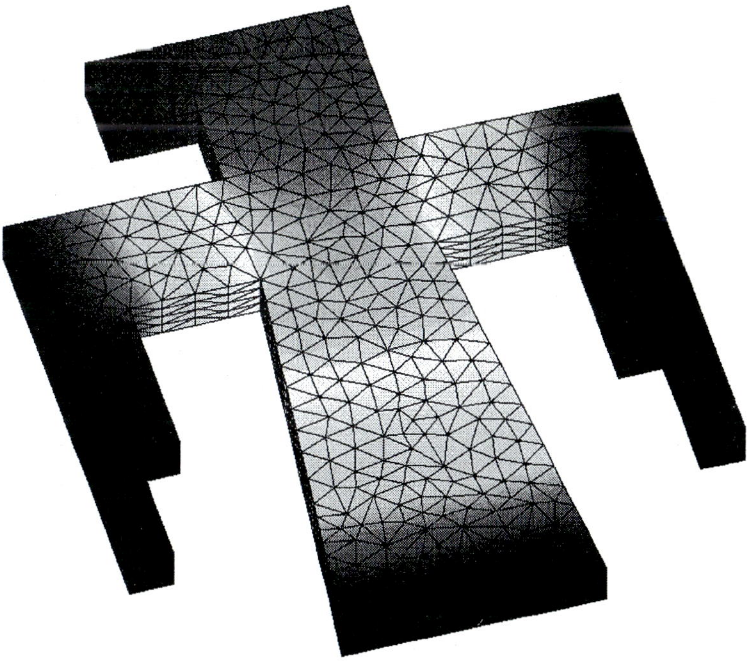

Fig. 2. Solution of Heat Conduction Equation on the Sandia Thunderbird

From such information, it is possible to automatically obtain a map from cell and node numbers to global equation and degree of freedom numbers, and to get the distribution of equations over processors. Additionally, the graph of the global matrix can be determined automatically. The terms appearing in the variational equation are examined to identify the appropriate master elements. The user need never worry about low-level details such as elements, degree of freedom and equation numbering maps, and processor maps; everything is worked out automatically by Sundance in an internal preprocessing step.

When Sundance is ready to assemble a stiffness matrix and load vector, it communicates with low-level linear algebra objects through the TSF [4] interface. There are many high-quality numerical linear algebra packages in use, and the use of the TSF interface allows third-party packages to be used in Sundance via TSF. Specification of the actual linear algebra representation to be used is done by providing a `TSFVectorType` object to the `StaticLinearProblem` constructor. This object knows how to build a vector space given a mesh and set of functions, and the vector space in turn knows how to build a vector of the appropriate type. Because Sundance does the element/node to equation mapping internally, it is sufficient to use a rather simple interface that allows initialization of the matrix data structure from the matrix graph, and then allows insertion of blocks of matrix elements.

The map-building preprocessing step and matrix assembly step are both fully parallelized. Parallelization of a Sundance calculation is simply a matter of using a partitioned mesh and a parallelized matrix-vector system. The same Sundance symbolic problem specification will work in both serial and parallel.

The guiding design principle behind Sundance is that the user should be able to specify a finite element analysis problem in terms of a symbolic statement of a variational form and boundary conditions, augmented by high-level discretization specifiers such as basis function type and quadrature method. The user should be shielded from all details of low-level bookkeeping of elements, nodes, and degrees of freedom. A sophisticated user will want fine control over discretization, but that control should be done through high-level directives.

Figure 2 shows results of a solver based on the above code fragment run on a three-dimensional figure of modest geometrical complexity. The mesh was generated by extruding a 2D mesh created with the Triangle mesh generator. The specification of the equation and boundary conditions involved fewer then 20 lines of Sundance code.

3 Flexibility of the Symbolic Interface

The power of the symbolic interface quickly becomes apparent when we attempt to change the problem or formulation. In this section, we show several illustrative examples.

We begin by modifying the heat diffusion problem above to include a convective term. The steady-state convection-diffusion equation is

$$\nabla^2 T = \mathrm{Pe}\,(\mathbf{u}\cdot\nabla)\,T, \tag{3}$$

where the Peclet number Pe is a dimensionless number indicating the relative importance of convection and diffusion. We can write a Galerkin convection-diffusion code with not much more work than was needed for the heat diffusion problem,

```
Expr T = new Unknown(Lagrange(1));
Expr varT = new Variation(Lagrange(1));

Expr eqn = -(grad*varT)*(grad*T) + Peclet*varT*U*(grad*T);
```

where the last line gives the convection-diffusion operator after integration by parts.

Depending on the problem under consideration, the velocity field `U` might be assumed known, in which case it can be represented as a Sundance expression, or it might be unknown, in which case we would set up a nonlinear problem to solve for both `T` and `U`.

Continuing with the convection-diffusion example, we can quickly write a code that will obtain the sensitivity to changes in a design parameter. As a trivial example, we can find the sensitivity of the above convection-diffusion problem to the Peclet number, i.e.

$$S = \frac{\partial T}{\partial \mathrm{Pe}}. \tag{4}$$

An equation for S follows from differentiating Equation 3 with respect to Pe,

$$\nabla^2 S = \mathrm{Pe}\,(\mathbf{u} \cdot \nabla)\, S + (\mathbf{u} \cdot \nabla)\, T. \tag{5}$$

Equation 5 is easily written in Sundance as

```
Expr S = new Unknown(Lagrange(1));
Expr varS = new Variation(Lagrange(1));

Expr sensEqn = -(grad*varS)*(grad*S) - Peclet*varS*U*(grad*S)
               - varS*U*(grad*T);
```

Running this code along with the original problem yields the results in Figure 3. Again, only a few additional lines of Sundance code were required to compute a quantity that would have been very difficult to compute with a standard code.

Work is in progress to automate the derivation of sensitivity equations, Jacobians, or Hessians given a PDE. This is the driver behind the development of Sundance the ability to quickly code the operators needed in a PDE application or a PDE-constrained optimization problem.

We can also use this symbolic flexibility to change the finite-element formulation. For example, in a first-order least-squares formulation [1] of the heat diffusion equation, we write a first order system

$$\nabla \cdot \mathbf{q} = f \tag{6}$$

$$\nabla T = \mathbf{q} \tag{7}$$

and minimize the functional

$$J(\mathbf{q}, T) = \|\nabla \cdot \mathbf{q} - f\|^2 + \|\nabla T - \mathbf{q}\|^2 + \text{b.c.}, \tag{8}$$

where the boundary conditions may be imposed with a similar least-squares condition or with a restriction on the function space of variations. The resulting variational form can be coded quickly in Sundance as shown in Figure 4. Alternate boundary condition imposition methods, e.g. least-squares or essential, can be represented and experimented with easily in similar Sundance code.

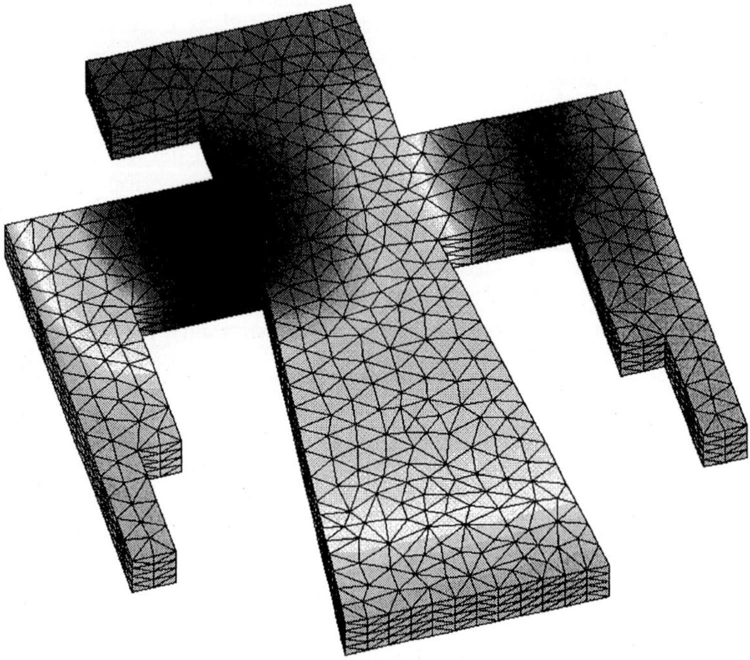

Fig. 3. Sensitivity to Peclet number on the Sandia Thunderbird. The underlying velocity field is a simple potential flow, also calculated by Sundance.

```
Expr T = new Unknown(Lagrange(2));
Expr varT = new Variation(Lagrange(2));
Expr Qx = new Unknown(Lagrange(2));
Expr varQx = new Variation(Lagrange(2));
Expr Qy = new Unknown(Lagrange(2));
Expr varQy = new Variation(Lagrange(2));
Expr Q = List(Qx, Qy);

Expr dx = new Derivative(0);
Expr dy = new Derivative(1);
Expr grad = List(dx, dy);

Expr foslsPoisson = (varQ - grad*varT)*(Q-grad*T)
               + (grad*varQ)*(grad*Q - f);
```

Fig. 4. Fragments of Sundance code for least-squares solution of the heat diffusion equation

4 A PDE-constrained optimization example

We now show a simple example of how to use Sundance to set up an optimization problem with a PDE constraint. Consider the Poisson equation with source terms parameterized with a design variable α,

$$\nabla^2 u = \sum_k \alpha_k \sin k\pi x. \tag{9}$$

A simple optimization problem is to choose α such that the state function u is a good fit to a target function $\hat{u}$. This target-fitting problem can be posed as a least-squares problem with objective function

$$f(\alpha) = \frac{1}{2}\int_\Omega (u(\alpha) - \hat{u})^2 + \frac{R}{2}\sum_k \alpha_k^2 \tag{10}$$

where R sets the control cost. As written, we could solve this problem with a pattern search method in which we solve 9 for u at each function evaluation. Alternatively, we can let the states become independent variables, but impose equation 9 as a constraint. In that case, we have a Lagrangian

$$L(\alpha, u, \lambda) = \frac{1}{2}\int_\Omega (u - \hat{u})^2 + \frac{R}{2}\sum_k \alpha_k^2 - \int_\Omega \nabla u \cdot \nabla \lambda - \sum_k \alpha_k \int_\Omega \lambda \sin k\pi x \tag{11}$$

where λ is a Lagrange multiplier. The necessary condition for solving the optimization problem is that the variations of the Lagrangian with respect to α, u, and λ are all zero.

This example is a quadratic program with an equality constraint, and the solution is obtained with a single linear solve of the KKT system. The KKT system is indefinite and is most efficiently solved using a block Schur complement method.

Note that in this problem, the state variable u and Lagrange multiplier λ are unknown functions defined with a finite-element basis. However, the design parameters are unknown "global" parameters, defined independently of the mesh. In Sundance, mesh-based unknowns are `UnknownFunction` expression subtypes and global unknowns are `UnknownParameter` expression subtypes. To optimize performance in parallel, Sundance imposes the restriction that all global unknowns must appear in a separate block from any meshed unknowns; that block is then replicated across processors while blocks containing meshed unknowns are distributed. Matrix blocks mapping between the global unknown space and a meshed unknown space are implemented with multivectors, in which each row (or column, depending on the orientation of the block) is a distributed vector. The specification of the unknowns and block structure for this problem is done with the following Sundance code:

```
Expr u = new UnknownFunction(new Lagrange(2));
Expr v = u.variation();

Expr lambda = new UnknownFunction(new Lagrange(2));
Expr mu = lambda.variation();

Expr alpha1 = new UnknownParameter();
Expr alpha2 = new UnknownParameter();
Expr alpha3 = new UnknownParameter();
Expr alpha = List(alpha1, alpha2, alpha3);
Expr beta = alpha.variation();

TSFVectorType petra = new PetraVectorType();
TSFVectorType dense = new DenseSerialVectorType();

TSFArray<Block> unks = tuple(Block(alpha, dense),
             Block(u, lambda, petra));

TSFArray<Block> vars = tuple(Block(beta, dense),
             Block(mu, v, petra));
```

Once the unknowns have been specified, we can write out the objective function and Lagrangian in symbolic form:

```
Expr objectiveFunction = 0.5*Integral(pow(u-target, 2.0))
             + 0.5*alpha*alpha;
Expr lagrangian = objectiveFunction
             - Integral((dx*u)*(dx*lambda))
             - Integral(lambda*forcing);
```

The equation set can be obtained by taking symbolic variations of the Lagrangian.

```
Expr eqn = lagrangian.variation(List(u, lambda, alpha));
```

We will solve the system using a Schur complement solver, using TSF's block manipulation capabilities. The user-level code to specify a Schur complement solver for a 2 by 2 block system is

```
TSFPreconditionerFactory prec = new ILUKPreconditionerFactory(1)
TSFLinearSolver innerSolver = new BICGSTABSolver(1.0e-12, 1000);
TSFLinearSolver outerSolver = new BICGSTABSolver(1.0e-10, 1000);

TSFLinearSolver solver =
             new SchurComplementSolver(innerSolver, outerSolver)
```

5 Summary of Current State and Future Plans

In its present state, Sundance provides a high-level problem description system for formulating and solving linear, elliptic PDEs in parallel. Nonlinear or parabolic problems can be posed and solved by looping over a sequence of linear elliptic problems. Field variables, global or distributed design variables, objective functions, and constraints can be represented as symbolic objects, so that it is possible to represent a complete nonlinear programming problem – PDE constraints, design variable constraints, and objective function – using a high-level symbolic description.

Low-level objects such as basis functions, quadrature rules, and cell types are implemented as inheritors of base classes for those objects, so that new types can be added as extensions. Currently supported are Lagrange and serendipity interpolation, Gauss-Legendre and nodal quadrature, and point, line, triangle, quad, and tetrahedron cells.

Current work on Sundance includes improvements to its symbolic system, incorporation of automatic differentiation, and application to a number of practical design problems underway at Sandia. Longer-term plans include enabling Sundance to use the extensive SIERRA family of finite-element support routines, including h-adaptive meshing and dynamic load balancing.

We believe that the Sundance in its current form will prove to be a useful tool for prototyping algorithms and models and that when coupled with the more powerful SIERRA finite-element framework, it will prove to be an efficient means of joining the latest optimization algorithms with high-performance, massively parallel PDE codes.

6 Acknowledgements

Thanks are due to Bart van Bloemen Waanders, Ray Tuminaro, Juan Meza, and especially Paul Boggs for encouragement and advice. Dawn Chamberlain, Monica Martinez-Canales, and Chris Newman were hardy alpha users and provided invaluable feedback and numerous bug reports. Ray Tuminaro and Mike Heroux provided assistance with the linking of the Aztec and Trilinos solver libraries.

References

1. P. B. Bochev and M. D. Gunzberger. Finite element methods of least squares type. *SIAM Review*, 40:789–837, 1998.
2. J. Coplien. *Advanced C++: Programming Styles and Idioms.* Addison-Wesley, New York, NY, 1992.
3. M. S. Gockenbach and W. W. Symes. An overview of hcl 1.0. *ACM Transactions on Mathematical Software*, 25:191–212, 1999.
4. K. Long and M.A. Heroux. The trilinos solver framework. In O. Marques and T. Drummond, editors, *2002 ACTS Workshop Proceedings, Berkeley, CA, Sept 3–6, 2002*, 2002.

Color Plates

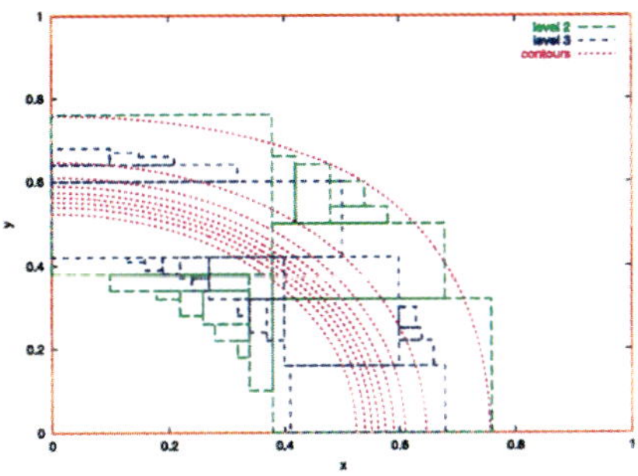

Plate 1-a. AMR with 50×50, 3 refinement levels and 2 buffer zones. 9 contours between 1.1 and 1.9 are plotted.

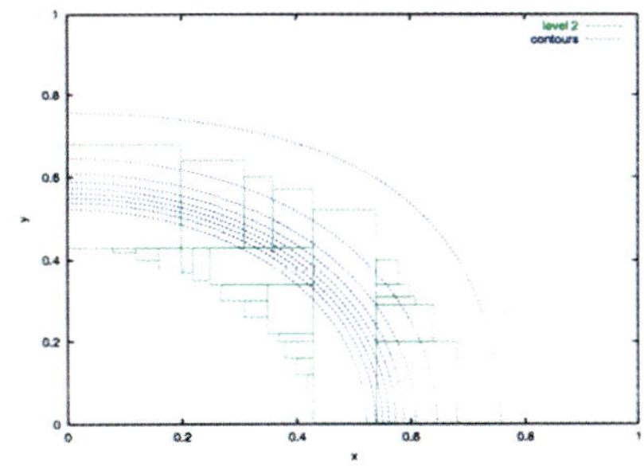

Plate 1-b. AMR with 100×100, 2 refinement levels and 2 buffer zones. 9 contours between 1.1 and 1.9 are plotted.

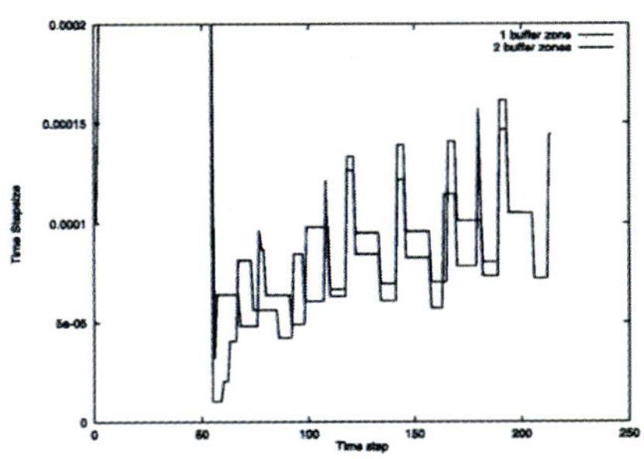

Plate 1-c. The warm restart can cause the DAE solver to reduce the time step after each refinement. Three-level refinement and a 50×50 base grid is used.

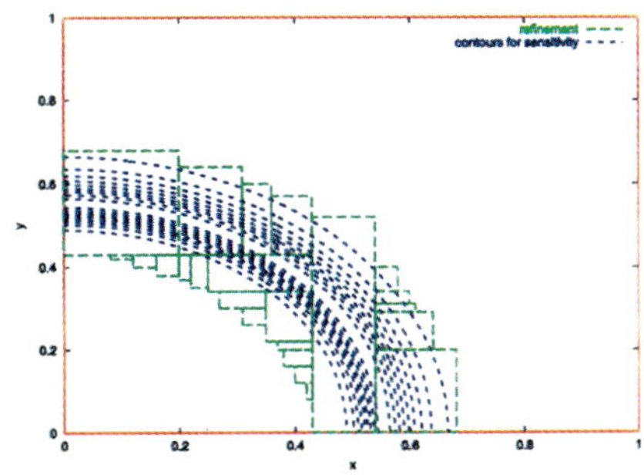

Plate 1-d. Contour plots for the sensitivity with respect to R. 9 contours between 2 and 18 are used. The contours increase from both sides to the middle of the refinement.

Plate 1. Results for reaction-diffusion problem (2) (see Fig. 5.1 on page 129).

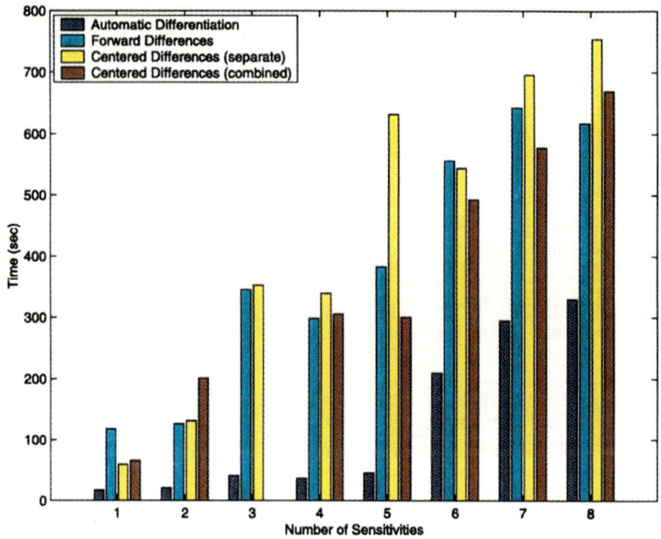

Plate 2. Comparison of SensPVODE performance (total time to solution) for various derivative computation strategies. Results are the average of three runs on 16 processors of a Linux cluster (see Fig. 5 on page 143).

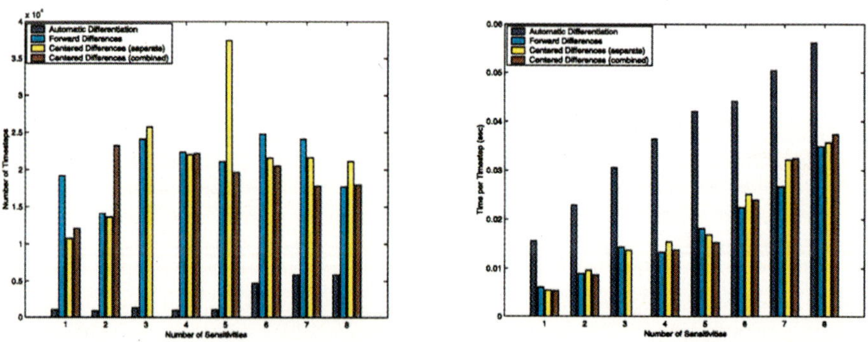

Plate 3. Number of timesteps and time per timestep for various derivative computation strategies in SensPVODE. Results are the average of three runs on 16 processors of a Linux cluster (see Fig. 6 on page 144).

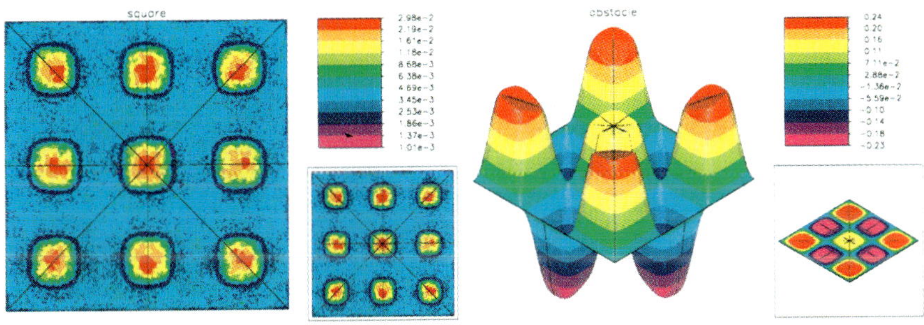

Plate 4. Elliptic variational inequality. On the left is an plot of the element size in the adaptive mesh with $n = 64000$ vertices. On the right is the corresponding solution (see Fig. 1 on page 229).

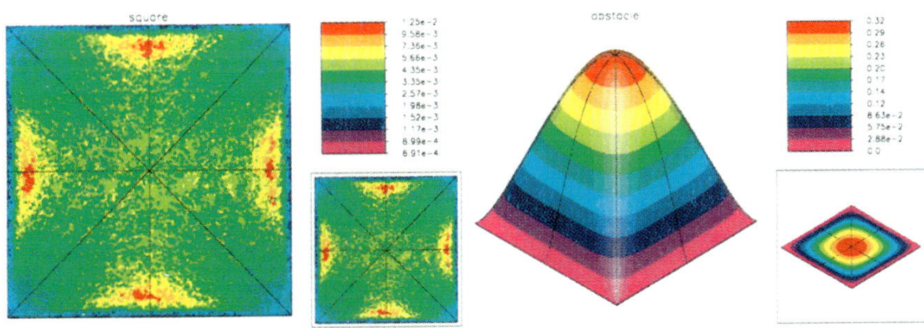

Plate 5. Elastic-plastic torsion problem. On the left is an plot of the element size in the adaptive mesh with $n = 64000$ vertices. On the right is the corresponding solution (see Fig. 2 on page 230).

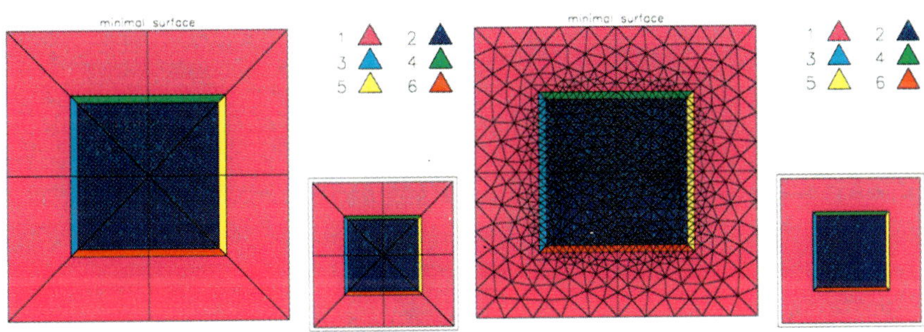

Plate 6. Minimal surface problem: On the left, we illustrate the skeleton corresponding to the definition of u_L. On the right, we show the corresponding initial mesh with 685 mesh points (see Fig. 3 on page 232).

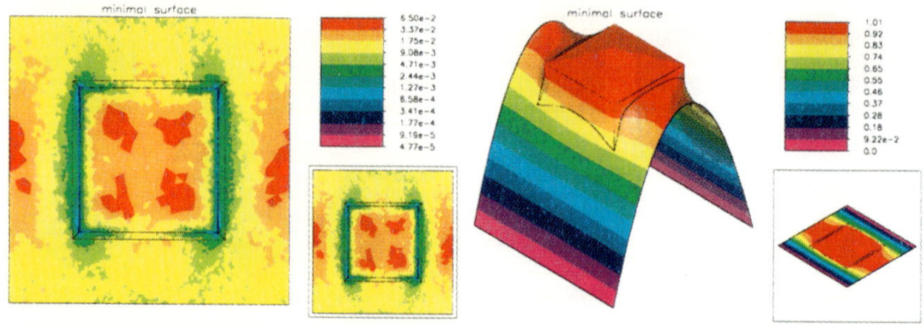

Plate 7. Minimal surface problem. On the left is an plot of the element size in the adaptive mesh with $n = 64000$ vertices. On the right is the corresponding solution (see Fig. 4 on page 232).

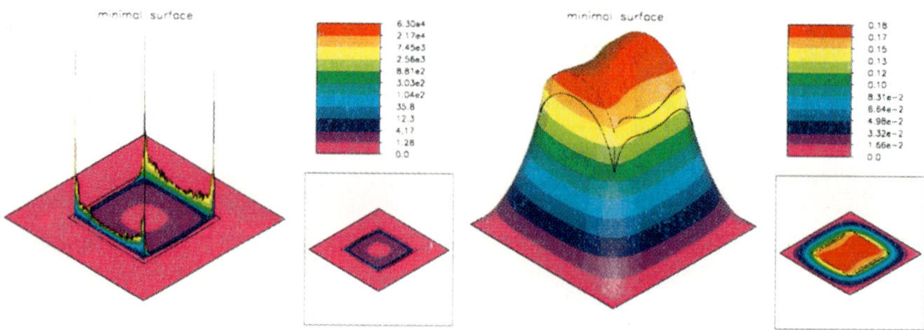

Plate 8. Minimal surface problem. On the left is a plot of the Lagrange multipliers for the mesh with $n = 64000$ vertices. On the right is a plot of the smoothed Lagrange multipliers (see Fig. 5 on page 233).

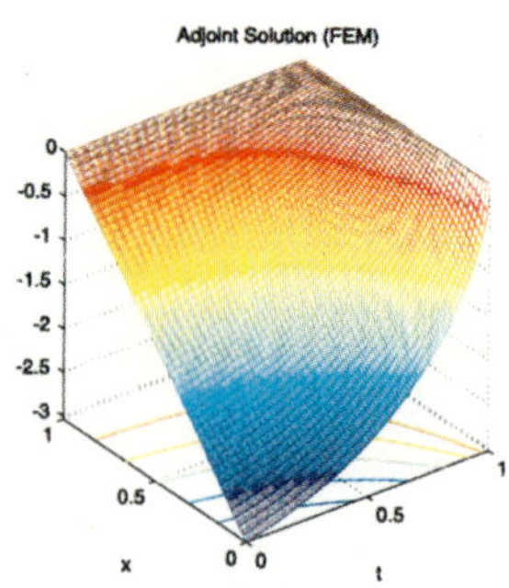

Plate 9. FEM Adjoint Solution in Iteration $k = 0$ (see Fig. 3 on page 278).

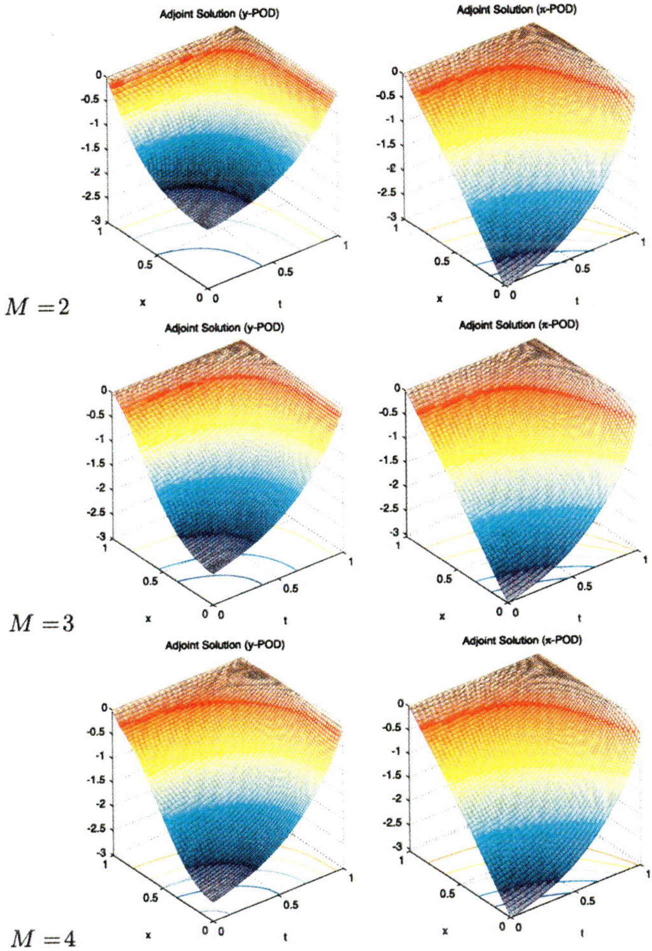

Plate 10. POD Based Adjoint Solutions in Iteration $k = 0$ (see Fig. 4 on page 279).

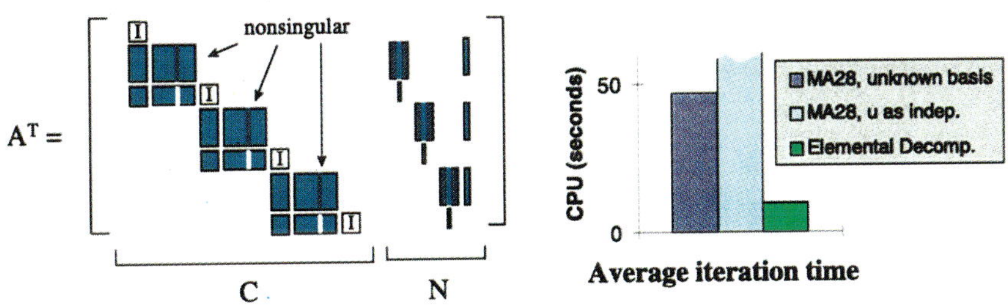

Plate 11. Tennessee Eastman example: Tailored linear algebra for DAE optimization. Note: associate the legond entries from top to bottom with left to right in the bar graph (see Fig. 4 on page 325).

Editorial Policy

§1. Volumes in the following three categories will be published in LNCSE:

i) Research monographs
ii) Lecture and seminar notes
iii) Conference proceedings

Those considering a book which might be suitable for the series are strongly advised to contact the publisher or the series editors at an early stage.

§2. Categories i) and ii). These categories will be emphasized by Lecture Notes in Computational Science and Engineering. **Submissions by interdisciplinary teams of authors are encouraged.** The goal is to report new developments – quickly, informally, and in a way that will make them accessible to non-specialists. In the evaluation of submissions timeliness of the work is an important criterion. Texts should be well-rounded, well-written and reasonably self-contained. In most cases the work will contain results of others as well as those of the author(s). In each case the author(s) should provide sufficient motivation, examples, and applications. In this respect, Ph.D. theses will usually be deemed unsuitable for the Lecture Notes series. Proposals for volumes in these categories should be submitted either to one of the series editors or to Springer-Verlag, Heidelberg, and will be refereed. A provisional judgment on the acceptability of a project can be based on partial information about the work: a detailed outline describing the contents of each chapter, the estimated length, a bibliography, and one or two sample chapters – or a first draft. A final decision whether to accept will rest on an evaluation of the completed work which should include

- at least 100 pages of text;
- a table of contents;
- an informative introduction perhaps with some historical remarks which should be accessible to readers unfamiliar with the topic treated;
- a subject index.

§3. Category iii). Conference proceedings will be considered for publication provided that they are both of exceptional interest and devoted to a single topic. One (or more) expert participants will act as the scientific editor(s) of the volume. They select the papers which are suitable for inclusion and have them individually refereed as for a journal. Papers not closely related to the central topic are to be excluded. Organizers should contact Lecture Notes in Computational Science and Engineering at the planning stage.

In exceptional cases some other multi-author-volumes may be considered in this category.

§4. Format. Only works in English are considered. They should be submitted in camera-ready form according to Springer-Verlag's specifications. Electronic material can be included if appropriate. Please contact the publisher. Technical instructions and/or TEX macros are available via http://www.springer.de/math/authors/help-momu.html. The macros can also be sent on request.

General Remarks

Lecture Notes are printed by photo-offset from the master-copy delivered in camera-ready form by the authors. For this purpose Springer-Verlag provides technical instructions for the preparation of manuscripts. See also *Editorial Policy.*

Careful preparation of manuscripts will help keep production time short and ensure a satisfactory appearance of the finished book.

The following terms and conditions hold:

Categories i), ii), and iii):
Authors receive 50 free copies of their book. No royalty is paid. Commitment to publish is made by letter of intent rather than by signing a formal contract. Springer-Verlag secures the copyright for each volume.

For conference proceedings, editors receive a total of 50 free copies of their volume for distribution to the contributing authors.

All categories:
Authors are entitled to purchase further copies of their book and other Springer mathematics books for their personal use, at a discount of 33,3 % directly from Springer-Verlag.

Addresses:

Timothy J. Barth
NASA Ames Research Center
NAS Division
Moffett Field, CA 94035, USA
e-mail: barth@nas.nasa.gov

Michael Griebel
Institut für Angewandte Mathematik
der Universität Bonn
Wegelerstr. 6
D-53115 Bonn, Germany
e-mail: griebel@iam.uni-bonn.de

David E. Keyes
Computer Science Department
Old Dominion University
Norfolk, VA 23529–0162, USA
e-mail: keyes@cs.odu.edu

Risto M. Nieminen
Laboratory of Physics
Helsinki University of Technology
02150 Espoo, Finland
e-mail: rni@fyslab.hut.fi

Dirk Roose
Department of Computer Science
Katholieke Universiteit Leuven
Celestijnenlaan 200A
3001 Leuven-Heverlee, Belgium
e-mail: dirk.roose@cs.kuleuven.ac.be

Tamar Schlick
Department of Chemistry
Courant Institute of Mathematical Sciences
New York University
and Howard Hughes Medical Institute
251 Mercer Street
New York, NY 10012, USA
e-mail: schlick@nyu.edu

Springer-Verlag, Mathematics Editorial IV
Tiergartenstrasse 17
D-69121 Heidelberg, Germany
Tel.: *49 (6221) 487-8185
e-mail: peters@springer.de
http://www.springer.de/math/peters.html

Lecture Notes in Computational Science and Engineering

Vol. 1 D. Funaro, *Spectral Elements for Transport-Dominated Equations.* 1997. X, 211 pp. Softcover. ISBN 3-540-62649-2

Vol. 2 H. P. Langtangen, *Computational Partial Differential Equations.* Numerical Methods and Diffpack Programming. 1999. XXIII, 682 pp. Hardcover. ISBN 3-540-65274-4

Vol. 3 W. Hackbusch, G. Wittum (eds.), *Multigrid Methods V.* Proceedings of the Fifth European Multigrid Conference held in Stuttgart, Germany, October 1-4, 1996. 1998. VIII, 334 pp. Softcover. ISBN 3-540-63133-X

Vol. 4 P. Deuflhard, J. Hermans, B. Leimkuhler, A. E. Mark, S. Reich, R. D. Skeel (eds.), *Computational Molecular Dynamics: Challenges, Methods, Ideas.* Proceedings of the 2nd International Symposium on Algorithms for Macromolecular Modelling, Berlin, May 21-24, 1997. 1998. XI, 489 pp. Softcover. ISBN 3-540-63242-5

Vol. 5 D. Kröner, M. Ohlberger, C. Rohde (eds.), *An Introduction to Recent Developments in Theory and Numerics for Conservation Laws.* Proceedings of the International School on Theory and Numerics for Conservation Laws, Freiburg / Littenweiler, October 20-24, 1997. 1998. VII, 285 pp. Softcover. ISBN 3-540-65081-4

Vol. 6 S. Turek, *Efficient Solvers for Incompressible Flow Problems.* An Algorithmic and Computational Approach. 1999. XVII, 352 pp, with CD-ROM. Hardcover. ISBN 3-540-65433-X

Vol. 7 R. von Schwerin, ***M**ulti **B**ody **S**ystem **SIM**ulation.* Numerical Methods, Algorithms, and Software. 1999. XX, 338 pp. Softcover. ISBN 3-540-65662-6

Vol. 8 H.-J. Bungartz, F. Durst, C. Zenger (eds.), *High Performance Scientific and Engineering Computing.* Proceedings of the International FORTWIHR Conference on HPSEC, Munich, March 16-18, 1998. 1999. X, 471 pp. Softcover. 3-540-65730-4

Vol. 9 T. J. Barth, H. Deconinck (eds.), *High-Order Methods for Computational Physics.* 1999. VII, 582 pp. Hardcover. 3-540-65893-9

Vol. 10 H. P. Langtangen, A. M. Bruaset, E. Quak (eds.), *Advances in Software Tools for Scientific Computing.* 2000. X, 357 pp. Softcover. 3-540-66557-9

Vol. 11 B. Cockburn, G. E. Karniadakis, C.-W. Shu (eds.), *Discontinuous Galerkin Methods.* Theory, Computation and Applications. 2000. XI, 470 pp. Hardcover. 3-540-66787-3

Vol. 12 U. van Rienen, *Numerical Methods in Computational Electrodynamics.* Linear Systems in Practical Applications. 2000. XIII, 375 pp. Softcover. 3-540-67629-5

Vol. 13 B. Engquist, L. Johnsson, M. Hammill, F. Short (eds.), *Simulation and Visualization on the Grid.* Parallelldatorcentrum Seventh Annual Conference, Stockholm, December 1999, Proceedings. 2000. XIII, 301 pp. Softcover. 3-540-67264-8

Vol. 14 E. Dick, K. Riemslagh, J. Vierendeels (eds.), *Multigrid Methods VI.* Proceedings of the Sixth European Multigrid Conference Held in Gent, Belgium, September 27-30, 1999. 2000. IX, 293 pp. Softcover. 3-540-67157-9

Vol. 15 A. Frommer, T. Lippert, B. Medeke, K. Schilling (eds.), *Numerical Challenges in Lattice Quantum Chromodynamics.* Joint Interdisciplinary Workshop of John von Neumann Institute for Computing, Jülich and Institute of Applied Computer Science, Wuppertal University, August 1999. 2000. VIII, 184 pp. Softcover. 3-540-67732-1

Vol. 16 J. Lang, *Adaptive Multilevel Solution of Nonlinear Parabolic PDE Systems.* Theory, Algorithm, and Applications. 2001. XII, 157 pp. Softcover. 3-540-67900-6

Vol. 17 B. I. Wohlmuth, *Discretization Methods and Iterative Solvers Based on Domain Decomposition.* 2001. X, 197 pp. Softcover. 3-540-41083-X

Vol. 18 U. van Rienen, M. Günther, D. Hecht (eds.), *Scientific Computing in Electrical Engineering.* Proceedings of the 3rd International Workshop, August 20-23, 2000, Warnemünde, Germany. 2001. XII, 428 pp. Softcover. 3-540-42173-4

Vol. 19 I. Babuška, P. G. Ciarlet, T. Miyoshi (eds.), *Mathematical Modeling and Numerical Simulation in Continuum Mechanics.* Proceedings of the International Symposium on Mathematical Modeling and Numerical Simulation in Continuum Mechanics, September 29 - October 3, 2000, Yamaguchi, Japan. 2002. VIII, 301 pp. Softcover. 3-540-42399-0

Vol. 20 T. J. Barth, T. Chan, R. Haimes (eds.), *Multiscale and Multiresolution Methods.* Theory and Applications. 2002. X, 389 pp. Softcover. 3-540-42420-2

Vol. 21 M. Breuer, F. Durst, C. Zenger (eds.), *High Performance Scientific and Engineering Computing.* Proceedings of the 3rd International FORTWIHR Conference on HPSEC, Erlangen, March 12-14, 2001. 2002. XIII, 408 pp. Softcover. 3-540-42946-8

Vol. 22 K. Urban, *Wavelets in Numerical Simulation.* Problem Adapted Construction and Applications. 2002. XV, 181 pp. Softcover. 3-540-43055-5

Vol. 23 L. F. Pavarino, A. Toselli (eds.), *Recent Developments in Domain Decomposition Methods.* 2002. XII, 243 pp. Softcover. 3-540-43413-5

Vol. 24 T. Schlick, H. H. Gan (eds.), *Computational Methods for Macromolecules: Challenges and Applications.* Proceedings of the 3rd International Workshop on Algorithms for Macromolecular Modeling, New York, October 12-14, 2000. 2002. IX, 504 pp. Softcover. 3-540-43756-8

Vol. 25 T. J. Barth, H. Deconinck (eds.), *Error Estimation and Adaptive Discretization Methods in Computational Fluid Dynamics.* 2003. VII, 344 pp. Hardcover. 3-540-43758-4

Vol. 26 M. Griebel, M. A. Schweitzer (eds.), *Meshfree Methods for Partial Differential Equations.* 2003. IX, 466 pp. Softcover. 3-540-43891-2

Vol. 27 S. Müller, *Adaptive Multiscale Schemes for Conservation Laws.* 2003. XIV, 181 pp. Softcover. 3-540-44325-8

Vol. 28 C. Carstensen, S. Funken, W. Hackbusch, R. H. W. Hoppe, P. Monk (eds.), *Computational Electromagnetics.* Proceedings of the GAMM Workshop on "Computational Electromagnetics", Kiel, Germany, January 26-28, 2001. 2003. X, 209 pp. Softcover. 3-540-44392-4

Vol. 29 M. A. Schweitzer, *A Parallel Multilevel Partition of Unity Method for Elliptic Partial Differential Equations.* 2003. V, 194 pp. Softcover. 3-540-00351-7

Vol. 30 T. Biegler, O. Ghattas, M. Heinkenschloss, B. van Bloemen Waanders (eds.), *Large-Scale PDE-Constrained Optimization.* 2003. VI, 349 pp. Softcover. 3-540-05045-0

Vol. 31 M. Ainsworth, P. Davies, D. Duncan, P. Martin, B. Rynne (eds.) *Topics in Computational Wave Propagation.* Direct and Inverse Problems. 2003. VIII, 399 pp. Softcover. 3-540-00744-X

Vol. 32 H. Emmerich, B. Nestler, M. Schreckenberg (eds.) *Interface and Transport Dynamics.* Computational Modelling. 2003. XV, 432 pp. Hardcover. 3-540-40367-1

Vol. 33 H. P. Langtangen, A. Tveito (eds.) *Advanced Topics in Computational Partial Differential Equations.* Numerical Methods and Diffpack Programming. 2003. XIX, 659 pp. Softcover. 3-540-01438-1

Texts in Computational Science and Engineering

Vol. 1 H. P. Langtangen, *Computational Partial Differential Equations.* Numerical Methods and Diffpack Programming. 2nd Edition 2003. XXVI, 855 pp. Hardcover. ISBN 3-540-43416-X

Vol. 2 A. Quarteroni, F. Saleri, *Scientific Computing with MATLAB.* 2003. IX, 257 pp. Hardcover. ISBN 3-540-44363-0

For further information on these books please have a look at our mathematics catalogue at the following URL: `http://www.springer.de/math/index.html`